Vietnam Tourism

Policies and Practices

Vietnam Tourism

Policies and Practices

Huong T. Bui

Ritsumeikan Asia Pacific University (APU), Japan

Giang T. Phi

VinUniversity, Vietnam

Long H. Pham

Vietnam National University (Ha Noi), Vietnam

Huong H. Do

Thai Binh Duong University, Vietnam

Andrew Le

International College of Management Sydney, Australia

Binh Nghiem-Phu

University of Hyōgo, Japan

CABI

CABI is a trading name of CAB International

CABI
Nosworthy Way
Wallingford
Oxfordshire OX10 8DE
UK

Tel: +44 (0)1491 832111
Fax: +44 (0)1491 833508
E-mail: info@cabi.org
Website: www.cabi.org

CABI
WeWork
One Lincoln St
24th Floor
Boston, MA 02111
USA

T: +1 (617)682-9015
E-mail: cabi-nao@cabi.org

A catalogue record for this book is available from the British Library, London, UK.

Library of Congress Cataloging-in-Publication Data

Names: Bui, Huong T., author.
Title: Vietnam Tourism: policies and practice
Huong T. Bui, Ritsumeikan Asia Pacific University (APU), Japan,
Giang T. Phi, VinUniversity, Vietnam,
Long H. Pham, Vietnam National University, Vietnam,
Huong H. Do, Thai Binh Duong University, Vietnam,
Andrew Le, International College of Management Sydney, Australia,
Binh Nghiem-Phu, University of Hyogo, Japan.

Description: Boston, MA : CABI is a trading name of CAB International, [2022] | Includes bibliographical references and index. | Summary: 'Vietnam has experienced rapid growth within its tourism industry during the past decade. This growth is part of Vietnam's transitional economy in which a wide range of forms of tourism have emerged'-- Provided by publisher.
Identifiers: LCCN 2021038917 (print) | LCCN 2021038918 (ebook) | ISBN 9781789242782 (Hardback) | ISBN 9781789242799 (eBook) | ISBN 9781789242805 (ePub)
Subjects: LCSH: Tourism--Vietnam. | Tourism--Social aspects--Vietnam. | Economic development--Vietnam.
Classification: LCC G155.V49 B85 2022 (print) | LCC G155.V49 (ebook) | DDC 338.4/791597--dc23
LC record available at https://lccn.loc.gov/2021038917
LC ebook record available at https://lccn.loc.gov/2021038918

References to Internet websites (URLs) were accurate at the time of writing.

ISBN-13: 9781789242782 (hardback)
9781789242799 (ePDF)
9781789242805 (ePub)

DOI: 10.1079/9781789242782.0000

Commissioning Editor: Claire Parfitt
Editorial Assistant: Lauren Davies
Production Editor: James Bishop

Typeset by Exeter Premedia Services Pvt Ltd, Chennai, India
Printed and bound in the USA by Integrated Books International, Dulles, Virginia

Contents

Authors

Huong T. Bui is a professor at the College of Asia Pacific Studies, Ritsumeikan Asia Pacific University (APU), Japan. She holds a PhD in tourism management from Griffith University (Australia). Prior to her career in academia, Huong worked in the tourism industry and was a consultant for tourism development projects in South-east Asia for the World Bank, Japan International Cooperation Agency (JICA) and the Netherlands Development Organization (SNV). She has received research grants from the Japan Society for Promotion of Sciences (JSPS) on dark tourism, heritage tourism, and disaster risk management for the tourism sector in Asia. Her research expertise revolves around conflict resolution, crisis management, sustainable resource management and adaptive resilience of tourism destinations. She has published 40 journal articles in leading tourism journals and edited two books on tourism in Asia: *Tourism and Development in Southeast Asia* (2020, Routledge) and *Nature-based Tourism in Asia Mountain's Protected Area* (2021, Springer). Email: huongbui@apu.ac.jp

Giang T. Phi is an assistant professor in tourism management and innovation/entrepreneurship at the College of Business and Management, VinUniversity, Hanoi, Vietnam. She holds a PhD in tourism management from Griffith University, Australia, and is an expert in tourism, social entrepreneurship, design thinking and sustainable development. The first edition of her edited book is *Critical Issues in Tourism Co-Creation* (2021, Routledge). Besides serving on the advisory board of Community-based Tourism Travel and Consulting social enterprise since 2014, Giang has engaged in high-profile consulting projects for the European Commission and Tourism Accommodation Australia. Having lived and worked in four countries across three continents, Giang now focuses on bringing her global experience to drive positive changes through impactful teaching, research and community services in Vietnam. Email: giang.ptl@vinuni.edu.vn

Long H. Pham is an associate professor and Dean of the Faculty of Tourism Studies, University of Social Sciences and Humanities, Vietnam National University (Hanoi). He is also a visiting professor at Rikkyo University, Japan. He graduated from the Vietnam National University (Vietnam), University of Utara (Malaysia), and holds a PhD from Rikkyo University in Japan. His research areas include ecotourism, community-based tourism, sustainable tourism development, and tourism policies and governance in South-east Asia. He is a leading consultant in sustainable tourism, community-based tourism and ecotourism in protected areas in Vietnam for international organizations such as the British Council Vietnam, Korea International Cooperation Agency (KOICA), German Development Agency (GIZ), United National Development Programme (UNDP), United Nations Educational,

Scientific and Cultural Organization (UNESCO), United States Agency for International Development (USAID), International Labour Organisation (ILO) and JICA. Long Pham has recently received a research grant from the National Foundation for Science and Technology (NAFOSTED), Vietnam, on value co-creation and destination brand equity. He is a senior policy adviser in tourism planning and development in Vietnam. Email: phamhonglong@gmail.com

Huong H. Do is a lecturer at Thai Binh Duong University, Nha Trang, Vietnam. She was previously Head of the Tourism Department at Hoa Sen University and founding Dean, School of Tourism at the former Hoa Sen College, Ho Chi Minh city. This is also where she devoted her career to develop a theory-practice-balanced curriculum in tourism and hospitality higher education programmes. She earned her PhD in tourism management (Griffith University), her MSc in strategic tourism management (SKEMA Business School), and her MBA in international business (Asian Institute of Technology). Her research focuses on tourist behaviour, destination management and sustainable tourism. Huong was also a national consultant for local tourism destinations and ecotourism development in protected areas in Vietnam funded by local governments and the Asian Development Bank. Email: dhhuongtourism@gmail.com

Andrew Le is a senior lecturer and Postgraduate Program Manager (Tourism, Hospitality, and Events) at the International College of Management Sydney (ICMS), Australia. Previously, he worked as a lecturer and course convenor at several business schools at Southern Cross University, Griffith University and Hanoi University. Andrew holds a PhD from Griffith University (Australia). His research interests cover tourism and hospitality, and sustainable development. Andrew is a reviewer and publishes in these areas in top-tier journals and with publishers such as Springer, CABI and the International Union for Conservation of Nature (IUCN). Additionally, he has worked as a team leader and senior consultant for international development projects funded by Europe Aid, Asian Development Bank (ADB), UNESCO, the Dutch Organisation for Internationalisation in Education (NUFFIC), Spanish Agency for International Development Cooperation (AECID), SNV and USAID. Email: ale@icms.edu.au

Binh Nghiem-Phu is an assistant professor at the School of Economics and Management, University of Hyōgo, Japan. Before that, he taught in other educational institutions in Vietnam and Japan, such as the Thuong Mai University (Vietnam) and Fukushima College – National Institute of Technology (Japan), as a full-time lecturer. The majority of his research is directed towards understanding consumers' perceptions and evaluations of the characteristics and images of products, services, organizations and places. He adopts the approaches of applied psychology theories to the implementation of marketing and management activities. His research papers or academic services can be found in or are with the *Asia Pacific Journal of Marketing and Logistics*, the *Journal of Air Transport Management* and *Tourism and Hospitality Research*. Email: binhnghiem@gmail.com

Guest Contributors

Quynh N. Bui is a lecturer of the Faculty of Tourism Studies, University of Social Sciences and Humanities, Vietnam National University in Hanoi. Upon completing a Master's in social sciences in marketing at the University of East Anglia (UK), she is enrolling in a PhD in tourism economics at the National Economics University, Vietnam. Her research interests include tourism and development, responsible tourism, sustainability in tourism, personal and organizational behaviours in tourism, marketing in tourism and event tourism. She worked as a tourism specialist in tourism development, ecotourism and community-based tourism in Vietnam for international organizations such as GTZ, AECID and JICA. Email: buinhatquynh1512@gmail.com

He T. Bui is a master's student at Ritsumeikan Asia Pacific University (APU), Japan. Prior to enrolment in academia, she worked at the National Archives of Vietnam for 10 years. She has co-authored some thematic books on Vietnamese history in the colonial period, including *Educational Policies in Colonial Vietnam 1858–1945* (2016, IC Publisher) and *Communal Reform* (2018, IC Publisher). Her current research is about tourism history, with a focus on the French colonization of Indochina. Email: buihek54@gmail.com

Phuong-Anh Dang is a lecturer at the University of Social Science and Humanities, Vietnam National University, Hanoi. She holds a PhD in cultural studies at Vietnam National Institute of Culture and Art Studies. She completed her BA and MA at the Faculty of Tourism Studies, Vietnam National University in Hanoi (2005 and 2012). She has received several grants from Indian Technical and Economic Cooperation (ITEC) (India, 2014), NAFOSTED (Vietnam, 2019) and the International Convention of Asian Scholars (ICAS) (the Netherlands, 2021). Phuong Anh's research focuses on tourism, heritage and cultural studies, sustainable tourism development, and community engagement in planning and development. Email: dangphuonganh@vnu.edu.vn

Alexandru O. Dimache an assistant professor at the College of Asia Pacific Studies, Ritsumeikan Asia Pacific University (APU), Japan. He holds a PhD in Tourism Management from the Hong Kong Polytechnic University (Hong Kong), an Erasmus European Masters in Tourism Management (EMTM), and a BSc in Tourism Management from the University of Surrey (UK). Beyond academia, he worked as a consultant and guide for adventure travel companies in Romania and the UK. His experience and expertise span across dark tourism, heritage and museum studies, interpretation, identity and attachment, psychology, linguistics, and social entrepreneurship. Email: adimache@apu.ac.jp

Thuy T. Duong is a lecturer at the Department of Geography, Vietnam National University. She holds a PhD in resource management and environment from Vietnam National University, Hanoi. Thuy has more than 10 years of research in tourism geography. She serves as a key member of five leading national tourism projects and has published more than 20 journal articles in various domestic and international journals. She delivers courses in tourism geography at Vietnam National University and other universities. Email: duongthithuy@hus.edu.vn

Thomas E. Jones is a professor of Environment and Development Studies at Ritsumeikan Asia Pacific University (APU) in Kyushu, Japan. His research interests include nature-based tourism, protected area management and environmental policy. Dr Jones completed his PhD at the University of Tokyo and has conducted visitor surveys on Mount Fuji and in the Japanese Alps. His latest publication is the edited volume *Nature-based Tourism in Asia Mountain's Protected Area* (2021, Springer). Email: 110054tj@apu.a.jp

Thu G. Hoang is the Dean of the Faculty of Management and Tourism at Hanoi University. He received his PhD in industrial/organizational psychology from Portland State University in 2011. Dr Hoang's research focuses on evaluating the impact of various work environment factors (human resources practices, job design, leadership) on key organizational and individual outcomes (job attitudes, job performance and employee's well-being) in a wide range of business settings. As a practitioner, he also provides consulting services to organizations to address recruitment and selection, training and development, performance management, workplace motivation and reward systems, and quality of work life. Email: thuhg@hanu.edu.vn

Phuong M. Ngo is an officer of the Investment, Trade and Tourism Promotion Agency under Quang Tri People's Committee. She completed her Master's course at Ritsumeikan Asia Pacific University (APU), Japan. She published articles in *Indian and Asian Studies* and *Journal of Tourism and Adventure*. She is working in the promotion of the tourism sector in Quang Tri, Vietnam. Email: minhphuong.aepd@gmail.com

Tuan Q. Pham is an associate professor at Vietnam National University, Hanoi. He was the Dean of the Faculty of Geography from 2010 to 2020. He holds a PhD in natural resources and environment, an MSc in landscape ecology and environment, and a BA in geography. He is a member of the Scientific Council of the University of Science, Vietnam National University, and a member of the executive committee of the Vietnam Geography Association. Associate Professor Pham is a leading expert in Vietnam on natural resources and environment management and has been involved in various projects in the conservation of coastal zones, national parks and protected areas in Vietnam. He has been extensively involved in teaching, supervising graduate students, and publishing in domestic and international journals. Email: phamquangtuan@hus.edu.vn

Thanh D. Tran is an associate professor at the Saigon International University (www.siu.edu.vn), Vietnam. He graduated from the National University of Kyiv, Ukraine, and holds a PhD in geography from Vietnam National University (VNU), Hanoi. He was one of the founders of tourism geography major and tourism programmes and the Faculty of Tourism Studies, University of Social Sciences and Humanities, VNU, Hanoi. His research interests are sustainable tourism development, community-based tourism, ecotourism, geotourism, tourism and climate change. Dr Tran has published textbooks in tourism such as *Topography* (2005, VNU), *Tourism Geography* (2012, VNU) and *Introduction to Tourism* (2021 VNU), which many tourism training institutions in Vietnam reference. Email: tranducthanh@siu.edu.vn

Hang T.B. Tran is a senior lecturer at the Faculty of Hospitality and Tourism, Thuong Mai University, Vietnam. She is the Director of the Scientific Research Management Office at this university. Prior to the current position, she was the Head of Tourism Enterprises Management Department, Faculty of

Hospitality and Tourism of the Thuong Mai University. Dr Tran has supervised many master's and doctorate students. Her research and teaching relate to managerial issues in hospitality and tourism companies in Vietnam. She received research grants from the Ministry of Education and Training and the Ministry of Science and Technology. Her publications have been used as teaching materials in various universities in Vietnam. Email: tranbichhang@tmu.edu.vn

Le-Anh Trinh is the Head of the Department of Events Management, Faculty of Tourism Studies, University of Social Sciences and Humanities, Vietnam National University (VNU), Hanoi. He is also an inspiring lecturer and a well-known TV show host, master of ceremonies (MC) and event conductor. Le-Anh has been engaged in different forms of the arts and performances alongside his interests and expertise in tourism and development studies and the social and cultural issues surrounding festivals and events in Vietnam. In his capacity as MC, Le Anh has been involved in numerous major events; he has had a strong public influence in Vietnamese society. Currently, he is a contributor for major media channels such as VTV (Vietnam Television), VOV (Voice of Vietnam) and top newspapers in Vietnam on topics linked to tourism product development, implications and policy in cultural tourism and events. Email: anhtl@ussh.edu.vn

Hai Q. Truong is a professor at Vietnam National University, Hanoi and Vice Chairman of Vietnam Geography Association, member of the Executive Board of National Extraordinary Science and Technology Program on Publication of *Vietnam National Gazetteer*, member of the International Landscape Ecology Association, and member of the Southeast Asian Geography Association. After acquiring a PhD at the University of Sofia, Bulgaria, in 1991, Professor Dr Truong has collaborated with universities and research institutes around the world under the scope of various scientific research and exchange programmes, especially participating in the Fulbright Visiting Scholar Program in the USA for 1998–1999. The focus of his research includes, but is not limited to, tourism geography, tourism resource assessment, tourism planning, ecotourism and sustainable tourism. He has published more than 30 articles on tourism in prestigious national and international journals, chaired two national projects on tourism, and trained numerous PhD students. Email: haitq@vnu.edu.vn

List of Figures

List of Tables

1 Vietnam Tourism: a View from Within

Huong T. Bui[1]* and Giang T. Phi[2]
[1]Ritsumeikan Asia Pacific University (APU), Japan; [2]VinUniversity, Hanoi, Vietnam

Introduction

Vietnam Tourism: Policies and Practices is a unique work offering an emic view of tourism in Vietnam. Thirteen chapters in the book cover a broad range of tourism development issues with reference to the historical, political, cultural and economic background of the country. The compendium of work in this volume fills an existing gap in tourism publications, where Vietnam is often considered a relatively minor destination within the broader context of tourism in Southeast Asia and where much of the literature overly links tourism to the Vietnam War (1959–1975). More importantly, the research represented in this volume consists of original empirical studies conducted by Vietnamese authors and contributors over a long period of engagement with tourism in the country as government officers, development consultants, tourism researchers, educators and industry workers. The writers are thus able to access government policy papers, industry reports, research documentation, data, and fieldworks in the native language, Vietnamese. This, in turn, ensures the inclusion of relevant and up-to-date insights on tourism policies, structures and governance, various forms of tourism, and the industry's operational research from both theoretical and practical perspectives. The authors are also experienced with global and regional tourism trends and theories owing to their diverse backgrounds in teaching, research and consulting inside and outside of Vietnam. This work is the culmination of their knowledge and experience and their inspiration to embrace both local and global perspectives, representing the viewpoint of a new generation of researchers in the area of Vietnam tourism.

Current Issues in Vietnam Tourism

Vietnam is situated on the east of the Indochinese Peninsula, bordering Cambodia to the west and Laos and China to the north. The territory covers an area of some 327,500 km^2. Its long and narrow territory stretches 1700 km from north to south, and in the centre it spans a mere 50 km from east to west (see Fig. 1.1).

On the one hand, Vietnam possesses 3260 km of extensive coastline, stretching from the Gulf of Tonkin to the Gulf of Thailand with a plethora of fine beaches and archipelagos. On the other hand, mountainous areas cover three-quarters of the land surface, typified by hilly terrains. Vietnam thus boasts a high level of biodiversity. Vietnam ranks as the 16th most biodiverse country in the world (Mittermeier *et al.*, 2004). It is home to several of the world's most iconic species, including the Indochinese tiger (*Panthera tigris*) and the Asian elephant (*Elephas maximus*). Since 1992, four mammal

*Corresponding author: huongbui@apu.ac.jp

DOI: 10.1079/9781789242782.0001

Fig. 1.1. Administrative map of Vietnam. Map courtesy of Thuy T. Duong.

species have been discovered previously unknown to science (Timmins and Duckworth, 2001). Furthermore, 109 large mammals and about 850 bird species have been recorded in Vietnam, which is also home to between 9600 and 12,000 plant species. Vietnam is also known as one of the world's richest countries in agricultural biodiversity, with over 800 plant species cultivated in diversified agro-ecological production (USAID (United States Agency for International Development), 2013).

From a cultural perspective, with over 4000 years of civilization, Vietnam has a long line of cultural heritage ranging from historical sites, monuments, both French colonial and 'Indo-Chinese' architecture, war heritage, spiritual practices and intangible heritages.

Vietnam is also ethnically diverse with a myriad of languages, cultures and traditions belonging to 54 different ethnic groups, most of whom are ethnic minorities living in the high plateau and mountainous regions (Nguyen, 2016). The combination of Vietnam's rich historical, natural and cultural resources forms a solid basis for developing unique tourism products (Jansen-Verbeke and Go, 1995), yet there are arguably significant challenges related to the sustainable use of these resources.

Travel in Vietnam had existed since the feudal era to French colonization, but it was not until the last decade of the 20th century that tourism development accelerated upon the ending of major wars and conflicts (see Chapter 2, this volume). The centralized political and economic system as the main driver of post-war reconstruction left its traces in contemporary tourism governance, policy and planning. However, the market economy with the socialist orientation has largely shaped current tourism development (see Chapter 3, this volume). Tourism has been one of the first economic sectors of Vietnam to open up to foreign investment and privatization in the era after the *Đổi Mới* (Renovation) economic reforms were initiated in 1986. The tourism industry witnessed a transition from the state-owned enterprises in the early 1990s to the privatization of those in the early 2000s and the flourishing of the large domestic and international groups that dominate the hotel industry. The Law on Tourism (originating from the Tourism Ordinance of 1997, then becoming the Tourism Law in 2005, amended to its current form in 2017) highlights the importance of the operating environment in supporting tourism development (Semone *et al.*, 2011). Unlike some other countries in the region, where tourism has been encouraged but primarily left to develop organically, Vietnam has implemented a structured planning approach to the tourism sector, guided by the Master Plans (World Bank, 2019). This centralized planning approach to tourism development has been exemplified in Chapter 4 (this volume). The diversity of tourism products built on various natural and cultural resources of the country is presented from Chapters 5 to 9 (this volume), covering different forms of tourism such as marine and island tourism (Chapter 5), ecotourism (Chapter 6), community-based tourism (Chapter 7), heritage tourism (Chapter 8) and dark tourism at the former battlefields of the Vietnam War (Chapter 9). Due to its tendency to employ high shares of low-skilled, rural, and youth workers, tourism has also positively impacted poverty reduction and has facilitated some redistribution of income from richer to poorer localities in Vietnam (World Bank, 2019). The chapters covered in Part I and Part II of this volume reflect this social viewpoint of tourism development.

Since the beginning of the 21st century, Vietnam has been experiencing a tourism boom, establishing itself as one of South-east Asia's top tourist destinations while achieving record growth in international and domestic visitors. In 2019, Vietnam welcomed 18 million international tourists, a target reached 10 years earlier than planned. In the same year, domestic tourism grew exponentially, reaching 85 million trips and surpassing the target set for 2030 (VNAT, 2020). Tourism's direct and indirect contribution to Vietnam's gross domestic product (GDP) reached 9.2% in 2019, with tourism as the country's single largest service export. The focus of tourism development has also switched from primarily economic objectives to social and environmental sustainability. Among other things, the institutionalization of sustainable and responsible tourism development has subsequently impacted the operation and management of tourism-related businesses. Chapters 10–13 of this volume reflect the embracement of new values in the business operations of the tourism industry. In particular, the tour operators' perception and implementation of responsible tourism practices are presented in Chapter 10. Various questions concerning corporate social responsibility in the current situation of the coronavirus disease 2019 (COVID-19) pandemic are further addressed in Chapter 11. Moreover, the intertwining of policies and stakeholder engagement, reflective of the top-down governance of the event industry, are demonstrated in Chapter 12. Human resource management practices are discussed in Chapter 13.

Vietnam has achieved the most considerable overall improvement in destination competitiveness among its South-east Asian neighbours (i.e. Brunei, Myanmar, Cambodia, Timor-Leste, Indonesia, Laos, Malaysia, the Philippines, Singapore and Thailand) since 2015. According to *The Travel & Tourism Competitiveness Report*

2017 (WEF, 2017), Vietnam improved in the areas of ICT (information and communications technology) readiness, international openness, safety and security, as well as ground and port infrastructure. However, several areas remained relatively behind regional benchmarks, including tourist service, air transport infrastructure and prioritization of the tourism sector (WEF, 2017). As a whole, while the South-east Asian region performed very poorly in the dimension of environmental sustainability, Vietnam ranked 129th globally. This ranking raises critical issues, drawing immediate attention to policy making and implementation. Other concerns about current tourism development in Vietnam are visitor growth, declining visitor yield, and overcrowding pressures at some popular tourist destinations. These issues need to be addressed to ensure that future tourism growth is more sustainable, inclusive and geographically balanced (World Bank, 2019).

The unexpected factor of the global COVID-19 pandemic has had a tremendous impact on the tourism industry in Vietnam. In 2020–2021, there was a paucity of international tourist arrivals and the suspension of many investment and development projects in the tourism sector. Meanwhile, with a population of nearly 100 million, Vietnamese domestic tourism has grown to be a driving force for the tourism industry's recovery during and after the pandemic. The changes caused and accelerated by the pandemic offer an opportunity for those involved in Vietnam tourism to look back, revise and reconsider a more sustainable direction for future growth of the tourism industry in this country (see Chapters 3, 6, 8 and 11, this volume). Under these circumstances, a comprehensive assessment of Vietnam tourism from multiple perspectives, accommodating rapid changes in the operating environment of the industry, is needed and is one of the key contributions made by this book.

Current Research on Vietnam Tourism

Apart from partial demand arising from uncertainty and long-standing problems, there is also a need from tourism academia for a collection of empirical work on tourism in Vietnam published in English. Research on Vietnam tourism has proliferated in the past decades owing to a growing number of students and scholars with tourism backgrounds, along with the dramatic growth of the tourism industry over the past 30 years.

The previous three books on Vietnam tourism were written by American authors and often presented Vietnam in contrast and opposition to the American viewpoint, with excessive reference to the Vietnam War. For example, *Vietnam Tourism* by Arthur Berger (2005) deconstructs the author's perception and experiences of travelling in Vietnam. Being one of the first volumes of Vietnam tourism published in the early 2000s, the data and facts presented in the book have become outdated. The ethnographic approach that the author took limits the exploration of Vietnam tourism from a touristic viewpoint and, therefore, lacks a macro-level analysis of the country's tourism system. Another work, *Tours of Vietnam: War, Travel Guides and Memory* by Scott Laderman (2009), represents Vietnam tourism through the memory of the Vietnam War and is limited to American encounters. From this perspective, the work excessively discusses only one form of tourism: dark tourism experienced by an American tourist. Other modern forms of tourism in Vietnam, and the wide range of stakeholders involved in Vietnam tourism, are absent from the volume. The most recent volume is *Entrepreneurialism and Tourism in Contemporary Vietnam*, written by Jamie Gillen (2016). The author uses a cultural-economic framework to depict everyday life in Vietnam better and demonstrate that relationships are transactional in much more than monetary terms, even among competitors in the tourism industry. The book shifts away from presenting Vietnam as a part of the American history of foreign relations and examines modern Vietnam tourism from the production or supply-side perspective. In contrast, the consumption or demand-side perspective and the issues surrounding tourism policies and governance are neglected.

To a lesser extent, Vietnam has often been presented in collections, confined to one or several chapters of edited volumes on tourism in South-east Asian countries. For example, *Tourism in Southeast Asia: a New Direction* (Chon, 2000) discusses selected aspects of ecotourism, impacts of tourism on local communities, collaborative tourism marketing, and includes one chapter on Vietnam tourism (Mok and Lam, 2000). In the same year, the book *Tourism in South and Southeast Asia* (Hall and Page, 2000) debates tourism development in the region from a variety of perspectives (historical, social, cultural, economic, policy-based, ecological, planning, etc.), also featuring Vietnam tourism policy in the latter half of the 1990s (Cooper, 2000). *Tourism in Southeast Asia: Challenges and New Directions* (Hitchcock *et al.*, 2009) provides a wide range of themes as well as geographical areas of tourism development in the region, with a chapter on cultural politics in the border region of Vietnam and China (Chan, 2009), and Vietnam tourism business in the transitional economy (Bennett, 2009). *Heritage Tourism in Southeast Asia* (King *et al.*, 2010) covers issues in post-war Vietnam related to the tourist gaze at heritage sites (Johnson, 2010), heritage attractions in transition (Suntikul *et al.*, 2010a), and handicraft heritage (Hitchcock *et al.*, 2010). *Tourism and Monarchy in Southeast Asia* (Porananond and King, 2016) includes the topic of travel in monarchical Vietnam (Nguyen, 2016), and managing tourism in the former royal city of Huế (Bui, 2016). The most recent volume on *Tourism and Development in Southeast Asia* (Dolezal *et al.*, 2020) includes a chapter on the economic contribution of ecotourism in Vietnam (Pham and Bui, 2020).

Collections of books on broader Asian tourism have also included chapters about Vietnam, mainly from anthropological perspectives. *Asian Tourism: Growth and Changes* (Cochrane, 2008) is among the very few collections to introduce a historical perspective of tourism in the colonial period (DeWald, 2008). The collection of writing, *Domestic Tourism in Asia: Diversity and Divergence* (Singh, 2009), features domestic Vietnamese travelling for spiritual purposes such as meditation (Alneng, 2009). The book *Asia on Tour: Exploring the Rise of Asian Tourism* (Winter *et al.*, 2009) mentions the marginalization and preconception of domestic tourists in Vietnam in the practices of the travel agency industry (Gillen, 2009). Policies and governance of Vietnam tourism (Truong and Le, 2017) are discussed in *The Routledge Handbook of Tourism in Asia* (Hall and Page, 2017). The issue of managing a United Nations Educational, Scientific and Cultural Organization (UNESCO) World Heritage Site in Trang An (Bui *et al.*, 2018) was featured in the edited volume *Managing Asian Destinations* (Wang *et al.*, 2018).

The past decade has also witnessed a proliferation of publications on Vietnam tourism in various academic journals. Publications on Vietnam tourism in relation to development often focus on sustainability and community-based and pro-poor tourism (Phi *et al.*, 2017; Trupp *et al.*, 2020). Recent work on policies and governance outlines the evolution of policies for tourism development (Truong, 2013; Truong and Le, 2017), with a particular focus on pro-poor tourism (Truong *et al.*, 2014; Truong, 2017). In addition, there is a growing body of work on Vietnam tourism's impacts on society, the economy and the environment (Pham, 2012; Mai *et al.*, 2014; Hampton *et al.*, 2018) as well as work covering tourism social entreprises and social innovation (Phi, 2017; Phi *et al.*, 2016, 2017). Heritage tourism in Vietnam is another well-researched topic (Bui and Jolliffe, 2011a; Bui *et al.*, 2018; Bui and Dolezal, 2020), with much of the work examining the case of Hội An (Bui *et al.*, 2020; Hoang *et al.*, 2020; Thu *et al.*, 2020). Domestic tourism markets (Bui and Jolliffe, 2011b), along with differences within international and domestic markets, are addressed (Bui and Le, 2016). Meanwhile, dark tourism in Vietnam is researched through: (i) the different lenses of ideology and politics (Alneng, 2002); (ii) multifaceted practices of memory and knowledge production (Schwenkel, 2006); (iii) the visitor experience at battlefields (Upton *et al.*, 2018); (iv) migrant experiences of war sites (Le, 2014); (v) landscapes of war (Lema and Agrusa, 2013); (vi) wartime attractions (Henderson, 2000, 2007; Gillen, 2018); and (vii) interpretation of war sites (Ngo and Bui, 2019).

Ecotourism and nature-based tourism have been approached from topics such as governance and policies (Suntikul *et al.*, 2010b), partnership between private and public sectors (Ly and Xiao, 2016), destination planning and management (Buckley, 2004), gender and development (Tran and Walter, 2014), revenue diversification (Pham and Bui, 2020), carrying capacity (Ly and Nguyen, 2017) and changing landscape of the fragile environment resulted from tourism (Bui and Pham,

2021). Research on marine and island tourism emphasizes sustainability in Marine Protected Areas (MPAs) (Lindsey and Holmes, 2002; Mai and Smith, 2015; Le and Ta, 2020; Pham-Do and Pham, 2020), tourism planning (Mai and Smith, 2018), governance (Khuu *et al.*, 2021a, b) and solid waste treatment (Tsai *et al.*, 2021). Besides, there is also research on heritage and tourism development on some islands such as Con Dao (Hayward and Tran, 2014). However, research on industry operations, hotels, events and human resource development has been relatively under-researched. Although the analysis is comprehensive and varied, the scope of a single article or book chapter has limits due to issues of length and focus. A holistic approach to the phenomena of tourism in contemporary Vietnam is the central point and advantage of this volume, covering the existing shortage of comprehensive research on tourism in Vietnam.

Organization of the Book

Vietnam Tourism: Policies and Practices provides a comprehensive review of tourism development in Vietnam. Part I of the volume includes four chapters, introducing and outlining the history of Vietnam tourism and the role and function of the public sector in tourism policy making, planning and governance of the industry. Part II, with five chapters, offers analysis and assessment of various types of tourism, including marine and island tourism, ecotourism, community-based tourism, heritage, and dark tourism. Part III of the book comprises four chapters that examine the practices of hotels, tour operators, events, and the general issues of human resource development and management of the industry. A summary of the parts and their chapters is outlined below.

Part I. Tourism Policies and Practices Across Time and Space

Chapter 1 by Bui and Phi defines the scope of the book with reference to the political, social and economic situation of the country. By reviewing existing work on Vietnam tourism published in English, the authors highlight the issues of minorities and fragmentation with regards to current writing on the topic. This justifies the importance and significance of compiling a compendium of current work on policies and practices of tourism development from various levels of administration across different types and forms of tourism and encompassing diverse segments of the tourism industry.

Chapter 2 looks at tourism development from the Vietnamese feudal time to French colonization, and during the Indochina War and Vietnam War when the country was divided into North and South Vietnam prior to the reunification in 1975. Bui and Bui analyse the historical context of the major forces shaping the country's production and consumption of tourism that have long-lasting impacts on modern tourism. The approach to tourism history, reaching back to the monarchical and colonial periods, is distinguished from previous writing that excessively centres on tourism post-1975.

Chapter 3 by Bui *et al.* reviews tourism policies and governance in contemporary Vietnam. In post-war Vietnam (from 1975 onwards), tourism has changed its mission, from the fulfilment of a political role under the principle of a centrally planned economy to a bridging and supporting role in a transitional period from the centrally planned economy to the market economy. The industry has continued to transform into a strategic economic sector, serving as a springboard for other sectors to grow towards full international integration with high competitiveness. Behind the transitional roles and functions of the tourism industry is the process of institutionalization of tourism, created through laws and regulations. The authors address some critical issues of policy making and implementation such as over-optimistic, quantity-driven planning for tourism development as well as challenges to overcoming the dramatic impacts of the COVID-19 pandemic on the industry.

Chapter 4 exemplifies the process of tourism planning of Đắk Lắk province in the Central Highlands. The authors (Duong *et al.*) highlight the centralized planning approach to resource assessment for tourism product development. A systematic analysis is applied to reorient tourism development by promoting alternative forms of tourism such as ecotourism, cultural, agricultural and wellness forms of tourism. The findings indicate that tourism resources have been vastly underestimated, and the tourism development has not matched the potential of the resources in the region yet.

Part II. Types and Forms of Tourism

Chapter 5, written by Do and Phi, focuses on marine and island tourism, one of the four strategic tourism products. The authors systematically review tourism resources and infrastructure for marine and island tourism in Vietnam and then analyse stakeholder roles and their involvement in policy formulation and implementation. Through an in-depth analysis of two case studies on Côn Đảo and Phú Quốc Island, the authors compare two different approaches to marine and island tourism and address current issues as well as challenges for the future of sustainable development.

Chapter 6 is a compilation of policy papers on ecotourism by Pham *et al.*, providing readers with an overview of ecotourism in forest-protected areas in Vietnam, from theoretical concepts and historical evolution of forest conservation to the current structure of governance and management of forest resources. The work projects ecotourism in connection to economic and socio-cultural perspectives of the country as well as in relation to the behaviours of ecotourists. The review synthesizes existing studies published through various channels both in and outside of the country, along with numerous governmental reports and research conducted by non-governmental organizations (NGOs) on ecotourism, protected areas and forest conservation.

Chapter 7 defines community-based tourism (CBT) as a form of alternative tourism that seeks to engage the host community throughout the planning, development and management of tourism to reduce negative impacts and maximize the positive benefits of tourism for locals. The two authors, Phi and Pham, draw from conversations with CBT stakeholders, participant observation and secondary data to identify challenges facing CBT in the areas of marketing, capacity development and regulations. Concerning opportunities, there is an increase in cross-sectoral collaboration, digitalization in tourism and tourism policies in the (post-) COVID-19 period.

Chapter 8 examines the symbiosis of heritage conservation and tourism development at a UNESCO (United Nations Educational, Scientific and Cultural Organization) cultural heritage site, one of the most popular touristic sites in Vietnam. Bui, Le and Jones present findings derived from a longitudinal study over 5 years spanning before and during the COVID-19 pandemic (from 2016 to 2020). Although Hội An has dramatically benefited from mass tourism, the excessive increase in visitor numbers and concentration in the core heritage zone of the city has negatively impacted the economic wealth of the residents of the ancient town. COVID-19 halted the development of tourism, which attracts attention to the overdependence of the city on tourism, and offers an opportunity for administrators to rethink and revise Hội An's existing tourism policies.

Chapter 9 discusses the current practices of dark tourism conducted in the former demilitarized zone (DMZ), evoking the confrontation of two ideologies between Vietnamese and American troops in the Vietnam War. The two authors, Ngo, Bui and Dimache, examine the structure of the dark tourism industry, issues of governance, promotion and production of dark tourism and the involvement of tourists and tour guides in the historic sites in the DMZ of Quảng Trị province. The analysis of construction and narration at former battlefields offers an in-depth understanding of dark tourism at former war sites in post-war Vietnam.

Part III. The Tourism Industry Practices

Chapter 10 addresses the challenges to implementing responsible tourism among tour operators. The authors of the chapter, Bui *et al.*, propose rethinking responsible tourism via a behavioural approach. Findings outline fundamental factors affecting the intention to implement responsible tourism, including management policies, group dynamics, perceived social pressure, and cultural philosophy on the implementation of responsible tourism. The discussion opens up a discussion on responsible tourism when current practices are highly driven by external social conditions; they do not come from within the awareness of the tourism business.

Chapter 11, written by Nghiem-Phu and Tran, examines how hotel businesses in Vietnam overcome the challenges caused by COVID-19 by transforming lodgings into quarantine facilities. Analysing press and user-generated content, it was found that both positive and negative images were attached to the quarantine hotels. The findings have demonstrated that corporate social responsibility (CSR) could be implemented to sustain business in the current circumstances while preparing for brand strengthening for recovery.

Chapter 12 analyses policies and stakeholders' engagement in the planning and management of festivals and cultural events. The authors, Dang *et al.*, utilize a case-study approach with four vignettes: (i) Gióng Festival (Hà Nội city); (ii) Thành Tuyên Festival (Tuyên Quang province); (iii) Tịch Điền Festival (Hà Nam province); and (iv) the Tea Festival (Thái Nguyên province). Findings from semi-structured interviews with different stakeholders reveal a range of endogenous and exogenous drivers presented on vertical axes for stakeholder engagement. Important findings from this study relate to the process of changing roles of the local community, local authorities and the central state as the events evolve over time.

Chapter 13 on human resource management practices in the Vietnam tourism industry is compiled by Hoang, Do and Le. The authors analyse current practices on recruitment, selection, performance appraisal and labour performance, training and development, and compensation in tourism and its related industries. The outcomes of the analysis are a range of recommendations for the industry in relation to job analysis and design tools to enhance the effectiveness of human resource management across different sectors of the tourism industry.

A Synthesis

The collection of work presented in this book examines Vietnam tourism from both production and consumption sides, both subject to the intervention of the governance of the public sector. The work presented in this book is original and reflects a collection of fieldwork and empirical research conducted by the authors and contributors in their decades of involvement in Vietnam tourism. The authors and contributors have uniquely drawn from their professional and consulting experience to provide insights on current tourism practices, complementing their academic viewpoints. Written and published during a time of uncertainty and rapid change caused by a global health crisis, the COVID-19 pandemic has severely impacted the tourism industry. The contents of this volume update the situation of the past 2 years (2020–2021) regarding the current policies and support the industry and cover how tourism stakeholders reinvent themselves for building resilience as they face constant changes in the markets.

This volume lays the foundation for future research and study of the tourism phenomena in Vietnam and hopes to inspire new thinking that contributes to leading Vietnam tourism towards long-term resilience and success. The book also seeks to become a reliable source of reference for students, researchers, industry practitioners and policy makers interested in modern tourism in the Vietnamese context, with implications for the broader tourism industry in South-east Asia and tourism in developing countries.

References

Alneng, V. (2002) 'What the Fuck is a Vietnam?': touristic phantasms and the popcolonization of (the) Vietnam (War). *Critique of Anthropology* 22(4), 461–489.

Alneng, V. (2009) Zen and the art of tourism maintenance: a meditation on so-called proto-tourism in Vietnam. In: Singh, S. (ed.) *Domestic Tourism in Asia*. Earthscan, London, pp. 59–78.

Bennett, J. (2009) The development of private tourism business activity in the transitional Vietnamese economy. In: Hitchcock, M., King, V. and Parnwell, M. (eds) *Tourism in Southeast Asia: Challenges and New Directions*. NIAS, Copenhagen, pp. 146–164.

Berger, A.A. (2005) *Vietnam Tourism*. The Haworth Hospitality Press, New York.

Buckley, R. (2004) Ecotourism planning and destination management in Vietnam. In: Diamantis, D. (ed.) *Ecotourism: Management and Assessment*. Thompson, London, pp. 313–322.

Bui, H.T. and Dolezal, C. (2020) The tourism-development nexus in Southeast Asia: history and current issues. In: Dolezal, C., Trupp, A. and Bui, H.T. (eds) *Tourism and Development in Southeast Asia*. Routledge, London, pp. 23–40.

Bui, H.T. and Jolliffe, L. (2011a) Heritage and aspects of nation: Vietnam's Ho Chi Minh Museum. In: Frew, E. and White, L. (eds) *Tourism and National Identities*. Routledge, London, pp. 164–175.

Bui, H.T. and Jolliffe, L. (2011b) Vietnamese domestic tourism: an investigation of travel motivations. *Austrian Journal of South-East Asian Studies* 4(1), 10–29.

Bui, H.T. and Le, T.-A. (2016) Tourist satisfaction and destination image of Vietnam's Ha Long Bay. *Asia Pacific Journal of Tourism Research* 21(7), 795–810. DOI: 10.1080/10941665.2015.1075564.

Bui, H.T. and Pham, L.H. (2021) Governance and management of protected areas in Vietnam: nature-based tourism in mountain areas. In: Jones, T., Bui, H. and Appollo, M. (eds) *Nature-based Tourism in Asia Mountain Protected Areas – a Trans-regional Review of Peaks and Parks*. Springer, Cham, Switzerland, pp. 173–196.

Bui, H.T., Le, A. and Ngo, D. (2018) Managing UNESCO World Heritage in Vietnam: visitor evaluation of heritage mission and management of Trang An Landscape Complex. In: Wang, Y., Shakeela, A., Kwek, A. and Khoo-Lattimore, C. (eds) *Managing Asian Destinations*. Springer Nature Singapore, Singapore, pp. 89–106.

Bui, H.T., Jones, T.E., Weaver, D.B. and Le, A. (2020) The adaptive resilience of living cultural heritage in a tourism destination. *Journal of Sustainable Tourism* 28(7), 1022–1040. DOI: 10.1080/09669582.2020.1717503.

Bui, T.T. (2016) Management of world cultural heritage for sustainable tourism in Hue Royal Capital, Vietnam. In: Porananond, P. and King, V.T. (eds) *Tourism and Monarchy in Southeast Asia*. Cambridge Scholars Publishing, Newcastle upon Tyne, UK, pp. 103–117.

Chan, Y.W. (2009) Cultural and gender politics in China-Vietnam border tourism. In: Hitchcock, M., King, V. and Parnwell, M. (eds) *Tourism in Southeast Asia: Challenges and New Directions*. NIAS, Copenhagen, pp. 206–221.

Chon, K. (ed.) (2000) *Tourism in Southeast Asia: a New Direction*. Haworth Hospitality Press, New York.

Cochrane, J. (2008) *Asian Tourism: Growth and Change*. Elsevier, Amsterdam.

Cooper, M. (2000) Tourism in Vietnam: Doi Moi and the realities of tourism in the 1990s. In: Hall, C.M. and Page, S. (eds) *Tourism in South and Southeast Asia: Issues and Cases*. Routledge, London, pp. 167–177.

DeWald, E. (2008) The development of tourism in French colonial Vietnam, 1918–1940. In: Cochrane, J. (ed.) *Asian Tourism: Growth and Change*. Elsevier, Amsterdam, pp. 221–231.

Dolezal, C., Trupp, A. and Bui, H.T. (eds) (2020) *Tourism and Development in Southeast Asia*. Routledge, London.

Gillen, J. (2009) Disruptions of dialectic and a stereotypical response: the case of the Ho Chi Minh city, Vietnam, tourism industry. In: Winter, T., Teo, P. and Chang, T. (eds) *Asia on Tour – Exploring the Rise of Asian Tourism*. Routledge, London, pp. 196–206.

Gillen, J. (2016) *Entrepreneurialism and Tourism in Contemporary Vietnam*. Lexington Books, London.

Gillen, J. (2018) It begins with the bombs: operationalizing violence at a Vietnamese dark tourism site. *Cultural Geographies* 25(4), 519–536.

Hall, C.M. and Page, S. (eds) (2017) *The Routledge Handbook of Tourism in Asia*. Routledge, London.

Hall, C.M. and Page, S. (eds) (2000) *Tourism in South and Southeast Asia: Issues and Cases*. Butterworth Heinemann, Oxford.

Hampton, M.P., Jeyacheya, J. and Long, P.H. (2018) Can tourism promote inclusive growth? Supply chains, ownership and employment in Ha Long Bay, Vietnam. *The Journal of Development Studies* 54(2), 359–376.

Hayward, P. and Tran, G.T.H. (2014) At the edge: heritage and tourism development in Vietnam's Con Dao archipelago. *Journal of Marine and Island Cultures* 3(2), 113–124.

Henderson, J.C. (2000) War as a tourist attraction: the case of Vietnam. *International Journal of Tourism Research* 2(4), 269–280.

Henderson, J.C. (2007) Communism, heritage and tourism in East Asia. *International Journal of Heritage Studies* 13(3), 240–254.

Hitchcock, M., King, V.T. and Parnwell, M. (2009) Tourism in Southeast Asia revisited. In: Hitchcock, M., King, V.T. and Parnwell, M. (eds) *Tourism in Southeast Asia: Challenges and New Directions*. University of Hawaii Press, Honolulu, Hawaii, pp. 1–42.

Hitchcock, M., Nguyen, H. and Wesner, S. (2010) Handicraft heritage and development in Hai Duong. In: Hitchcock, M., King, V. and Parnwell, M. (eds) *Heritage Tourism in Southeast Asia*. NIAS, Copenhagen, pp. 221–235.

Hoang, T.D., Brown, G. and Kim, A.K.J. (2020) Measuring resident place attachment in a World Cultural Heritage tourism context: the case of Hoi An (Vietnam). *Current Issues in Tourism* 23(16), 2059–2075.

Jansen-Verbeke, M. and Go, F. (1995) Tourism development in Vietnam. *Tourism Management* 16(4), 315–321. DOI: 10.1016/0261-5177(95)97356-U.

Johnson, M. (2010) Aspring to the 'tourist gaze': selling the past, longing for the future at the Wolrd Heritage Site of Hue, Vietnam. In: Hitchcock, M., King, V. and Parnwell, M. (eds) *Heritage Tourism in Southeast Asia*. NIAS, Copenhagen, pp. 173–201.

Khuu, D.T., Jones, P.J. and Ekins, P. (2021a) A governance analysis of Con Dao National Park, Vietnam. *Marine Policy* 127, 103986.

Khuu, D.T., Jones, P.J. and Ekins, P. (2021b) Governance analysis of Nha Trang Bay and Cu Lao Cham Marine Protected Areas, Vietnam. *Marine Policy* 127, 104330.

King, V., Hitchcock, M. and Parnwell, M. (2010) *Heritage Tourism in Southeast Asia*. University of Hawaii Press, Honolulu, Hawaii.

Laderman, S. (2009) *Tours of Vietnam: War, Travel Guides and Memory*. Duke University Press, Durham, North Carolina.

Le, C.C. and Ta, C. (2020) The sustainability of marine tourism development in the South Central Coast, Vietnam. *Tourism Planning & Development* (advanced online). Available at: https://doi.org/10.1080/21568316.2020.1837226

Le, K.L. (2014) Cu Chi tunnels: Vietnamese transmigrant's perspective. *Annals of Tourism Research* 46(4), 75–88. DOI: 10.1016/j.annals.2014.02.007.

Lema, J. and Agrusa, J. (2013) Revisiting the war landscape of Vietnam and tourism. In: Butler, R. and Suntikul, W. (eds) *Tourism and War*. Routledge, London, pp. 245–253.

Lindsey, G. and Holmes, A. (2002) Tourist support for marine protection in Nha Trang, Viet Nam. *Journal of Environmental Planning and Management* 45(4), 461–480.

Ly, T.P. and Nguyen, T.H.H. (2017) Application of carrying capacity management in Vietnamese national parks. *Asia Pacific Journal of Tourism Research* 22(10), 1005–1020.

Ly, T.P. and Xiao, H. (2016) An innovative model of park governance: evidence from Vietnam. *Journal of Ecotourism* 15(2), 99–121.

Mai, N.T.T., Rahtz, D.R. and Shultz, C.J. (2014) Tourism as catalyst for quality of life in transitioning subsistence marketplaces: perspectives from Ha Long, Vietnam. *Journal of Macromarketing* 34(1), 28–44.

Mai, T. and Smith, C. (2015) Addressing the threats to tourism sustainability using systems thinking: a case study of Cat Ba Island, Vietnam. *Journal of Sustainable Tourism* 23(10), 1504–1528. DOI: 10.1080/09669582.2015.1045514.

Mai, T. and Smith, C. (2018) Scenario-based planning for tourism development using system dynamic modelling: a case study of Cat Ba Island, Vietnam. *Tourism Management* 68, 336–354.

Mittermeier, R.A., Robles Gil, P., Hoffmann, M., Pilgrim, J.D. and Brooks, T.M. (2004) *Hotspots Revisited: Earth's Biologically Richest and Most Endangered Ecoregions*. CEMEX, Mexico City.

Mok, C. and Lam, T. (2000) Vietnam's tourism industry: its potential and challenges. In: Chon, K. (ed.) *Tourism in Southeast Asia: a New Direction*. Haworth Hospitality Press, New York, pp. 157–162.

Ngo, P.M. and Bui, H.T. (2019) Contested interpretation of Vietnam War heritage: tour guides' mediating roles. *Journal of Tourism & Adventure* 2(1), 61–84.

Nguyen, H.P. (2016) Tourism and monarchy in Vietnam. In: Porananond, P. and King, V.T. (eds) *Tourism and Monarchy in Southeast Asia*. Cambridge Scholars Publishing, Newcastle upon Tyne, UK, pp. 118–141.

Pham, L. (2012) Tourism impacts and support for tourism development in Ha Long Bay, Vietnam: an examination of residents' perceptions. *Asia Social Science* 8(8), 28–39.

Pham, L.H. and Bui, H.T. (2020) Ecotourism and sustainable development in Vietnam's protected areas. In: Trupp, A., Dolezal, C. and Bui, H.T. (eds) *Tourism and Development in Southeast Asia*. Routledge, London, pp. 59–72.

Pham-Do, K.H. and Pham, T.T.T. (2020) Tourism in marine protected areas: a view from Nha Trang Bay, Vietnam. *Tourism Management Perspectives* 33, 100623.

Phi, G. (2017) Social entrepreneurship in Vietnam. In: Scott, N. (ed.) *Managing Growth and Sustainable Tourism Governance in Asia and the Pacific*. World Tourism Organization (UNWTO), Madrid.
Phi, G., Whitford, M. and Reid, S. (2016) What's in the black box? Evaluating anti-poverty tourism interventions utilising theory of change. *Current Issues in Tourism* 21(17), 1930–1945.
Phi, G., Whitford, M. and Dredge, D. (2017) Knowledge dynamics in the tourism-social entrepreneurship nexus. In: Sheldon, P. and Daniele, R. (eds) *Tourism and Social Entrepreneurship*. Springer, New York, pp. 155–172.
Porananond, P. and King, V. (eds) (2016) *Tourism and Monarchy in Southeast Asia*. Cambridge Scholars Publishing, Newcastle upon Tyne, UK.
Schwenkel, C. (2006) Recombinant history: transnational practices of memory and knowledge production in contemporary Vietnam. *Cultural Anthropology* 21(1), 3–30.
Semone, P., Laws, E., Ruhanen, L., Wang, Z. and Noel, S. (2011) Governance approaches in new tourist destination countries: introducing Tourism Law in Cambodia, Laos and Vietnam. In: Laws, E., Richins, H., Agrusa, J. and Noel, S. (eds) *Tourist Destination Governance: Practice, Theory and Issues*. CAB International, Wallingford, UK, pp. 25–39.
Singh, S. (2009) *Domestic Tourism in Asia: Diversity and Divergence*. Earthscan, London.
Suntikul, W., Butler, R. and Airey, D. (2010a) Vietnam's heritage attractions in transition. In: Hitchcock, M., King, V. and Parnwell, M. (eds) *Heritage Tourism in Southeast Asia*. NIAS, Copenhagen, pp. 202–220.
Suntikul, W., Butler, R. and Airey, D. (2010b) Implications of political change on national park operations: Doi Moi and tourism to Vietnam's national parks. *Journal of Ecotourism* 9(3), 201–218.
Thu, T.T., Ryan, C. and Hung, D.B. (2020) Heritage, education and processes of change in Vietnamese rural tourism: a case study from Hoi An. *Journal of Vacation Marketing* 26(3), 378–394.
Timmins, R.J. and Duckworth, J.W. (2001) Priorities for mammal conservation in the ROA. In: Baltzer, M.C., Nguyen Thi Dao. and Shore, R.G. (eds) *Towards a Vision for Biodiversity Conservation in the Forests of the Lower Mekong Ecoregion Complex*. WWF Indochina, Hanoi.
Tran, L. and Walter, P. (2014) Ecotourism, gender and development in Northern Vietnam. *Annals of Tourism Research* 44, 116–130.
Truong, D.V. (2017) Poverty in tourist paradise? A review of pro-poor tourism in South and South-East Asia. In: Hall, C.M. and Page, S. (eds) *Routledge Handbook of Tourism in Asia*. Routledge, London, pp. 121–138.
Truong, D.V., Hall, C.M. and Garry, T. (2014) Tourism and poverty alleviation: perceptions and experiences of poor people in Sapa, Vietnam. *Journal of Sustainable Tourism* 23(7), 1071–1089.
Truong, V.D. (2013) Tourism policy development in Vietnam: a pro-poor perspective. *Journal of Policy Research in Tourism, Leisure and Events* 5(1), 28–45.
Truong, V.D. and Le, A. (2017) The evolution of tourism policy in Vietnam, 1960–2015. In: Hall, C.M. and Page, S. (eds) *The Routledge Handbook of Tourism in Asia*. Routledge, London, pp. 191–204.
Trupp, A., Dolezal, C. and Bui, H.T. (2020) Mapping tourism, sustainability, and development in Southeast Asia. In: Dolezal, C., Trupp, A. and Bui, H.T. (eds) *Tourism and Development in Southeast Asia*. Routledge, London, pp. 2–22.
Tsai, F.M., Bui, T.D., Tseng, M.L., Lim, M.K. and Tan, R.R. (2021) Sustainable solid-waste management in coastal and marine tourism cities in Vietnam: a hierarchical-level approach. *Resources, Conservation and Recycling* 168, 105266.
Upton, A., Schänzel, H. and Lück, M. (2018) Reflections of battlefield tourist experiences associated with Vietnam War sites: an analysis of travel blogs. *Journal of Heritage Tourism* 13(3), 197–210.
USAID (United States Agency for International Development) (2013) *Vietnam Tropical Forest and Biodiversity Assessment: US Foreign Assistance Act, Section 118/119 Report*. August. USAID, Hanoi.
VNAT (Vietnam National Administration of Tourism) (2020) *Vietnam Tourism: Annual Report of 2019*. Labour Publishing House, Hanoi.
Wang, Y., Shakeela, A., Kwek, A. and Khoo-Lattimore, C. (2018) *Managing Asian Destinations*. Springer Nature Singapore, Singapore.
WEF (World Economic Forum) (2017) *The Travel & Tourism Competitiveness Report 2017: Paving the way for a more sustainable and inclusive future*. WEF, Geneva, Switzerland.
Winter, T., Teo, P. and Chang, T.C. (eds) (2009) *Asia on Tour: Exploring the Rise of Asian Tourism*. Routledge, London.
World Bank (2019) *Taking Stock: Recent Economic Development of Vietnam (Special Focus: Vietnam's Tourism Developments)*. World Bank Group, Hanoi.

2 The History and Evolution of Vietnam Tourism

Huong T. Bui* and He T. Bui
Ritsumeikan Asia Pacific University (APU), Japan

Introduction

The early days of Vietnam tourism date back to the feudal period, while modern tourism infrastructure and administration were established during French colonial rule. Overshadowed by nearly 20 years of the Vietnam War and economic struggle in the post-war period, tourism development was halted until its resumption in the early 1990s. Literature on tourism in Vietnam often centres only on the post-unification period (Truong and Le, 2017), which has largely left out the foundation of the system's development, infrastructure and touristic resources. This chapter offers a comprehensive view of the evolution of tourism in Vietnam against the backdrop of the long and complicated history of the country. The history of Vietnam tourism divides into four major periods, tourism in: (i) the monarchical period (from the 10th century to the 18th century); (ii) French colonial Vietnam (from the 19th to the early 20th century); (iii) post-colonial divided Vietnam (from the mid-1950s to the mid-1970s); and (iv) in contemporary Vietnam (from 1975 to the present). This chapter covers the first three periods, while tourism development in the post-1975 period is discussed in Chapter 3 of the volume.

The Monarchical Period

Feudal Vietnam established its monarchy after regaining its autonomy from feudal China with the territory of Đại Việt, along with Champa, Funan (until the 7th century), and Water Chenla (from the 8th to 19th centuries), leaving trails of heritage remaining as present-day tourism destinations. The Đại Việt monarchy began in 938, located in Northern Vietnam bordered by China to the north, Champa to the south and Laos to the west. The nation developed firmly and gradually annexed the kingdoms of Champa and Water Chenla and established a united Vietnam in the early 19th century (VNSSC (Vietnam Social Sciences Committee), 1981).

The establishment of cultural heritage sites

Over ten centuries of development, the Đại Việt's monarchical regime has its capital cities in Hoa Lư (Ninh Bình province), to Thăng Long (present-day Hà Nội), Thanh Hóa and Huế, leaving remnants of outstanding architectural work, such as palaces, citadels, pagodas, and landscapes throughout the north and central regions of the country, including the complex of artistic and religious architecture of Tràng

*Corresponding author: huongbui@apu.ac.jp

 Vietnam Tourism: Policies and Practices (eds Huong T. Bui *et al.*)
DOI: 10.1079/9781789242782.0002

An Landscape Complex in Ninh Bình (UNESCO World Heritage Site (WHS) designated in 2014), the Central Sector of the Imperial Citadel of Thăng Long (WHS designated in 2010), and the Citadel of the Hồ Dynasty in Thanh Hóa province (WHS designated in 2011). The solid stone rampart walled town in the Citadel is the largest and the most profound historical architectural creation of the Vietnamese monarchy and among very few examples of stone architecture remaining in South-east Asia. The last dynasty in the history of the Vietnamese monarchy was the Nguyen (1802–1945), which contributed a series of architectural works to the Complex of Huế Monuments (WHS designation in 1999), and its associated performances, the Vietnamese Court Music (UNESCO Masterpiece of the Oral and Intangible Heritage of Humanity in 2003).

Champa, an ancient kingdom in the centre of Vietnam, was made up of several independent kingdoms, linked by commercial, matrimonial and political ties, namely Amaravati (Đà Nẵng, Quảng Nam, and Quảng Ngãi), Vijaya (Bình Định), Kauthara (Khánh Hòa) and Panduranga (Ninh Thuận, and Bình Thuận). They shared various common cultural traits such as language and religious beliefs (Schweyer, 2011), classified as the Sa Huỳnh culture in official Vietnamese historical narrative (Hung, 2015). Evidence on the origins of Champa and its royal power date back to the end of the 5th century. The dynasty was based in Trà Kiệu, known as Sinhapura 'the Lion City' which arose from the 6th century and remained until the 8th century. The 10th century was the golden age of Champa, reflected in the monuments and sculptures. Around 1025, Đại Việt colonization of territories to the south began, known as 'the March to the South (*Nam Tiến*)', followed by a military campaign against central Champa (Amaravati) in 1044, and a final attack in 1471 at Vijaya (present-day Bình Định). From that time onwards, Champa was limited to the southern province of Kauthara (present-day Khánh Hòa) and Panduranga (present-day Ninh Thuận and Bình Thuận). The reduction of Champa to the southern provinces, the advance of Islam, and new contacts with peninsular Malaysia forced Champa to adapt to the situation (Schweyer, 2011). From the 15th to the 18th centuries, the kingdom of Champa gradually weakened and was annexed by Đại Việt. Early research of Champa by colonial researchers was mostly interested in the Indian aspects of Cham art, architecture and religion (Lockhart, 2011). The Mỹ Sơn Sanctuary (Quảng Nam province) reflects typical Champa royal architecture designated by UNESCO as a WHS in 1999. The Champa Kingdom left its trace in numerous terracotta towers constructed for worship by the Cham ethnic community, such as Po Nagar Tower (Khánh Hòa), Po Klaung Girai Tower (Ninh Thuận) and Po Rome Tower (Ninh Thuận). Archeological evidence suggests that the Cham ethnics were mainly a maritime-oriented people, and Champa was a coastal kingdom (Lockhart, 2011). Hội An, an ancient town located in the coastal plain of Quảng Nam province, was the place where the Cham ethnic group earned their living through fishing and trade between the 2nd and 15th centuries. In the 9th and 10th centuries, Hội An became an important commercial port, and Vietnamese settlers in the 16th and 17th centuries engaged in farming, fishing and handicraft production there. Hội An's identity, its sense of a market economy, and embrace of diversity further flourished during the 17th to 18th centuries, the 'Age of Commerce', when the rulers were dependent upon and eager for foreign trade (Reid, 1993). In 1999, Hội An Ancient Town was designated as a UNESCO WHS under criterion (*ii*), the fusion of culture over time; and criterion (*v*) the well-preserved culture of an Asian trading port.

Funan was an ancient state which developed into a mighty empire from the 1st to the 7th centuries, stretching from northern coastal areas bordering India to the south of Vietnam. Funan was established on the foundations of the Óc Eo culture and included the entire southern delta region. By the 7th century, Funan was annexed by the Chenla dynasty of the Khmer Kingdom. Chenla was then established and divided into two, Land Chenla and Water Chenla, a part of which was uncultivated with sparse population and merged into present-day Vietnam (Ninh, 2006). There are some traces of Funan and Chenla that remain today, such as archeological sites of royal religious relicts and burial tombs in An Giang, Đồng Tháp and Tiền Giang, mainly discovered by French scientists between 1931 and 1943 and later through excavations in the late 1980s by Vietnamese archeologists. The distinctiveness of the Óc Eo culture in the relicts of Funan-Chenla Kingdom form some of the major heritage attractions of southern Vietnam (Hung, 2015).

Characteristics of tourism in the monarchical period

In the monarchy, emperors and the mandarin class travelled on missions assigned by the court, such as royal processions around the country or travelling abroad for diplomatic purposes. Vietnam's feudal dynasties frequently sent envoys to China dedicating a tribune or paying homage, and Chinese envoys also regularly went to Vietnam to enthrone the king or place requirements upon the Vietnamese court. Other forms of travel included those of the noble class on sightseeing tours to enjoy the landscape. Travel activities of emperors were often reflected in their writings, such as poems about the beauty of natural sceneries of visited places. Similar patterns of literary writing on the natural beauty of the country were recorded in numerous texts by feudal mandarins visualizing the routes, places, landscapes and cultural lives of those places (Hung, 2015). In the late 19th and early 20th century, the last emperor of the Nguyễn Dynasty, King Bảo Đại, under the protectorate of the French, developed a series of resorts as his summer palaces and residences across major seaside and mountain areas, modelling the French style.

Religious tourism existed in the form of travel by Chinese and Indian monks to Vietnam or by Vietnamese monks around the country. Historically the most important is the pilgrimage journey of King Trần Nhân Tông (1278–1293) with the establishment of a monastery for a new branch of Buddhism – Trúc Lâm Yên Tử. Christian missionaries also left numerous notes on travelling. Alexandre de Rhodes (1994), after a 3-year mission in Tonkin, wrote the *Report of Tonkin* in 1650. Borri (1998) wrote about a 5-year mission in the *Report of Cohinchine* in 1631. These documents give detailed accounts of the whole country in the 17th century. Writings by Vietnamese Catholic priests recorded the spread of Christianity in Vietnam, stories of religious life, and journeys of foreign and Vietnamese priests (Hien and Duong, 1981).

Commercial tourism refers to the travel of traders from China, India, Champa, Chenla, Siam, Cambodia, Java, Malaysia and from such countries of the West as England, France, America and the Netherlands from the 17th century. Several seaports and river ports were established during the maritime trading period throughout the country, including Vân Đồn (Quảng Ninh), Phố Hiến (Hưng Yên), Thăng Long (Hà Nội), Hội An (Quảng Nam) and Gia Định (Hồ Chí Minh city) (Hung, 2015).

In short, tourism during the monarchical period in Vietnam mainly made by emperors, nobles or officials on duty or for diplomatic trips, as well as activities of the merchant class in international and domestic trade and commerce. Travel for missionaries was also noted. The infrastructure and significant architecture of the period established an important foundation for tourism to be developed, particularly in shaping the identity of the country as a cultural destination.

Tourism in Colonial Vietnam

The French colonization of Vietnam started in 1858, and by 1884, all three regions of Nam Kỳ (Cochinchina), Trung Kỳ (Annam) and Bắc Kỳ (Tonkin) were conquered, forming three *pays* (countries in French) under different political regimes. The French consolidated their power in the region by creating the Indochinese Union in 1887 headed by a governor-general, gradually incorporating Vietnam, Cambodia, Laos and Chinese Kouang-Tchéou-Wan. The French Indochinese Union's various territories did not carry the same political weight. There was only one colony, Cochinchina, the economically dominant territory. Tonkin was the most powerful as the seat of the General Government of Indochina. Annam and Cambodia retained their royal power that had no political function. Laos, a country with less political weight and more strategic value, served as a buffer zone. Demographically, Cochinchina was the largest centre of European settlement, followed by Tonkin, Annam, Cambodia and Laos (Demay, 2014).

Tourism was a major and indispensable player in colonial policies in four ways. First, tourism helped produce a country's identity and its portrayal, presenting the colony in a fun and entertaining way. Second, tourism worked to master the territory in spatial terms, primarily through the construction of transportation networks and accommodation. Third, tourism contributed to the development of society by increasing the number of colonists and developing social ties and shared values. Finally, tourism was a component of a country's economic activity, bringing foreign currency to the colony (Demay, 2014). Tourism increasingly influenced political and economic changes in colonial Vietnamese society. From 1898 to 1940, tourism in Indochina developed immensely, divided into three periods: (i) 1898–1920: the transfer of tourism practices and the building of hill stations; (ii) 1920–1930: the development tourism of Indochina; and (iii) 1930–1940: promoting tourism abroad and within the colony (Demay, 2014). The last period (1930–1940) was a turning point for the travel culture of Vietnam (DeWald, 2008).

Heritage conservation

The development of tourist activities was coupled with a preservation policy with a twofold aim: preserving sites for historical and environmental reasons while also safeguarding a range of tourist sites (archeological, historical, natural and hunting sites). The preservation of monuments served a dual purpose: (i) it ensured the suitability of tourism operations; and (ii) helped to legitimize French colonization by providing a scientific sanction (Demay, 2014). Monumental archeology, increasingly linked to tourism, allowed the state to appear as the guardian of a generalized and local tradition. The old sacred sites were to be incorporated into the map of the colony, along with their ancient prestige (Anderson, 2006).

Transportation

The railway networks

The first railroad was constructed in 1885, linking Sài Gòn to Mỹ Tho. However, it was not until 1898 that Governor-General Doumer launched a comprehensive railroad network programme. It comprised a dorsal line (the *Transindochinois*) stretching from Lạng Sơn to Mỹ Tho and three transversal lines throughout the country. The northern section from the Chinese border to Tourane (present-day Đà Nẵng), constructed between 1902 and 1928, served Hà Nội, Nam Định, Vinh, Huế and Tourane (present-day Đà Nẵng) (see Fig. 2.1). The southern section completed in 1913 connected Mỹ Tho with Nha Trang. The last section from Tourane to Nha Trang was put into service in 1936, allowing the direct travel from Hà Nội to Sài Gòn by bridging the extant gap between the two previous sections (Gourdon, 1931). By facilitating passengers' travel, the *Transindochinois* also brought potential customers to those future tourist hubs (Maignan, 1936).

Road networks

In 1912, Governor-General Sarraut initiated the construction of an extensive road network. By 1926, the road network covered more than 3000 km. Colonial Road No.1, running from the Chinese border to the Siamese (present-day Thailand) border, followed the ancient Mandarin Road opened by the Vietnamese monarchy in previous centuries, linking the four capital cities of Tonkin, Annam, Cochinchina and Cambodia. Non-colonial roads were developed in every '*pays*' of the colony, reaching remote areas especially the hinterland of Annam (Gouvernement général de l'Indochine, 1926). The development of road networks boosted car transportation. In 1931, there were 20,000 cars in Indochina; a quarter belonged to transportation companies (Gourdon, 1931). In many cities in Annam, there were private companies, either European or native, that offered car rental services (Bureau du Tourisme de Hue, n.d.).

Water lines

Regular service of mail steamers and mixed liners connected Indochina to the outside world via major ports from Hải Phòng in the north to Sài Gòn in the south (Gouvernement général de l'Indochine, 1926). Ships of the French companies *Messageries Maritimes* and *Chargeurs Réunis* linked Marseille, Sài Gòn and Hải Phòng and also reached China and Japan. Other regional lines connected Tonkin with Hong Kong, Sài Gòn with Bangkok, Java and Swatow,

Fig. 2.1. The map of *Transindochinois* railroads in 1925. From I2898, Vietnam National Archives Center No.1.

and commercial lines joined Indochina to the USA, the Philippines, Tahiti and New Caledonia (Gourdon, 1931). In 1931, it took: (i) 24 days to travel from Marseille to Saigon via Djibouti, Colombo, and Singapore; (ii) 13 days from Yokohama via Kobe, Shanghai, and Hong Kong; (iii) 4 days from Batavia; and (iv) 27 days from San Francisco (Sylvestre de Sacy, 1931). The major ports of Sài Gòn, Đà Nẵng and Hải Phòng offered services to many smaller coastal ports running the length of the country (DeWald, 2008).

Air transportation

In 1928, an office of air travel was established by the General Government of Indochina, responsible for improving infrastructure for air travel (Agence économique de l'Indochine, 1937). In 1929, the first travel by seaplane between Saigon and Angkor was carried out in 4 hours, marking the inauguration of air travel within Indochina (Bontoux, 1929). Until 1937, with the exception of Haut-Laos, civil airfields (98 in total) were located throughout the colony: 30 in Annam, 21 in Tonkin, 18 in Laos, 16 in Cambodia, 11 in Cochinchina, and two in Kouang-Tchéo-Wan (Agence économique de l'Indochine, 1937). Outside the colony, the Compagnie Air-Orient, which operated a flight route from Marseille to Beirut, prolonged this line to Sài Gòn via Iraq, Persia, India, Burma and Siam (Gourdon, 1931). In 1933, the French pilot René Lefèvre finished his journey from Paris to Sài Gòn in 10 days and 7 hours on board of a Mauboussin plane.

Until the first half of the 20th century, Indochina was equipped with modern transport infrastructure, which linked all countries of the colony with each other and connected the colony to the outside world. Tourism benefited from this vast infrastructure, and thousands of workers migrated from the north to the south and vice-versa using this form of commuting (Goscha, 1996).

Development of resorts

Resorts constituted a cornerstone of the colonization process, addressing economic, health and strategic needs simultaneously. Accustomed to temperate weather, the tropical climate had debilitating effects on Europeans; they found it harmful and dangerous to their health. Therefore, as soon as they arrived in Indochina, the French military authority was interested in finding appropriate places in the colony which could provide the troops with an environment similar to the European climate zone so that they could be reinvigorated without returning to France, the metropole (Glaillard, 1919). To maintain their presence, colonizers developed new areas conducive to recreating the metropole's environment and climate. Resorts were linked with the emerging communication network (Demay, 2014).

Hill stations

Hill stations were European inventions to maintain health for military personnel negatively affected by the tropical weather. The Indochinese government adopted the Indian models for establishing hill stations. As a result of special instructions given by the Ministry of Colonies, the Governor-General of Indochina required a military delegation to find an appropriate place in the Tam Đảo mountain range in Tonkin to create a sanatorium in 1904 (Coronnat, 1904). In search of a sanatorium at a greater elevation, near a railroad, and on a larger plateau which allowed for the construction of necessary buildings, the residents of Lao Kay (present-day Lào Cai) proposed the plateau Sa Pa (Tourrès, 1909). In addition to Tam Đảo and Sa Pa, Ba Vì in Tonkin was studied as a location for establishing a hill station in the 1940s. However, the project was soon abandoned due to a lack of financial resources and political turbulence (Tucat, 1942).

In Annam, following the exploration of the Lang-Biang plateau (Đà Lạt) by a French doctor, Yersin, the city of Đà Lạt was developed into a European summer capital. A comprehensive planning map for Đà Lạt was made in 1899 (Glaillard, 1919). However, it was not until 1916 that construction was initiated, mainly driven by Governor-General Roume (Gaide, 1930). Đà Lạt underwent successive modifications in terms of urban planning and transportation (see Fig. 2.2) and was frequented by visitors from throughout Indochina, even during the Vietnam War (Cindy, 2013). At the turn of the century, Bà Nà (in present-day Đà Nẵng) was discovered by Captain Debay, who had been dispatched by

Fig. 2.2. Artificial lake and golf club in Đà Lạt (date unknown). From GGI158, Vietnam National Archives Center No.1.

Doumer to search for a sanatorium near Đà Nẵng or Huế. The project, however, was delayed until 1916, following exploration led by the Director of Health in Annam. Construction of buildings started in 1919 (Gaide, 1930). Almost two decades later, the Bạch Mã sanatorium was instituted and established in 1932. The site continued to be developed until 1945 (Fife, 2009).

Seaside resorts

While high mountains required careful investigation before any attempt to develop a hill station, the coastal area appeared to be a feasible solution in the short term. Unlike hill stations, few seaside resorts were created by state demand, with only three falling under this category: (i) Cap Saint-Jacques (the present-day Vũng Tàu); (ii) Sầm Sơn (as a seaside sanatorium in Thanh Hóa); and (iii) Kep (in Cambodia). Cap Saint-Jacques near Saigon was the first seaside resort created in Indochina. In 1894, the area became an urban centre, and public works were carried out to turn it into an important port and city. A sanatorium was constructed, offering a low price to patients and convalescents (Picanon, 1896). In Tonkin, Đồ Sơn (Hải Phòng) was a proposed location for development from the turn of the 20th century on. However, tourists rapidly left Đồ Sơn for Sầm Sơn, which gained more attention from the government. Tourists interested in spending their holidays in a seaside resort had many other choices, for example Hòn Gai in Quảng Yên, Cửa Lò (Vinh), Cửa Tùng (Quảng Trị), Thuận An and Lăng Cô (Huế), Mỹ Khê (Đà Nẵng) and Nha Trang (Gaide, 1930). The proliferation of seaside resorts was due to their easy accessibility and low-cost planning (Demay, 2014).

In short, hill stations and seaside resorts were selected via very different processes due to the geo-climatic characteristics of each environment. There were, however, certain common criteria for selecting potential sites: (i) accessibility; (ii) size; (iii) safety; and (iv) the presence of clean drinking water. Lower temperatures, due to altitude, favoured the mountain environment for the construction of hill stations. Authorities thus focused greater attention on them, especially since their construction costs were very high. Seaside resorts were ignored and deemed secondary (Demay, 2014).

Accommodation

In 1922, Governor-General Maurice Long wanted to equip the colony with comfortable

Fig. 2.3. Lang-Biang Palace, Đà Lạt (date unknown). From GGI158, Vietnam National Archives Center No.1.

hotels by searching for private investment. His initiative was unsuccessful due to a lack of faith in the future of tourism in Indochina (Agence économique de l'Indochine, 1930). In 1924, the *Société des Grands Hôtels Indochinois* (SGHI) was created to manage hotel properties. In Saigon, the most ancient and famous hotel was the Continental Palace, with 140 rooms. In Đà Lạt, the Grand Hôtel, the Lang-Biang Palace (see Fig. 2.3) and the Hôtel du Parc could receive between 300 and 350 guests (Agence économique de l'Indochine, 1930). In Bà Nà, there were many villas, chalets and a hotel (Agence économique de l'Indochine, 1924). One could find well-furbished hotels and bungalows along the Colonial Road No.1, in Tuy Hòa, Quảng Ngãi, Đồng Hới, Phan Thiết, Nha Trang and Quy Nhơn (Agence économique de l'Indochine, 1933). Many hotels in Tonkin provided hot water during wintertime. In Indochinese hotels, one could find spacious rooms, very open and ventilated, and equipped with completely private bathrooms. Good French cuisine, high-quality wines and liquors were served. Most large hotels in Indochina accepted travellers' cheques and coupons from leading travel agencies (Office Central du Tourisme Indochinois, n.d.). In 1938, there were more than 150 hotels in Indochina, with most of the largest ones being located in Vietnam. State-owned hotels were given to concessionaires under contract for operation with or without subsidy (Office Central du Tourisme Indochinois, 1938a).

Tourism governance and promotion

The governance of colonial tourism

French historians wrote about the emergence of a Paris-centred imperial tourism lobby for Indochinese tourism (Demay, 2014). This lobby was composed of metropolitan shipping companies (the *Messageries Maritimes* and the *Chargeurs Réunis*) and French tourism advocacy groups such as the *Touring Club de France*, joining the hotel industry in Indochina to form the first official tourism bodies. Inspired by tourism development in France and other countries in Western Europe and North America, this grouping advocated for tourism development in colonial states partly for their business interests. As early as 1917, the Governor-General of Indochina, Albert Sarraut, had given his support to tourism's development, and later as Minister of Colonies, he incorporated this new sector into his plan for the French colonial empire (Demay, 2014). A consensus was thus achieved, one that concentrated on the development of an internationally oriented tourism marketplace. In 1923, a Central Tourism Office (the *Comité Central du Tourisme*) was established and responsible for giving recommendations on tourism planning and tour programmes. The governor-general was the President, supported by Chiefs of Local administrations, the Director of Finance, the Inspector general of Public Works, the Director of Economic Services, the Director of the *Ecole Française d'Extrême-Orient*, the Chief Architect, and representatives of *Syndicats d'Initiative* and three members assigned by the governor-general representing the press, the hospitality industry and shipping companies (Agence économique de l'Indochine, 1923). Under this strategic consensus, the metropolitan tourism lobby and their colonial offspring (the railways, automobile transportation companies and hotel firms) took the lion's share of state assistance.

The *Office Central du Tourisme* was subordinate to the Department of Economic Services, created in 1923 and renamed as the *Office Indochinois du Tourisme* in 1928 and then as the *Office Central du Tourisme Indochinois* (OCTI) in 1935 (Office Central du Tourisme Indochinois, 1938a). This agency was in charge of supervising tourist activities in the colony, disseminating information intended to promote sites and monuments of the country, programing itineraries, and carrying out all activities in favour of tourism development in Indochina. From 1935 to 1937, information on hotels, transportation networks and tourist sites, among other things, were recorded and published in books, posters and advertisements in French, English and Dutch and distributed in many Asian cities such as Hong Kong, Singapore, Penang and Kuala Lumpur, among others (Office Central du Tourisme Indochinois, 1938a). In 1937, 10,000 leaflets entitled 'French Indochina', 10,000 tourist maps (in French, English and Dutch), 30,000 leaflets of Angkor accompanied with maps, 10,000 brochures of *La Route Mandarine* in French, and English, and 20,000 maps of Sài Gòn had been issued. The OCTI also held or participated in many exhibitions and fairs in cities such as Batavia, Sài Gòn, Hà Nội, Hải Phòng and Phnom Penh and exchanged correspondence with foreign partners (Office Central du Tourisme Indochinois, 1938a).

Other agencies

Syndicats d'Initiative (Tourism Information Office) were non-profit organizations created to promote tourism in Indochina. These offices were established in Sa Pa, Tam Đảo and Hà Nội and were grouped under the name of *Union des Syndicats Touristiques du Nord-Indochine* (Federation of Tourist Unions of North Indochina) to promote tourism of the Tonkin throughout Indochina. The Union comprised four special commissions in charge of advertising, hospitality, sites and transportation, respectively. Besides distributing information intended to make the Indochinese colony famous worldwide, the Union acted as a bridge between different tourist organizations inside and outside the colony, involved public authorities and transport providers in its projects as well as organizing tours inside Indochina. Đà Lạt had its own tourist information office established in 1936 (Office Central du Tourisme Indochinois, 1938a) (see Fig. 2.4).

The *Bureau du Tourisme* was founded in 1923 under the patronage of the *Syndicat d'Initiative de l'Indochine* and was responsible for advertising colonial tourism (see Fig. 2.5). In 1923, the Bureau had collected comprehensive documentation of voyages in Indochina and provided useful information, and organized tours in every area of the colony. However, it did not have the budget to conduct advertising activities. In May 1923, the Director of the Bureau, with financial help from the *Syndicat d'Initiative* and the *Automobile Club de Cochinchine et du Cambodge* issued the first number of the *Revue du*

Fig. 2.4. The cover page of the brochure *Dalat, Indochine Française* (*Syndicat d'Initiative de Dalat*, date unknown). From GGI158, Vietnam National Archives Center No.1.

Tourisme Indochinois. A year later, 30,000 copies of the guidebook had been distributed in the Far East, France and its colonies, in the UK, the USA and Australia. Two thousand copies of the first guidebook edited by the Bureau were sold (Syndicat d'Initiative, 1924). In the early 1920s, state agencies were used to collect information and played the role of an oft-neglected adviser. At the end of the 1920s, private organizations came together to create a travel agency. In 1936, the arrival of the Thomas Cook & Son agency reflected the growth of Indochina as a destination (Demay, 2014).

Tourism exhibitions

The Department of Economic Services of Indochina was in charge of organizing and preparing fairs and exhibitions held both within and outside Indochina (Direction des Affaires économiques de l'Indochine, 1925). In the first four decades of the 20th century, the

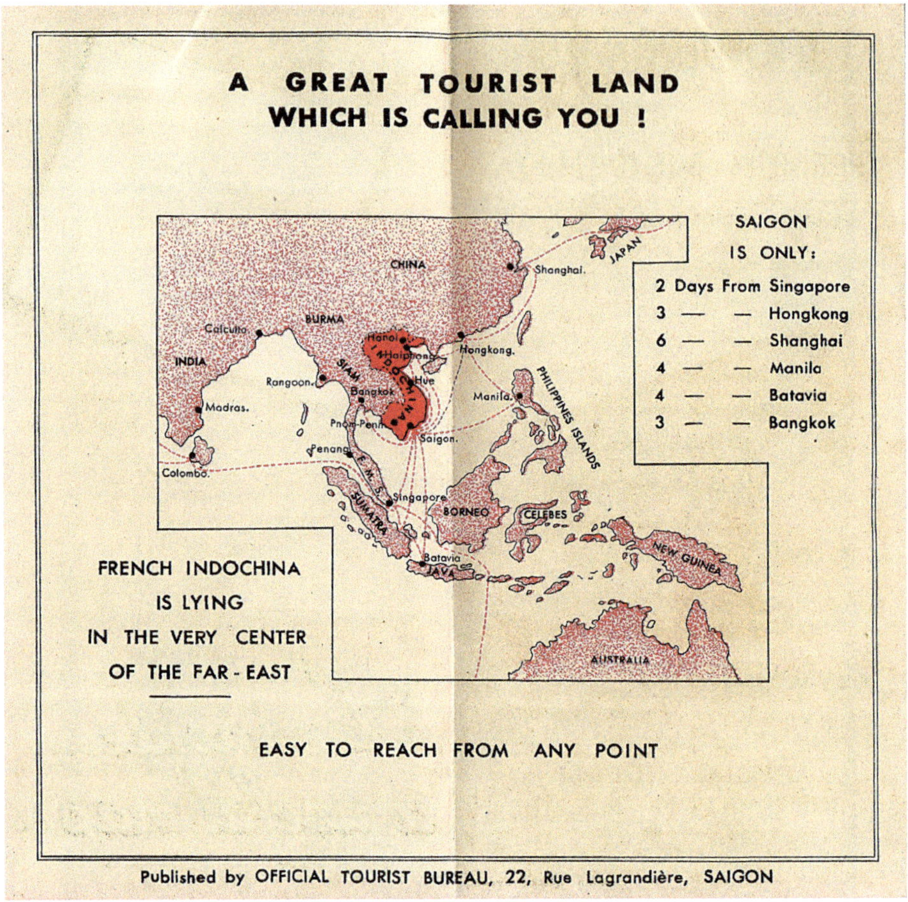

Fig. 2.5. A brochure by the *Bureau du Tourisme*. From TĐNK5948, Vietnam National Archives Center No.2.

Indochinese colony had participated in dozens of exhibitions in Asian, European and American countries, both as participant and organizer. For the international exhibition taking place in San Francisco in 1939, the colonial government built a pavilion that housed three sections: (i) tourism; (ii) hunting; and (iii) crafts (Gouverneur général de l'Indochine, 1938). The tourism section also experienced exceptional success, visited daily by American guests who showed a strong desire to go on a trip to Indochina. Charming vendors dressed in Vietnamese costumes constituted a complementary aspect of their participation which was an advertisement in favour of Indochina. This advertisement was reinforced by journal articles, radio talks, conferences, and initial contact between exhibition organizers and leaders of shipping companies and travel agencies (Sans, 1939).

However, prior to this, the colonial exhibition held in Paris in 1931 was an international event that had been the first of its kind in which Indochina participated in an extremely brilliant way (Aubert, 1929, 20 June). The exhibition would be a demonstration of the colonial effort of France in every field. Therefore, all services in the colony were required to contribute to the success of Indochinese participation in this exhibition. The Indochinese section was expected to include a central palace that displayed general services of the colony and some particular pavilions, which showed the five '*pays*' of the Union and the territory of Kouang-Tchéou-Wan (Aubert, 1929, 20 June). General exhibitions staged all forms of administrative, social, intellectual and economic activities of the colony, thus providing an overview of its development and offering a comprehensive description of the French work in Indochina. On the other hand, local exhibitions showed the life and civilization of each part of this Far East section of the French colonial empire (Gouvernement général de l'Indochine, *c.*1930).

Guide books and documentaries

The first guide book, *Guide du Voyageur*, that received official support from Governor-General Paul Doumer was published in 1902. Five years later, Governor-General Paul Beau collaborated with Madrolle to bring out the guidebook *Tonkin du Sud* (Fife, 2009). In 1936, the resident-superior in Annam ordered provincial leaders to prepare a travel brochure sold at low prices at tourism bureaus, hotels, bungalows, guest houses and railway stations (Résident supérieur en Annam, 1936). Travel agencies also contributed to promoting Indochina. In 1937–1938, the *Compagnie des Wagons-Lits* suggested a campaign that mainly included disseminating a brochure and a leaflet containing comprehensive documentation about Indochina, including technical details relating to the visit to the colony in all agencies of Wagons-Lits/Cook, and on board ships and planes coming to Indochina. Besides, posters showing the beauty and tourist attractions of Indochina were expected to be displayed permanently in large agencies of Wagons-Lits/Cook in Europe, Africa, Asia and America (Compagnie internationale des Wagons-Lits and des Grands express européens, 1937). Despite the efforts of public and private organizations to promote tourism in Indochina, the colony remained relatively unknown in the Far East and the USA (Inspecteur général des Travaux publics, 1935). In a report dated 25 March 1935, its author highlighted the non-existence of French activities intended to promote the Indochinese colony in the Dutch Indies. French books could not compete with their German and Italian counterparts, while French airlines were known for accidents and shipping routes for boat fires (Chauvet, 1935).

Tourist market and behaviour

In the early stages of colonization in Vietnam, participants in the tourism market were mainly French and from other European countries. Rondet-Saint's report in 1913 stated that Indochina had great potential for the tourism industry, and the colony was located in a region where tourist intensity was considerable, although there were only 150 tourists who arrived in Indochina (Rondet-Saint, 1913). His report gained the attention of the Ministry of Colonies, especially of the Committee of Colonial Tourism, who published a book on Indochina and made contact with travel agencies in 1914 in search of cooperation activities. In the same year, the Committee received an answer from the American company, EMS Hall, showing interest in organizing tours to Indochina (Tran, 2010). In 1935, the *Compagnie des Wagons-Lits* proposed to operate dining cars and bunks on some rail networks (Inspecteur général des Travaux publics, 1935).

In 1935, Alcan, delegate of the *Automobile Club Sud Indochinois* in Malaysia, recommended that the administration takes into consideration two types of foreign visitors: (i) those of grand cruises; and (ii) inhabitants of neighbouring countries. The former generally came in large numbers to visit Angkor. Due to a lousy organization and high price, they often made the round trip via Aranya in Siam (present-day Thailand) instead of Sài Gòn or via the coast of the Gulf of Siam. From January to April 1935, four cruises had visited Malaysia, Java and the Siamese coastline and sent between 300 and 500 passengers to Angkor, including *The RMS Asturias* (British), *The Empress of Britain* (Canadian), *The Resolute* (German-American) and *The SS Arandora Star* (British). The latter were citizens of Malaysia who were discouraged by the price of transportation between Sài Gòn and Singapore (Alcan, 1935).

Smaller ventures entered the tourism trade market without the direct and indirect support of the colonial tourism lobby. However, they did not possess the resources to compete effectively for the interests of international tourists. The prevailing wisdom about tourism's cosmopolitanism and its modernity excluded most of these enterprises from the market for international tourists. This left many businesses, especially Asian-run ones, to concentrate on the interests of other travellers, especially domestic tourists. As early as 1925, a committee of the Tourism Promotion Board observed the potential for growth in domestic tourism markets (Demay, 2014); the Board, however, continued to focus on the promise of the international tourism market. This left considerable room for smaller businesses to cater to domestic tourists. Such businesses publicized their services (mainly package tours) through advertising in Vietnamese and French newspapers and magazines.

The tourism industry gradually encouraged the participation of Vietnamese enterprises. In 1931, *Khanh Ky et Cie* organized a tour to France on the occasion of the colonial exhibition. After the success of this tour, *Khanh Ky et Cie* opened an office in Paris and cooperated with Thomas Cook & Son to provide tours to Paris in May and November 1937 (Tran, 2010). A letter from the OCTI dated 30 June 1938 provided a list of sightseeing tours carried out by the OCTI, in which the name of many Vietnamese service providers appeared. The OCTI was optimistic about domestic tourism and the emergence of native travel agencies. However, it preferred the expansion of several companies that had gained the public's trust by offering good service to the creation of native travel agencies in large numbers (Office Central du Tourisme Indochinois, 1938b).

By the late 1930s, however, the growth of the domestic market led to questions not only about the international focus of the state's agenda but also about heavy-handed state intervention in achieving that focus (Demay, 2014). Tourism between the two world wars was in two directions: the perceived exigencies of colonial development propelled the state's international tourism agenda, while domestic tourism grew in its shadow. It was the advent of World War II that, ironically, led the colonial state to actively advocate for the growth of domestic tourism. As the flow of international tourists dwindled after 1940 and the war prevented any plan for achieving economic growth through tourism, the wartime government in occupied France – the Vichy regime – continued to pursue tourism in Indochina for didactic reasons: the loftier goal of popularizing the state's ideology of work, family and nation. For example, in spite of the austere economic conditions, the colonial state continued to fund hill-station holidays for state employees, who benefited from this ideology (DeWald, 2008).

In summary, the geo-historical analysis of tourism in French Indochina demonstrates colonial authorities used tourism to contribute to the welfare of Indochinese society, especially the colonial community, to spur the colony's economic development and contribute to the manufacture of an Indochinese identity (Demay, 2014).

Post-colonial Tourism in North and South Vietnam (1954–1975)

Decolonization in 1954 was accompanied by the departure of most of the colonial community and the withdrawal of French administration from Northern Vietnam. It was followed a few years later by the outbreak of the Vietnam War (1959–1975), also known as the Second Indochina War. Having been divided after 1954 and following different political ideologies, North and South Vietnam embarked on a 20-year conflict. The Vietnam War was complicated by the involvement of American forces and alliances on both sides from 1965 to 1972. Both North and South Vietnam governments, during wartime, did not seriously comprehend nor sustain the development of the tourism industry.

Tourism in South Vietnam

Following the 1954 Geneva Accords, Vietnam was divided into two countries, and the Republic of Vietnam (RoV), known colloquially as South Vietnam, transformed from a French colony to a US-backed state. The National Tourist Office carried on from the colonial administration to the RoV. Tourism came to assume a notable role in the Sài Gòn government's cultural diplomacy campaign. Tourism carried the potential to generate positive sentiments about the southern state that would translate into popular support and diplomatic goodwill, serving both the Americans and their clients. The growth of tourism in the late 1950s in South Vietnam coincided with Washington's increased interest in Asia (Laderman, 2009). Travel was a political activity recognized by both civilian and military officials. The year 1957 recorded the arrival of 13,250 international tourists, which had nearly doubled in 1960 to 24,265 and surpassed 72,000 in 1970, mainly from the USA (Laderman, 2009). Destinations in the south, such as Đà Lạt, Nha Trang and Huế were reachable from Sài Gòn by air transport and established road networks. Pan American Airways (commonly known as Pan Am) was the only US airline getting tourists to Vietnam. The intensification of the war in the late 1960s, the withdrawal of US troops in 1973, and the collapse of the RoV in 1975 marked the short life of tourism in South Vietnam.

Tourism in North Vietnam

Under the political agenda of North Vietnam, international tourism served a political purpose. Thus total international tourist arrivals were very limited, numbering only 6130 in 1960, growing modestly to 36,910 in 1975 (Thanh, 2005). The Vietnam Tourist Company was set up in 1960 in the north, and it was placed under the Ministry of Foreign Affairs and managed by the Ministry of Public Security (Thanh, 2005). Although some tourist sites were already in place (e.g. Hà Nội, Tam Đảo), economic benefits were subordinated to national security. The Vietnam War ended in the mid-1970s, leaving severe damage to the economy, which was also worsened by the economic sanctions that were imposed by the American government. As a result, the development of tourism was not on any economic agenda. The Vietnam National Administration of Tourism was established in 1978 to manage all tourism activities nationwide. Foreign visitors were still few in number (e.g. 41,110 in 1980) (Thanh, 2005), and these mainly consisted of those coming from the then Soviet Union. It was not until 1986 when the *Đổi Mới* programme was adopted that Vietnam shifted from a centrally planned economy to a market-oriented one, where tourism was gradually recognized as an economic sector in its own right (Truong and Le, 2017).

Conclusions

This chapter reviews the evolution of tourism in Vietnam from the feudal age through the colonial period starting from the late 19th century and early 20th century. A brief period of post-colonial tourism in the 20 years of the Vietnam War was ended by the reunification in 1975. The form of tourism in the feudal time was noted for diplomatic missions marking the early days of international travel for business purposes. Numerous cultural destinations remain from this time associated with different cultural and commercial activities. French colonial administrators introduced tourism for the well-being of colonists and as an economic vehicle to achieve the colonial civilization mission. The establishment of hill stations, seaside resorts, commercial hotels and agencies, and the establishment of travel markets and related agencies laid the foundations for modern tourism development. After the withdrawal of the French from Vietnam, the legacy of colonial tourism was carried on in South Vietnam until the end of the Vietnam War. Though tourism was not a target for development in both the North and South due to the Vietnam War escalation, travel activities for political purposes did exist to limited extents in the divided country. The destinations and the infrastructure have continued to inform present-day tourism development established after 1975 outlined in other chapters in the book.

Acknowledgements

The work presented in this chapter is supported by Japan Society for the Promotion of Science (JSPS) research grant number 18H03462.

References

Agence économique de l'Indochine (1923) *Communiqué No.64 de la Presse indochinoise*. Département Droit, économie, politique (JO-60850), 30 September. National Library of France, Paris.

Agence économique de l'Indochine (1924) *Communiqué No.82 de la Presse indochinoise*. Département Droit, économie, politique (JO-60850), 31 July. National Library of France, Paris.

Agence économique de l'Indochine (1930) Le progrès de l'industrie hôtelière en Indochine française. *Bulletin de l'Agence économique de l'Indochine* 25, 91–95.

Agence économique de l'Indochine (1933) Le tourisme en Indochine. *Bulletin de l'Agence économique de l'Indochine* 61, 295–302.

Agence économique de l'Indochine (1937) L'Aviation en Extrême-Orient: l'œuvre aéronautique accomplie en Indochine. *Bulletin de l'Agence économique de l'Indochine* 94, 33–37.

Alcan (1935) *Lettre la Direction des Affaires économiques et administratives*. Résidence supérieure au Tonkin fonds (RST72710-01). National Archives Center No.1, Hanoi.

Anderson, B.R.O.G. (2006) *Imagined Communities: Reflections on the Origin and Spread of Nationalism*, revised edn. Verso, London.

Aubert (1929, 20 June) *Note de service*. Gouvernement général de l'Indochine fonds (GGI448). National Archives Center No.1, Hanoi.

Bontoux, H. (1929) *L'inauguration du tourisme aérien en Indochine. De la rivière de Saigon aux douves d'Angkor-Vat*. Les Editions d'Extrême-Asie, Saigon.

Borri, C. (1998) *Cochinchina in 1621 (Xứ Đàng Trong năm 1621)*. Tổng Hợp, Ho Chi Minh City.

Bureau du Tourisme de Hue (n.d.) *Notice touristique sur l'Annam*. Imperimerie d'Extrême-Orient, Hanoi.

Chauvet, L.H. (1935) *Lettre au Ministre des Affaires étrangères au sujet de la propagande étrangère et française aux Indes Néerlandaises*. Gouvernement général de l'Indochine fonds (GGI5760), 25 March. National Archives Center No.1, Hanoi.

Cindy, A.N. (2013) Where people and places meet: travel and the spatial identities of Indochina, France and Hue in 1920s–1940s Vietnamese print. Master's thesis, Michigan State University, Michigan. Available at: https://d.lib.msu.edu/etd/1835 (accessed 7 September 2021).

Compagnie internationale des Wagons-Lits and des Grands express européens (1937) *Lettre au Gouverneur général de l'Indochine au sujet du tourisme en Indochine*. Résidence supérieure au Tonkin fonds (RST72710-02), 19 May. National Archives Center No.1, Hanoi.

Coronnat (1904) *Recherche d'un emplacement de sanatorium dans le massif de Tam Dao*. Résidence supérieure au Tonkin fonds (RST38969), 25 May. National Archives Center No.1, Hanoi.

de Rhodes, A. (1994) *Hisotry of Tonkin Kingdom (Lịch sử vương quốc Đàng Ngoài)*. Dai Ket, Ho Chi Minh City.

Demay, A. (2014) *Tourism and Colonization in Indochina (1898–1939)*. Cambridge Scholars Publishing, Newcastle upon Tyne, UK.

DeWald, E. (2008) The development of tourism in French colonial Vietnam, 1918–1940. In: Cochrane, J. (ed.) *Asian Tourism: Growth and Change*. Elsevier, Amsterdam, pp. 221–231.

Direction des Affaires économiques de l'Indochine (1925) *Rapport au Conseil de Gouvernement sur le service assuré par le 1er Bureau de la Direction des Affaires économiques au cours de l'intersesssion 1924–1925*. Gouvernement général de l'Indochine fonds (GGI4202). National Archives Center No.1, Hanoi.

Fife, L.R. (2009) Bạch Mã: historical archaeology at a French colonial hill station, Central Vietnam, 1930–1991. PhD thesis, University of New England, Armidale, Australia. Available at: https://rune.une.edu.au/web/handle/1959.11/9243 (accessed 7 September 2021).

Gaide (1930) *Les stations climatiques en Indochine*. Imperimerie d'Extrême-Orient, Hanoi.

Glaillard (1919) *Rapport concernant les sanatoria de l'Indochine*. Gouvernement général de l'Indochine fonds (GGI7805), 31 December. National Archives Center No.1, Hanoi.

Goscha, C.E. (1996) Récits de voyage viêtnamiens et prise de conscience indochinoise (*c.*1920–*c.*1945). In: Salmon, C. (ed.) *Récits de voyages des Asiatiques: genres, mentalités, conceptions de l'espace*. EFEO, Paris, pp. 253–279.

Gourdon, H. (1931) *L'Indochine*. Librairie Larousse, Paris.

Gouvernement général de l'Indochine (1926) Les travaux publics de l'Indochine. *Bulletin économique de l'Indochine française* 182, 166–320.

Gouvernement général de l'Indochine (*c.*1930) *Exposition générale*. Résidence supérieure au Tonkin fonds (RST75004). National Archives Center No.1, Hanoi.
Gouverneur général de l'Indochine (1938) *Lettre au Chef du Service géographique au sujet de l'exposition San Francisco*. Gouvernement général de l'Indochine fonds (GGI403), 9 June. National Archives Center No.1, Hanoi.
Hien, P.N. and Duong, N.H. (1981) *Record of secret story about Western Christianity (Tây Dương Gia TôKý Lục)*. Social Sciences, Hanoi.
Hung, N.P. (2015) Tourism and monarchy in Vietnam. *Journal of Social Sciences and Humanities* 1(3), 260–276.
Inspecteur général des Travaux publics (1935) *A.s. Cie des Wagons-Lits*. Résidence supérieure au Tonkin fonds (RST72710-01), 23 October. National Archives Center No.1, Hanoi.
Laderman, S. (2009) *Tours of Vietnam: War, Travel Guides and Memory*. Duke University Press, Durham, North Carolina.
Lockhart, B. (2011) Colonial and post-colonial construction of 'Champa'. In: Tran, P. and Lockhart, B. (eds) *Then Cham of Vietnam: History, Society and Art*. NUS Press, Singapore, pp. 1–53.
Maignan, A. (1936) *L'achèvement du Transindochinois*. Editions du Moniteur des Travaux publics, Paris.
Ninh, L. (2006) *Champa Kingdom (Vương Quốc Chămpa)*. Vietnam National University, Hanoi.
Office Central du Tourisme Indochinois (n.d.) *Les Hotels en Indochine*. Unnamed publisher, Saigon.
Office Central du Tourisme Indochinois (1938a) *Glimpses of a Great Tourist Country: French Indochina at the Beginning of 1938*. unnamed publisher, Saigon.
Office Central du Tourisme Indochinois (1938b) *Rapport de fin d'année du Bureau Officiel du Tourisme*. Résidence supérieure au Tonkin fonds (RST72710-02), 26 January. National Archives Center No.1, Hanoi.
Picanon, E. (1896) *Hospitalisation avec tarif réduit des familles de fonctionnaires locaux au sanatorium de Cap Saint-Jacques et celui de Yokohama*. Gouvernement général de l'Indochine fonds (GGI5652). National Archives Center No.1, Hanoi.
Reid, A. (1993) *Southeast Asia in the Early Modern Era: Trade, Power, and Belief*. Cornell University Press, Ithaca, New York.
Résident supérieur en Annam (1936) *Lettre au Gouverneur général de l'Indochine au sujet du tourisme*. Résidence supérieure au Tonkin fonds (RST72710-01), 29 June. National Archives Center No.1, Hanoi.
Rondet-Saint, M. (1913) *Organisation du tourisme en Indochine*. Résidence supérieure au Tonkin fonds (RST20859). National Archives Center No.1, Hanoi.
Sans, P. (1939) *L'Indochine la 'Golden Gate International Exposition' de San Francisco*. Gouvernement général de l'Indochine fonds (GGI407). National Archives Center No.1, Hanoi.
Schweyer, A.-V. (2011) *Ancient Vietnam: History, Art and Archeology*. River Books Co, Bangkok.
Sylvestre de Sacy, S. (1931) Le tourisme en Indochine. *Revue du Touring Club de France* 345, 40–46.
Syndicat d'Initiative (1924) Réunion du Syndicat d'Initiative. *Écho annamite* 4, 4.
Thanh, T.D. (2005) *Introduction to Tourism*, 4th edn. Vietnam National University, Hanoi.
Tourrès (1909) *Au sujet de la création d'un sanatorium sur le plateau de Chapa-Losuitong*. Résidence supérieure au Tonkin fonds (RST38422), 2 June. National Archives Center No.1, Hanoi.
Tran, V.N. (2010) Tourism in Vietnam in the early 20th century (Du lịch Việt Nam đầu thế kỷ XX). *Journal of Vietnam National University* 26, 164–173.
Truong, V.D. and Le, A. (2017) The evolution of tourism policy in Vietnam, 1960–2015. In: Hall, C.M. and Page, S. (eds) *The Routledge Handbook of Tourism in Asia*. Routledge, London, pp. 191–204.
Tucat, G. (1942) *Rapport sur l'aménagement du mont Bavi*. Résidence supérieure au Tonkin fonds (RST67366). National Archives Center No.1, Hanoi.
VNSSC (Vietnam Social Sciences Committee) (1981) *History of Vietnam (Lịch sử Việt Nam)*, 1st edn. VNSSC, Hanoi.

3 Tourism Policies and Governance

Huong T. Bui[1], Long H. Pham[2]*, Thanh D. Tran[3] and Binh Nghiem-Phu[4]
[1]Ritsumeikan Asia Pacific University (APU), Japan; [2]Vietnam National University, Hanoi, Vietnam; [3]Saigon International University, Vietnam; [4]University of Hyōgo, Japan

Introduction

Following up on the evolution of Vietnam tourism presented in Chapter 2 (this volume), the current chapter focuses on its development post-1975. An analysis of tourism policy and governance in Vietnam cannot set aside the evolution of the Marxist political economic doctrine regulating the socio-economic system of the country since 1975. The transition from the central plan to the market-driven economy has been reflected in tourism policy making and implementation. Owing to its high international integration, its exceptional vulnerability to changes in global trends while being well connected to domestic consumption and mobilization of resources for post-war development, tourism mirrors the notion of a transitional economy better than any other sector. Current trends towards environmental and cultural protection also clearly imprint on the policy of tourism development. Although previous work provides a fundamental understanding of tourism policies in contemporary Vietnam (Truong, 2013; Truong and Le, 2017), the constant changes in tourism's external operating environment, especially prior to and during the COVID-19 pandemic, demand a thorough reassessment of tourism policy and governance. Also, while a handful of writings on Vietnam tourism policies in the early 2000s (Cooper, 2000), and 2010s (Suntikul *et al.*, 2008; Semone *et al.*, 2011) approached tourism from an exogenous free-market principle, none of them set the policy viewpoint from a Marxist political economy angle, and this has resulted in a shortage of in-depth analyses. Understanding of policies, however, demands indigenous knowledge of history and culture as well as accessibility to legal data in the native language. Bridging these gaps, the current book and chapter provide tourism scholars and practitioners with a systematic review of tourism policy and governmental approaches in contemporary Vietnam.

Given the complexity of tourism policy, markets and governance in the past 45 years, a combination of different methods was adopted in this analysis. The authors utilize historical chronological tables, charts and graphs to illustrate the process of transformation and the institutional changes in tourism occurring during that process. Official statistics and regular surveys of tourism markets not only assist in monitoring tourism development but also illustrate changes in policies. A content analysis of various reports, legal documents and books highlights the focus of tourism policies in different periods. All these methods in combination present the evolution of post-war tourism development policy and governance along the

*Corresponding author: phamhonglong@gmail.com

 Vietnam Tourism: Policies and Practices (eds Huong T. Bui *et al.*)
DOI: 10.1079/9781789242782.0003

temporal dimension of 45 years, across different markets, and on the cross-scale interactions of administration.

Tourism Policy and State Governance

A political economy approach to public policy

Public policy is influenced by the economic and social characteristics of society, as well as other features of the political system (Hall and Jenkins, 2004). Policy is a consequence of the political environment, values and ideologies, the distribution of power, institutional frameworks, and of decision-making processes (Simeon, 1976). Tourism development itself is political in terms of the policy decisions that have to be made about public expenditure (e.g. on human capital and physical resource development), together with support and mediation on the sustainable use of resources (Burns and Novelli, 2007).

State intervention in tourism is affected by changing political ideologies. The tendency to privatize and commercialize functions that were once performed by a government which has been almost universal in Western nations since the late 1970s, in the Eastern Europe Communist bloc in the early 1990s, and in the socialist states in Asia (e.g. China, Vietnam) has substantially affected the nature of national government involvement in the tourism industry. In such changing political climates, four approaches to political economy applicable in tourism are: (i) Marxian; (ii) regulation theory; (iii) international comparative; and (iv) post-structural (Bramwell, 2011). Bianchi (2018) argues that there is shifting focus in the political economy of tourism from technical, empirically driven analyses of tourism's contribution to economic development to the various strands of development theory. In particular, theoretical advancement potentially lies in the application of cultural political economy and Marxian thinking about the political economy of tourism.

For Marxists, ownership of the means of production is generally assigned the central role. The centrally controlled economies are characterized by an oligarchic system of government (Bernholz and Breyer, 1993, cited in Bahre, 2007). The constitution of ownership determines the way in which property rights to economic goods are distributed. Accordingly, seller and buyers on the market are not only exchanging goods for money, they also exchange rights (e.g. the prerequisite for the rational use of resources) in each of the transactions involved (Bahre, 2007). In centrally planned economies, all areas of society and policy domains are subordinate to governmental doctrine (i.e. the 'Party'). In contrast, in a democratic system with a market economy, acceptance of individual interest is a constituent part of that system, and the society is legitimized by the ties between decisions concerning society as a whole and individual interests (Teuber, 2001, p. 65, cited in Bahre, 2007). Owing to the interdependence of the political and the socio-economic system, these two basic types (centrally controlled economy and market economy) can be combined with different basic political systems (dictatorship and democracy): (i) market economy and democracy; (ii) market economy and dictatorship; (iii) centrally controlled economy and democracy; and (iv) centrally controlled economy and dictatorship. In the long term, a centrally controlled economy cannot be conceived without a dictatorship. However, a market economy is also compatible with dictatorship, at least in the medium term (Wrobel, 1999, p. 63, cited in Bahre, 2007). Thus, the transition from a centrally controlled economy to a market economy in the medium term might be able to accommodate the principles of single-party rule in its theory of political economy.

The role taken by tourism in transitional economies (i.e. from centrally planned to market economies) in the early 1990s was rather sudden and unplanned under rational political change in Eastern Europe (Bahre, 2007). Consequently, the Marxism political economy principles in a transitional economy were required to cover a rapid reform of property rights within different subsectors of the Eastern German tourism industry, for example. Within this context, the tourism industry is a functional category of economic production, to which many industries contribute, and in which all economic sectors participate indirectly, through the demand for tourism goods (Bahre, 2007). Notably, in relation to the potential use of political economy to understand the state's activities in tourism, 'the field of tourism studies has grossly neglected this complex view of the state' (Dredge and Jenkins, 2007, p. 38). Under the principles of Marxist political economy in transition, these characteristics of the transformation process of the tourism industry in former communist states might be applicable to Vietnam. However, due to the slow and incremental transition of the political, economic system in the country, some unique features of tourism policy might appear to be in contrast to those found in the rational and fast economic transitions in Eastern Europe.

Modelling tourism governance

Hall (2011) identifies two broad meanings of governance as: (i) a description of contemporary state adaptation to its economic and political environment; and (ii) a denoting of the role of the state in the coordination of socio-economic systems by steering the socio-economic system, or by coordination and self-government. Alternatively, 'governance is the set of tasks such as decision making, enforcement of decisions, communication of rules and measurement of performance that allow these functions of a system to proceed' (Laws *et al.*, 2011, p. 1). Also, national and regional tourist organizations have undergone restructuring to: (i) reduce their planning, policy and development roles and increase their marketing and promotion functions; and (ii) engage in a greater range of partnerships, networks, and other collaborative relationships with stakeholders (Hall and Jenkins, 2004). Functional shifts in the types of governance have been analysed in four modes: hierarchies, markets, networks, and communities (Hall, 2011) (see Fig. 3.1).

Referring to the governance typology described in Fig. 3.1, owing to the role of the state in international relations, legislation and regulation, hierarchical governance remains significant. However, while the role of government and state intervention in tourism, as well as in environmental governance are dominant themes in the literature, there is little direct discussion on hierarchical governance (Hall, 2011). Research on the market as a governance mechanism concerns corporatization and privatization of tourism functions and has shifted the influence of neoliberal political philosophy. Among these research studies, the concept of networks, and public–private partnerships in particular, has received considerable attention in tourism policy

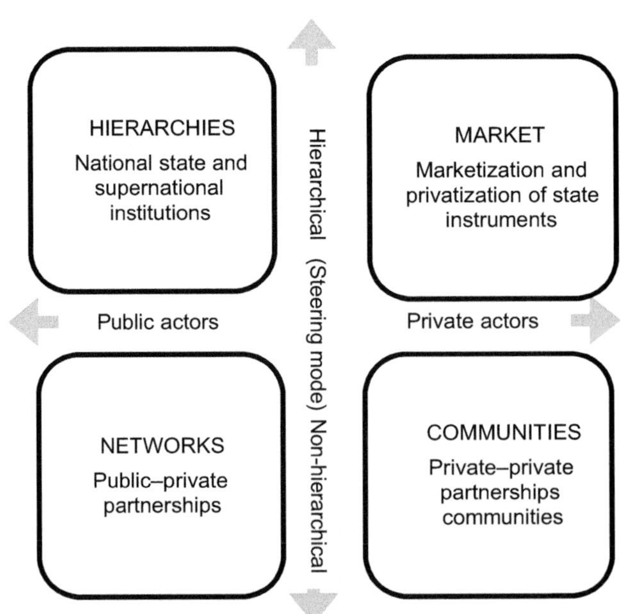

Fig. 3.1. The framework of governance typology. Adapted from Hall, 2011.

and planning as a 'middle way' or 'third way' between hierarchical and market approaches to tourism governance (Scott *et al.*, 2008). Networks have been proposed as a means to potentially integrate different policy perspectives. However, the extent to which networks may act to serve self-interest rather than the larger collective interest poses major challenges for their utility as a policy instrument (Dredge, 2006). The communities' approach is very much influenced by communitarianism, development and demands for more direct citizen involvement in governance. All three dimensions of governance as communities highlight the importance of public participation in public policy making (Hall, 2011).

To illustrate these theories and typologies, the authors analyse the changes and amendments to tourism development policies and the approach to state governance in Vietnam along with tourism market evolution, domestically and internationally. The findings are presented in the section below.

Analysis of the Changes in the Vietnam Tourism Policies and Markets

1976–1990: Fulfilment of political objectives

Upon the ending of the Vietnam War in 1975, the country entered a period of isolation from the West. Economic sanctions imposed by the USA and alliances from the mid-1970s to the early 1990s impeded the mobility of people and resources in and out of the country. International tourists were few. For example, in 1980, the country welcomed only 41,110 visitors (Tran, 2005). In 1960, the first state-owned Vietnam Tourism Incorporation (*Vietnamtourism*) was established under the Ministry of Foreign Trade to coordinate inbound and outbound travel for government officers and to manage existing resources such as hotels in North Vietnam. From 1969, this corporation operated under the command of the Prime Minister and then it was placed under the jurisdiction of the Ministry of the Interior. Shortly after 1975, this state-owned corporation took over and managed tourism facilities and infrastructure in South Vietnam. Gradually, it was dissolved into several regional state-owned companies in the late 1980s. International visitors in this period were foreign delegations, and experts largely from the socialist countries of the East European bloc coming by state invitation. The post-war economic struggle did not favour domestic tourism. There are no statistics of domestic travel prior to 1986 (Tran, 2005). By 1989, the country received 187,573 international arrivals and recorded 540,000 domestic trips (Tran, 2005). In the intermediate post-war period, tourism primarily enhanced the mutual understanding between the North and the South, and was used to fulfil political objectives, but was yet to be considered an economic activity in its own right (Truong, 2013).

In the post-war economic restructuring process, the state government played both roles of policy making and managing state-owned companies, the latter surviving on government subsidies. The centrally planned economy left no authority (and rights) for businesses to manage their own facilities, budgets and workforces. There were also no laws, legal frameworks and regulations to guide and regulate business operations (Tran, 2005). Apparently, tourism statistics were not the immediate concerns of the administrators.

The 1986 *Đổi Mới* programme shifted Vietnam from a centrally planned economy to a market-oriented one. Accordingly, tourism infrastructure and governance systems have gradually been established. The Vietnam Civil Aviation airline made its first international flight to Beijing and then Vientiane in 1976, to Bangkok in 1978, then gradually expanded its network to Hong Kong, Manila, Kuala Lumpur and Singapore (Truong and Le, 2017). The Vietnam National Administration of Tourism (VNAT), was established in 1978 and placed directly under the Prime Minister, performing three main responsibilities: (i) undertaking state management of tourism; (ii) managing tourism training institutions; and (iii) managing state-owned tourism businesses (Cooper, 1997; Tran, 2005).

The milestone for the encouragement of private sector and foreign investment was the promulgation of the *Law on Foreign Investment* in 1987. Good opportunities for tourism growth attracted transnational corporations to invest in the tourism sector (Jansen-Verbeke and Go, 1995). For instance, the Saigon Floating Hotel, which was opened in 1989 by Southern Pacific Hotels, was the first five-star hotel in Hồ Chí Minh city that offered international standards of services (Truong, 2013). In the North, the renovation of the historic Metropole Hotel to be the first five-star hotel managed by the Sofitel Group in Hà Nội was also initiated. By the end of this period, there were 45 hotels under foreign investment projects (Hobson *et al.*, 1994), indicating the transition from the state monopoly over tourism to the structure of a market economy under the command of the state government.

In 1990, *Vietnam's Tourism Year* campaign was launched to promote Vietnam's image to the world (Hobson *et al.*, 1994). However, the collapse of the Eastern Europe bloc in the late 1980s and early 1990s, meant that Vietnam tourism lost these markets, and the country was short of foreign currency to pay the debts incurred from prolonged wars and post-war reconstruction. In this situation, the Western European and North American markets brought an immediate solution in income from service exports. Tourist sites and facilities largely left from the colonial and imperial periods (in South Vietnam) were gradually renovated and upgraded in Hà Nội, Huế, Đà Nẵng, Hồ Chí Minh city, Đà Lạt, Vũng Tàu and Cần Thơ (Truong, 2013) to start welcoming foreign tourists.

1991–2010: Warming up the economic engine for taking off

The decision by the USA in January 1994 to lift the 20-year-old trade embargo, combined with Vietnam's efforts to develop an infrastructure for tourism brought significant opportunities for economic development. Tourism switched its role from a political means to a full economic mechanism for the generation of benefits. *Decree No. 46* issued by the central government in 1994 affirmed that tourism was a strategic component in the country's socio-economic development, industrialization and modernization (Truong and Le, 2017).

Between 1991 and 1994, the annual growth rate of international arrivals was constantly over 30% and hit a record high of 78.6% in reaching 1 million foreign visitors in 1994. An optimistic view of a tourism future, the *Master Plan for Tourism Development 1995–2010* set a target for the country to welcome 3.5–3.8 million foreign tourists by the end of 2000 (VNAT, 1994). The dilemma was that Vietnam did not develop clear actions to go with its strategic approaches (Cooper, 2000). Major problems were a lack of accurate and basic statistical information about visitors, a lack of investment funds, the high inflation rate, an inadequate system of business laws and policies to govern the business community, and poor public infrastructure (Lam and Mok, 1997). Unrealistic planning resulted in an over-supply of hotel capacity, exaggerated

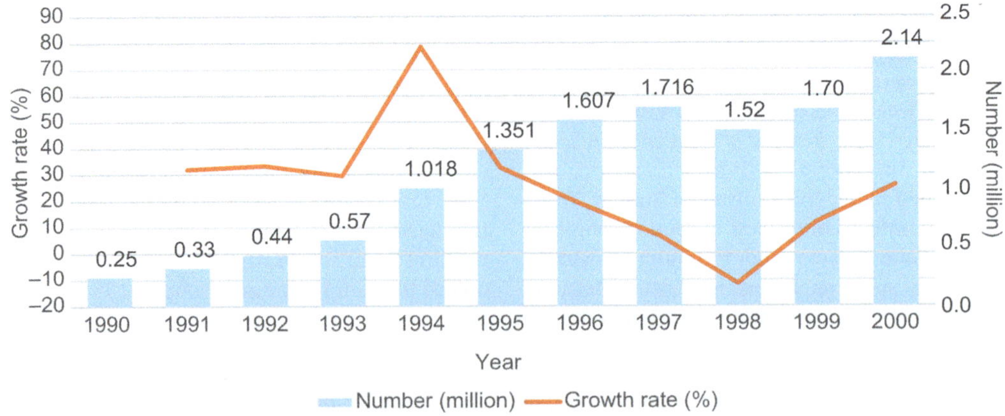

Fig. 3.2. International tourist arrivals (1990–2000). From General Statistics Office – VNAT, 2005.

by the Asian financial crisis in 1997 that led to a sharp decline in international arrivals (Cooper, 2000). By the year 2000, there were 2.1 million international visitors, only two-thirds of the targeted growth in numbers (see Fig. 3.2). The domestic market, however, was not seriously affected by the crisis. In 10 years, the domestic market increased tenfold, reaching 10 million by the end of the millennium, reflecting a booming post-war consumption trend starting from the 1990s (see Fig. 3.3).

Along with some state-owned enterprises there are various types of enterprises (e.g. collective, privately owned, family-run and foreign direct investment) participating in provision of tourism services. The barriers to private investment were gradually removed and foreign investment encouraged, with the issuance of the *Law on Private Companies* and the *Company Law* (1990), of the *Law on the Promotion of Domestic Investment* (1994), and the amendment of the *Law on Foreign Investment* (1992, 1996).

The *National Strategy for Tourism Development 2001–2010* (Government of Vietnam, 2002) set targets for upgrading and diversifying the tourism product, upgrading the quality of service, preservation and conservation of resources, and directed investment priorities. In 2006, tourism was incorporated as a key component of the *National Socio-economic Development Plan 2006–2010* that was approved at the 10th Party Congress (VNAT, 2005).

The growth rate of the domestic market remained high and stable throughout the decade. In 10 years, the domestic market increased by nearly 300%, reaching the size of 28 million trips in 2010 (Fig. 3.4). The international market fluctuated and was highly vulnerable to external shocks. The deep drop in international arrivals in

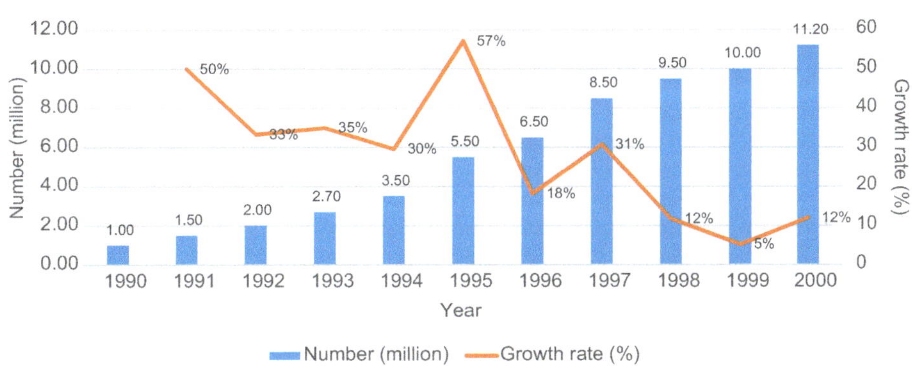

Fig. 3.3. Domestic trips (1990–2000). From General Statistics Office – VNAT, 2005.

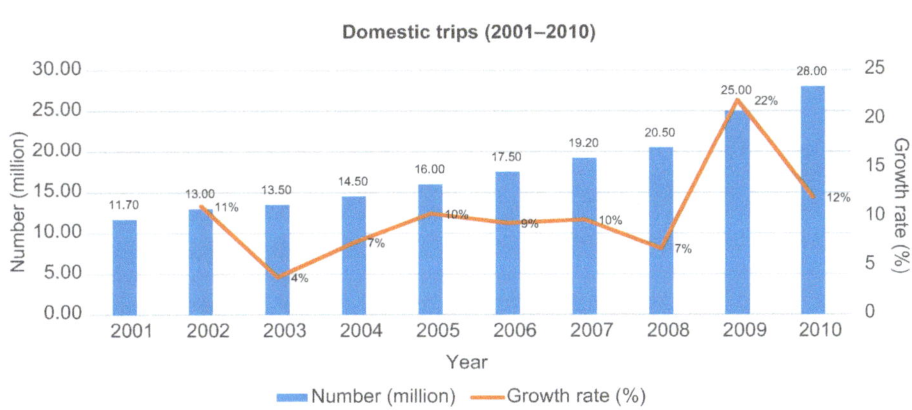

Fig. 3.4. Domestic trips (2001–2010). From General Statistics Office – VNAT, 2013.

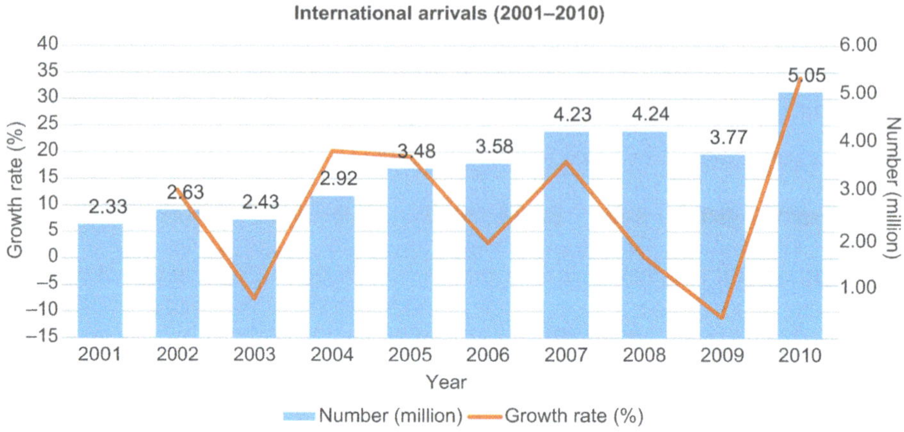

Fig. 3.5. International arrivals (2001–2010). From General Statistics Office – VNAT, 2013.

2003 was associated with the outbreak of severe acute respiratory syndrome (SARS) in East Asia, the economic recession brought the growth rate of international tourism to zero in 2008 and it declined by a further 11% in the following year. The market, however, bounced back in 2010, uplifting tourism to the new record high of 5.05 million in 2010 (Fig. 3.5). The market has also shifted from Europe and North America to East Asia, particularly the Mainland Chinese market, since 2005. Prior to 2005, the contribution of tourism to GDP was generally low (around 3%). Since 2005, owing to a change in the method of calculation by the implementation of the Tourism Satellite Account (TSA), the ratio of tourism to national GDP has been uplifted to 5% (see Fig. 3.6).

Overall, in the 20 years of development from 1990 to 2010, the first 10 years can be seen as a warming up period, the following 10 years, from 2001 to 2010 were the take-off period of Vietnam tourism signified by a dramatic increase in visitor arrivals, expansion of the domestic market and increased contribution to GDP.

2010–2020: Accelerating the economic and social dual engine

The *Master Plan for Tourism Development to 2020 with Vision to 2030* (VNAT, 2013) provided a roadmap for accelerating tourism development. Forecasts of inbound tourist numbers made in 2013 expected the country to welcome 10.5 million by 2020, 14 million by 2025, and 18 million by 2030. Domestic tourists would reach 47.5 million by 2020, 58 million in 2025 and

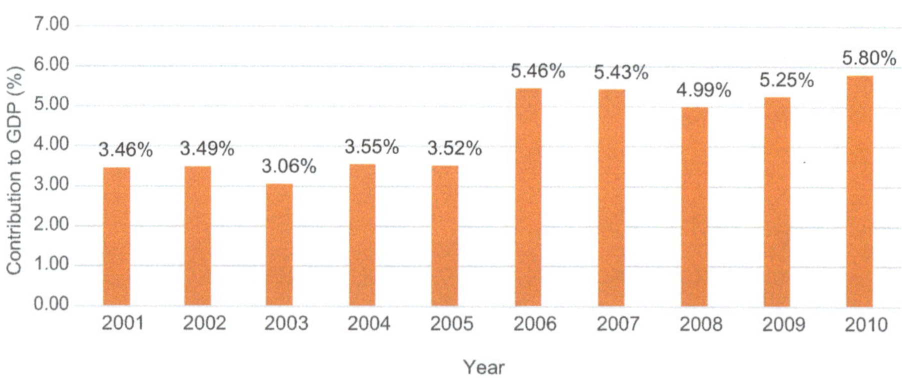

Fig. 3.6. Tourism contribution to GDP (2001–2010). From General Statistics Office – VNAT, 2013.

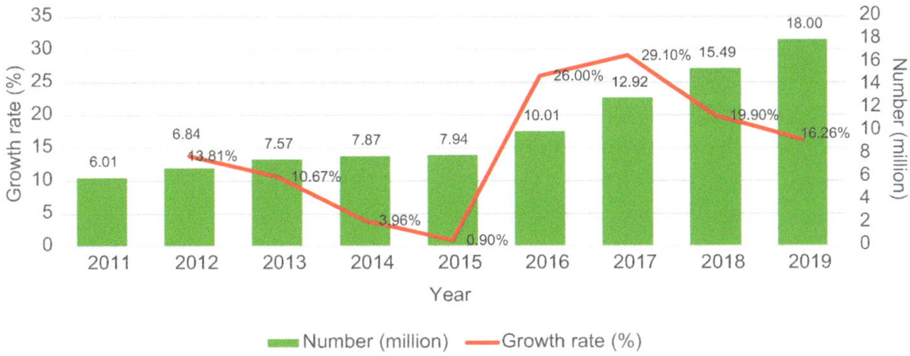

Fig. 3.7. International tourist arrivals (2011–2019). From General Statistics Office – VNAT, 2020.

71 million by 2030. Tourism receipts would reach US$18.5 billion by 2020, US$26.6 billion by 2025 and US$35.2 billion by 2030, double the figure of 2020. The contribution of tourism to GDP was set at 7.0% by 2020, 7.3% by 2025, and 7.5% by 2030 (VNAT, 2013). The actual performance of the tourist markets had surpassed all forecasted indicators. The international tourist arrivals reached 18 million in 2019 and domestic trips made a record height of 85 million in the same year (see Figs 3.7 and 3.8). In other words, Vietnam tourism has completed the objectives of development 10 years earlier than planned. Notably, both international and domestic markets accelerated since 2015. One of the reasons to top up the number of visitors was the change in the statistical method used in the calculation to include day visitors and those who do not use commercial accommodation, such as visiting friends and relatives (VFR) (VNAT, 2020).

The decade (2010–2020) has witnessed dramatic market transition. In the top ten international markets for Vietnam tourism, China and South Korea took the first and second positions. While the rest of the market showed steady growth, these two markets soared. In 10 years, the Chinese market increased by 500%, from 905,360 in 2010 to 5,806,425 in 2019, accounting for close to 30% of the market. The South Korean market also grew at an incredible rate close to 800% in 10 years, from 495,902 in 2010 to 4,290,802 (see Table 3.1). These two markets account for 50% of the market. While these growth rates are impressive, they show that Vietnam tourism is increasingly dependent on single markets.

When assessing market potential by propensity to spend and length of stay, the top ten international markets of Vietnam tourism can be divided into two groups. Visitors from

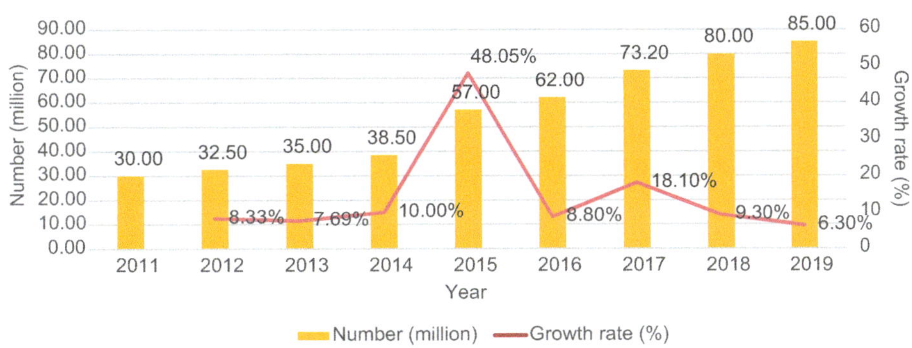

Fig. 3.8. Domestic trips (2011–2019). From General Statistics Office – VNAT, 2020.

Table 3.1. Top ten markets of Vietnam tourism (2008–2019). From General Statistics Office – VNAT, 2020.

	Year											
Market	2008	2009	2010	2011	2012	2013	2014	2015	2016	2017	2018	2019
China	643,344	518,948	905,360	1,416,804	1,428,693	1,907,794	1,947,236	1,780,918	2,696,848	4,008,253	4,966,468	5,806,425
South Korea	449,043	360,065	495,902	536,408	700,917	748,727	847,958	1,112,978	1,543,883	2,415,245	3,485,406	4,290,802
Japan	393,091	356,665	442,089	481,519	576,386	604,050	647,956	671,379	740,592	798,119	826,674	951,962
USA	414,851	403,027	430,993	439,872	443,826	432,228	443,776	491,249	552,644	614,117	687,226	746,171
Russia	49,004	66,203	82,751	101,631	174,287	298,126	364,873	338,843	433,987	574,164	606,637	646,524
Malaysia	174,545	165,623	211,337	233,132	299,041	339,510	332,994	346,584	407,574	480,456	540,119	606,206
Thailand	182,385	159,568	222,839	182,820	225,866	268,968	246,874	214,645	266,984	301,587	349,310	509,802
Australia	234,692	217,166	278,155	289,762	289,844	291,508	321,089	323,792	320,678	370,438	386,934	383,511
UK	107,091	115,492	139,152	156,290	170,346	184,663	202,256	212,798	254,841	283,537	298,114	315,084
France	182,068	172,959	199,351	211,444	219,721	209,946	213,745	211,636	240,808	255,369	279,659	287,655

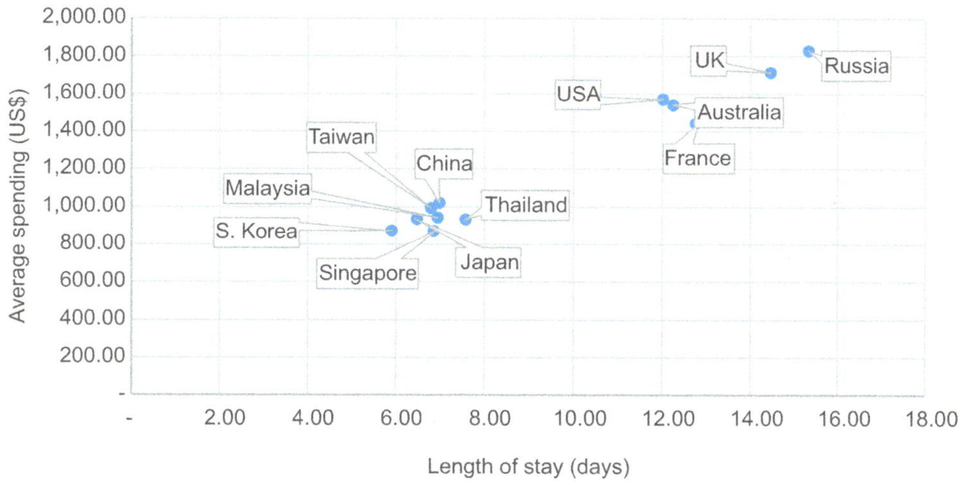

Fig. 3.9. Top 12 international markets in terms of length of stay (day) and spending (US$) (2019). From General Statistics Office – VNAT, 2020.

neighbouring Asian countries stay shorter periods and have lower average spending. While the number of visitors from China and South Korea has increased, visitors from these markets tend to have shorter stays than visitors from longer-haul markets such as Europe and the USA. The shift to North-east Asian markets therefore represents a decline in the overall economic yield of Vietnam's international visitor mix, as evidenced by the stagnant average daily spending of visitors between 2014 and 2017 (US$96), and a decline in average length of stay (from 11.1 days in 2013 to 10.2 days in 2017) (World Bank, 2019). On the other hand, the long-haul European and North American markets stay much longer, almost double the length of Asian market tourists and spend more (see Fig. 3.9).

The contribution of domestic tourism to overall tourism income demands further assessment. Vietnamese locals are spending more on their travel each year, from VND 2.78 million in 2015 to VND 3.93 million in 2019. In the same period, the spending of international visitors reduced from VND 24.85 million in 2015 to VND 23.39 million in 2019 (data from General Statistics Office, cited in VNAT, 2020). Domestic tourism in Vietnam reflects the rapid expansion of the middle class, who have a strong appetite for travel, and the improving affordability of air transport amid the growth of low-cost domestic air carriers (World Bank, 2019).

Tourism contribution to GDP has grown from 6.33% in 2015 to 9.20% in 2019 (Fig. 3.10), as a consequence of increasing both the number of tourists and their propensity to spend, particularly in the domestic market. The significant contribution to GDP confirms the strategic position of tourism in the national economy, shifting tourism from a support economic sector into a foundational sector that can facilitate and support the

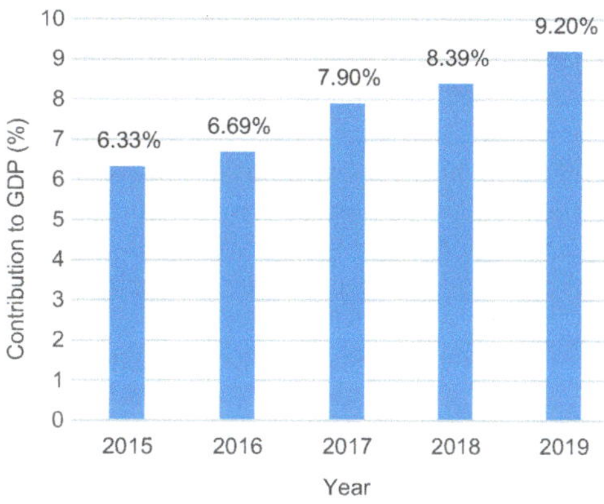

Fig. 3.10. Tourism contribution to GDP (2015–2019). From General Statistics Office – VNAT, 2020.

development of other related sectors such as retail, entertainment and restaurants. However, the tourism GDP multiplier value of 1.6 (e.g. US$1 of tourism spending creates an additional US$0.6 in income), is considerably smaller than elsewhere in developing South-east Asia, where it averaged around 2.4 in 2017, and lower than the global average for the tourism sector of 3.3. Similarly, the tourism spending multiplier impact on employment in Vietnam – around 1.7 – is lower compared with regional and global averages of 2.5 and 2.6, respectively. This points to a need to strengthen tourism sector linkages to the rest of the economy, enabling workers and firms in other sectors to indirectly benefit from Vietnam's growing tourism (World Bank, 2019).

The overheated development of tourism between 2015 and 2019 has already surpassed the highest growth scenario of 2020, reaching the target set for 2030. Vietnam welcomed 18 million international tourists 10 years earlier than planned. The domestic market has grown exponentially surpassing the target of 2030 and reaching 85 million trips in 2019. This demanded an adjustment to the Master Plan approved in 2018. The new *Master Plan for Việt Nam Tourism Development to 2030 with Vision to 2050* was launched in 2018 (VNAT, 2018). Major adjustments were made to the following points: (i) tourism as a strategic foundation sector that can facilitate the development of other economic sectors; (ii) fostering regional and international integration through tourism; and (iii) all levels of governance are responsible for, and resources are mobilized for, tourism development. More importantly, the new Master Plan sets targets to develop both international and domestic markets with an emphasis on increasing tourism export values. This is a significant shift in tourism policy from a quantity orientation (e.g. visitor numbers) to quality (e.g. spending and length of stay), together with other objectives such as sustainability, biodiversity, combating climate changes and fulfilling social objectives. The *Plan* sets targets to welcome 32 million international arrivals in 2025, 47 million in 2030 and the domestic market to reach 110 and 130 million, respectively. The contribution of tourism to GDP is expected to increase from 11.6% in 2025 to 13.9% in 2030. By 2050, Vietnam wishes to be a special destination with a strong reputation, brand, competiveness, and belonging in the top 20 of the worldwide tourism developed nations.

State Governance of Tourism

The governance of tourism in Vietnam shows distinctive characteristics of a centrally planned economy in transition to a market economy. The evolution of structure and functions of the national tourism organization, the VNAT, illustrates these characteristics.

Administrative structure

Established in 1979, the VNAT has accountability for 30 tourism companies and hundreds of hotels and restaurants on a national scale, taking over the entrepreneurial function from the Vietnam Tourism Incorporation. These companies had reformed into regional state-owned tourism companies, and remained until 1987, when VNAT became centrally responsible for the state governance of tourism, functioning as a national tourism organization with focus on destination promotion along with the primary governing tasks involved in tourism planning and education. In this period, tourism management was delegated from the state level to the provincial level, specifying the responsibilities of provinces and cities for managing their own tourism activities (Truong, 2013). Accordingly, a horizontal structure of tourism governance was established. Under the jurisdiction of VNAT, the Department of Commerce and Tourism was established at provincial levels (47 out of 64 provinces by 2007).

Since 2007, VNAT has become an organization under the jurisdiction of the Ministry of Culture, Sport and Tourism (MCST), advising and assisting the MCST on the state management and implementation of the law on tourism nationwide, and managing public services in tourism as per the law. Changing its structure from a single administrative division to a department in the MCST led to a major restructuring of the provincial administration, where tourism is placed within the Department of Culture and Information. The structure and functions of VNAT at the national level and the respective departments at the municipal level, in fact, have streamlined the administrative scale and scope of the organization. Accordingly, the VNAT is responsible for: (i) developing tourism laws; (ii) drafting resolutions of the National Assembly; (iii) drafting resolutions of the Standing Committee of the National Assembly; (iv) drafting resolutions and decrees of the government; and (v) drafting decisions and directives of the Prime Minister and legal documents on state management of tourism for the MCST to issue. VNAT also performs tasks related to the state management of tourism, including planning, destination promotion and branding along with tourism education. Functional units under the VNAT are the Department of Planning and Finance, the Department of Organization and Personnel, the Department of Travel, the Department of Hotel, the Department of Tourism Marketing, the Department of International Cooperation, and the General Department with three subunits, namely, the Institute of Tourism Development Research (ITDR), *Vietnam Tourism Review* and the Tourism Information Center.

Functions of VNAT

Legislative functions

As early as 1999, the Standing Committee of the National Assembly issued the *Ordinance on Tourism*. This first legal document stipulated four main areas: (i) tour operations (domestic and international); (ii) accommodation; (iii) transportation; and (iv) other supporting services (i.e. tour guiding). The four main areas of tourism business have expanded in the *Law on Tourism* (Government of Vietnam, 2005), creating a relatively complete legal corridor for tourism activities in the country. The latest modification of the *Tourism Law* was in 2017 by the National Assembly.

Planning functions

VNAT has drafted and monitored the implementation of major Master Plans for tourism development since the 1990s. These include the *Master Plan for Tourism Development 1995–2010*, the *National Strategy for Tourism Development 2001–2010*, the *Master Plan for Tourism Development to 2020 with Vision to 2030*, and the latest one the *Master Plan for Việt Nam Tourism to 2030 with Vision to 2050*. In accordance with these Master Plans being regularly developed at the national level by VNAT, specific provinces, municipalities and their encompassing tourist zones developed their own master plans. However, adherence to these master plans has often proven problematic, potentially undermining tourism development objectives and sustainability. Many of the plans remain on 'paper', have either not been adequately followed or have been modified during the course of their implementation. This has sometimes enabled investments which pose threats to the sustainability of tourism growth (World Bank, 2019).

Destination promotion functions

Since the first advertising campaign *Vietnam's Tourism Year* was launched in 1994 to promote Vietnam's image to the world, followed by the *Vietnam – a Destination for the New Millennium* in 2000, VNAT has been in charge of establishing new promotional campaigns every 5 years. The destination promotion and branding function of VNAT was reinforced with the establishment of the Department of Tourism Promotion in 2004, which has run three promotion campaigns: (i) *Welcome to Vietnam* (2004–2005); (ii) *Vietnam – the Hidden Charm* (2006–2011); and (iii) *Vietnam – the Timeless Charm* (since 2012). This indicates the emphasis on professionalism of tourism promotion under the responsibility of the national tourism organization.

Supporting associations

Since the early 2000s, the appearance of various associations supporting VNAT in dealing with operational issues and advisories has been part of the transition from a hierarchal model to a network model in tourism governance, in accordance to the description of the Fig. 3.1 by Hall (2011). The establishment of the Vietnam Tourism Association in 2002 marks this trend; the Association links, cooperates and supports members in business, market stabilization, quality improvement, product development and enhancing destination competitiveness. The Association also represents and protects the rights and interests of its members, representing 46 provinces and cities, and over 4000 enterprises including legal entities (travel agencies, transportation, travel agencies), collectives (agents, travel centres, clubs, forums), and individuals (guides, trainers, researchers) related to the tourism sector. The Vietnam Tourism Association has five affiliated associations (the Travel Association, the Hotel Association, the Tourism Training Association, the Tour Guide Association and the Chefs Association).

Challenges for Tourism Policy and Governance: the COVID-19 Pandemic

The COVID-19 pandemic has swamped the country since January 2020 with four waves of infection: (i) January–March 2020; (ii) July–December 2020; (iii) January–March 2021; and (iv) May 2021 onwards. Although Vietnam has been recognized as having an exceptionally good track record in controlling the pandemic, the tourism industry is one of the economic sectors suffering the most from this global crisis. The periodic suspension of inbound tourism since April 2020 has resulted in a very modest number of 3.8 million international tourists in 2020, a 73.8% decline from the figure of 2019. Domestic tourists, haunted by the fear of infection, had made only 42.5 million trips in 2020, down 41% from the previous year (Vietnam General Statistics Office, 2021). Revenue from tourism declined by 45.6% in Hà Nội, 68.2% in Đà Nẵng, 69.9% in Hồ Chí Minh city, 91.3% in Quảng Nam, and 95% in Khánh Hòa. Consequently, manpower was cut by 30.4% in the transportation and tourism sector, 29.9% in the accommodations sector, 17.4% in the entertainment sector, and 15.4% in the foods/restaurants sector (Vietnam General Statistics Office, 2021).

Several policies have been implemented to counter the impact of the pandemic on the industry. First, a series of tax, fee and interest reduction policies were introduced (CBRE, 2020). However, these have been proven to be ineffective as many businesses could not operate normally owing to the extremely strict social distancing and lock-down policy of the central and local governments. Second, promotional programmes for the stimulation of domestic tourism were implemented in May 2020, such as *Vietnamese travel in Vietnam*, and September 2020, such as *Vietnam tourism: Safe and attractive* (Nhan Dan, 2020). New tourist attractions were opened (e.g. VinWonders Phú Quốc) and hotel prices were halved (e.g. hotels in Sapa). Domestic airfares were reduced significantly (e.g. Bamboo Airways, Vietjet Air and Vietnam Airlines). These packages provided by a variety of stakeholders working in the industry resulted in a sharp increase in domestic demand in summer 2020 and brought some hope for Vietnam's tourism. None the less, they were quickly suspended due to the outbreak of the second wave of infection in major tourist cities. Recently (March 2021 onwards), discussion and even debate on the acceptance of the COVID-19 vaccination passport or certificate have been initiated. Driven by the successes of mass vaccination in certain countries (e.g. the UK, Israel and the USA), many people in the tourism industry and the public have advocated and supported an early reopening of the country's border to international tourists. Nevertheless, the government and a large proportion of the public are still hesitant and concerned about this move since it may fuel the domestic transmission of the virus, given the fact that the majority of the Vietnamese people have not been vaccinated (Viet Tuan, 2021). The widespread impact of the variant COVID-19 virus strains in Brazil, the UK, India and South Africa, for example, has also meant constraints on the reopening effort. The uncertainty of operation and demand from both domestic and international markets have pushed many tourism businesses to the edge of bankruptcy since the sector is dominated by small- and medium-sized enterprises.

The paucity of tourism in the wake of the pandemic, however, has driven the consideration of a new model for Vietnam's tourism development in the years to come. First, the market development strategy intensifies supporting the domestic market. The demand for domestic travel is waiting to boom when travel restrictions are removed. The booking rate for April and May 2021 has proven the intensity of the suppressed demand from Vietnamese travellers. The domestic market might assist businesses in recovery while the opening of international borders might take a year or two for recovery. Second, product development has been emphasized to meet the changes in market demand, and preference of travellers in the post-pandemic period. In particular, small-group tours and health and wellness tourism have come into the discussion. New product development encourages cooperation across regions and sectors to be upgraded to a new level. Finally, to deal with the emergence from the pandemic it has proved necessary to establish a crisis management fund to assist businesses in tourism, a fund that has been absent for many years (personal communication with tourism officers of VNAT, Hà Nội, 25 June 2021).

Discussion and Conclusions

The evolution of tourism policy and governance in post-war Vietnam since 1975 should be placed within a broader political economy guided by Marxist doctrine. For Marxists, ownership of the means of production generally is assigned to the central government, forming the basic principle of a centrally planned economy. Under this system, the sellers and the buyers exchange not only the ownership of goods but also the rights and duties determined by the existence of limited resources (Bahre, 2007). When properties are owned by the state, the rights to use them also belong to the state. The principle of the centrally controlled economy governed tourism policy in North Vietnam (1960–1975) and the immediate post-war period (1975–1990). The government played the ultimate role of operator/entrepreneur – owning and running state-owned tourism enterprises in these early times since the means of production (i.e. destinations, hotels, transport) were all owned by the government, and the government granted the rights to use the resources only to governmental organizations (i.e. VNAT) and state-owned enterprises and customers (i.e. government officers and delegations). When the national security issue in the post-war recovery period was prioritized, tourism policy placed emphasis on the fulfilment of its political role. This reflects changes in the jurisdiction under which the tourism industry was governed, from the Ministry of Foreign Trade to the Ministry of Interior, then directly under the Prime Minister, especially once foreign factors became involved in tourism production (i.e. visitors, investments etc.).

After *Đổi Mới* began in 1986 and lifting of the US trade embargo against Vietnam in 1994, the centrally planned economy has gradually been replaced by a market economy directed by one ruling party in the socialist state of Vietnam. Under this structure, the ownership of and rights to use factors of production are separated. Hence, private and foreign-invested businesses are now granted the rights to exchange goods and products in tourism, while the ownership of resources (e.g. land for hotel development) still belongs to the central state. From 1991 to 2010, tourism has taken off to become an important economic sector. The privatization of properties and deregulation of administration modified the functions of VNAT: (i) it switched from tourism operator to regulator – providing a legal framework for tourism business; (ii) it became the planner – leading the planning and development of tourist zones; (iii) it became coordinator – coordinating relevant ministries in tourism development; (iv) it became promoter – conducting promotion campaigns; and (v) it became educator – providing tourism education and training (Truong and Le, 2017).

In recent years, from 2010 onwards, as tourism has grown to be a spearhead industry with a significant contribution in terms of economic and social stability, policy and governance have increasingly adopted the principles of the market economy in combination with a social system moving towards democracy. This transition has led to an increasing openness to foreign visitors and investments, the delegation of power to lower levels, and encouragement of public participation in the planning and development of tourism. Examples include the establishment of the horizontal structure of tourism governance, and the appearance of models such as community-based tourism (see Chapter 7, this volume). Specifically, the increasingly important role of professional associations playing the coordinating and policy advisory roles to a national government organization (i.e. VNAT) has proven the existence of this trend. Against the backdrop of tourism governance typologies (Hall, 2011), systemic governance in Vietnam has shifted from the hierarchical mode to the market mode, while the early evidence of partnership and communities' governance is also observable. Functional change in tourism is another factor being noted in the last decade. The view of tourism as a secondary industry requiring intensive capital investment and generating revenue from foreign currency receipts has switched to a view of tourism as a tertiary industry that can connect to and lay the foundation for related industries to grow (refer to the *Master Plan for Việt Nam Tourism Development to 2030 with Vision to 2050* (VNAT, 2018).

Increasing concerns about economic, social and environmental sustainability in recent tourism policy debates are also evidenced. Conventionally, from the Western free-market economy viewpoint, socially sustainable forms of tourism, such as pro-poor and community-based tourism appear in Vietnam in the mid-2000s (Truong, 2013; 2014) along with the economy shifting to market-orientation mode and increasing democracy. Our research on tourism policy and governance from the Marxist viewpoint, however, challenges these contentions. We argue that the social function of tourism is an organic apparatus of the socialist system. The current practices of state-sponsored travel for government officers and workers in state-owned enterprises, public organizations and schools for the domestic market is an example (Michaud and Turner, 2006; Bui and Jolliffe, 2011). The same holds true for the concept of community-based pro-poor tourism, when the first village offering this type of travel was initiated in the 1960s in Lác Village in Hòa Bình (see Chapter 7, this volume). Perhaps, the philosophies underpinning NGOs and development aid focusing on disadvantaged groups in the society are congruent to the principles of the social function of tourism in the welfare socialist states of Eastern Europe (Bahre, 2007).

Another implication that emerges from our research centres is the debate on development-orientated tourism policy (Bianchi, 2018). The policy for the modernization of tourism facilities and resort development in peripheral areas has been criticized for being out of date and unsustainable (Bui and Dolezal, 2020). This policy prioritizes mass tourism at the expense of cultural and natural uniqueness, further worsening the social and economic fabric of the affected area, raising concerns about sustainability (World Bank, 2019). The shifting focus of the *Master Plan for Việt Nam Tourism Development to 2030 with Vision to 2050* has clearly articulated this issue and the goals the country needs to attain. The problems, however, are the ones of the capacity to implement and monitor this plan in the future.

This chapter has contributed to our further understanding of tourism policy and governance in post-war Vietnam spanning over 45 years of tourism development. It provides an in-depth analysis and explanation of the structure and functions of public administration of tourism in the country from the theoretical angle of the political economy. The study also comments on the industry's current challenges during the global health crisis in 2020 and 2021. The work, however, has several limitations that deserve further attention in future research. The authors centre the analysis on a limited number of secondary data sources, so it is unclear how effective the master plans have been in implementation. In particular, in-depth and highly specified data derived from the visitor spending survey and activities are missing from the analysis. Furthermore, expert opinions on the efficiency, problems and solutions should be considered in future research to overcome these limitations.

References

Bahre, H. (2007) Privatisation during market economy transformation as a motor of development. In: Burns, P. and Novelli, M. (eds) *Tourism and Politics: Global Frameworks and Local Realities*. Elsevier, Amsterdam, pp. 33–58.

Bianchi, R. (2018) The political economy of tourism development: a critical review. *Annals of Tourism Research* 70, 88–102.

Bramwell, B. (2011) Governance, the state and sustainable tourism: a political economy approach. *Journal of Sustainable Tourism* 19(4–5), 459–477.

Bui, H.T. and Dolezal, C. (2020) The tourism–development nexus in Southeast Asia: history and current issues. In: Dolezal, C., Trupp, A. and Bui, H.T. (eds) *Tourism and Development in Southeast Asia*. Routledge, London, pp. 23–40.

Bui, H.T. and Jolliffe, L. (2011) Vietnamese domestic tourism: an investigation of travel motivations. *Austrian Journal of South-East Asian Studies* 4(1), 10–29.

Burns, P. and Novelli, M. (2007) Tourism and politics: introduction. In: Burns, P. and Novelli, M. (eds) *Tourism and Politics: Global Frameworks and Local Realities on the Vietnam Hospitality Market*. Elsevier, Amsterdam, pp. 1–4.

CBRE (2020) *CBRE Hotels Insights: Implications Of COVID-19 On The Vietnam Hospitality Market*. CBRE, Ho Chi Minh City.

Cooper, M. (1997) Tourism planning and education in Vietnam: a profile 1995–2010. *Pacific Tourism Review* 1(1), 57–63.

Cooper, M. (2000) Tourism in Vietnam: Doi Moi and the realities of tourism in the 1990s. In: Hall, C.M. and Page, S. (eds) *Tourism in South and Southeast Asia: Issues and Cases*. Routledge, London, pp. 167–177.

Dredge, D. (2006) Policy networks and the local organisation of tourism. *Tourism Management* 27(2), 269–280.

Dredge, D. and Jenkins, J.M. (2007) *Tourism Planning and Policy*. Wiley, Milton, Queensland, Australia.

Government of Vietnam (2002) *National Strategy for Tourism Development 2001–2010*. Government of Vietnam, Hanoi.

Government of Vietnam (2005) *Law on Tourism*. National Political Publishing House, Hanoi.

Hall, C.M. (2011) A typology of governance and its implications for tourism policy analysis. *Journal of Sustainable Tourism* 19(4–5), 437–457.

Hall, C.M. and Jenkins, J. (2004) Tourism and public policy. In: Lew, A.A., Hall, C.M. and Williams, A. (eds) *A Companion to Tourism*. Blackwell Publishing, Malden, Massachusetts, pp. 525–540.

Hobson, P.J., Heung, V. and Chon, K.S. (1994) Vietnam's tourism industry: can it be kept afloat? *Cornell Hotel and Restaurant Administration Quarterly* 35(5), 42–48.

Jansen-Verbeke, M. and Go, F. (1995) Tourism development in Vietnam. *Tourism Management* 16(4), 315–321. DOI: 10.1016/0261-5177(95)97356-U.

Lam, C. and Mok, T. (1997) Hotel and tourism development in Vietnam. *Journal of Travel & Tourism Marketing* 7(1), 85–91. DOI: 10.1300/J073v07n01_06.

Laws, E., Agrusa, J., Scott, N. and Richins, H. (2011) Tourist destination governance: practice, theory and issues. In: Laws, E., Richins, H., Agrusa, J. and Noel, S. (eds) *Tourist Destination Governance: Practice, Theory and Issues*. CAB International, Wallingford, UK, pp. 1–16.

Michaud, J. and Turner, S. (2006) Contending visions of a hill-station in Vietnam. *Annals of Tourism Research* 33(3), 785–808.

Nhan Dan (2020) Promoting domestic travel with emphasis on safety. Available at: https://nhandan.vn/tin-tuc-du-lich/kich-cau-du-lich-noi-dia-giai-doan-2-chu-trong-an-toan-va-hap-dan-616426/ (accessed 25 September 2020).

Scott, N., Cooper, C. and Baggio, R. (2008) Destination networks: four Australian cases. *Annals of Tourism Research* 35(1), 169–188.

Semone, P., Laws, E., Ruhanen, L., Wang, Z. and Noel, S. (2011) Governance approaches in new tourist destination countries: introducing Tourism Law in Cambodia, Laos and Vietnam. In: Laws, E., Richins, H., Agrusa, J. and Noel, S. (eds) *Tourist Destination Governance: Practice, Theory and Issues*. CAB International, Wallingford, UK, pp. 25–39.

Simeon, R. (1976) Studying public policy. *Canadian Journal of Political Science/Revue canadienne de science politique* 9(4), 548–580.

Suntikul, W., Butler, R. and Airey, D. (2008) A periodization of the development of Vietnam's tourism accommodation since the open door policy. *Asia Pacific Journal of Tourism Research* 13(1), 67–80.

Tran, T.D. (2005) *Introduction to Tourism Sciences*. Vietnam National University, Hanoi.

Truong, V.D. (2013) Tourism policy development in Vietnam: a pro-poor perspective. *Journal of Policy Research in Tourism, Leisure and Events* 5(1), 28–45.

Truong, V.D. (2014) Pro-poor tourism: looking backward as we move forward. *Tourism Planning & Development* 11(2), 228–242.

Truong, V.D. and Le, A. (2017) The evolution of tourism policy in Vietnam, 1960–2015. In: Hall, C.M. and Page, S. (eds) *The Routledge Handbook of Tourism in Asia*. Routledge, London, pp. 191–204.

Vietnam General Statistics Office (2021) Vietnam tourism requires huge efforts and determination for 2021 recovery path. Available at: https://www.gso.gov.vn/du-lieu-va-so-lieu-thong-ke/2021/02/du-lich-viet-nam-2021-rat-can-quyet-tam-va-no-luc-de-vuot-kho/ (accessed 20 September 2021).

Viet Tuan (2021) Việt Nam readies vaccine passport entry for investors, tourists. *VnExpress*, 10 April. Available at: https://e.vnexpress.net/news/news/vietnam-readies-vaccine-passport-entry-for-investors-tourists-4260776.html (accessed 30 May 2021).
VNAT (Vietnam National Administration of Tourism) (1994) *Master Plan for Tourism Development 1995–2010*. VNAT, Hanoi.
VNAT (Vietnam National Administration of Tourism) (2005) 45 Years of Construction and Development of Việt Nam's Tourism Sector. Available at: https://vietnamtourism.gov.vn/index.php/items/1186 (accessed accessed 30 May 2021).
VNAT (Vietnam National Administration of Tourism) (2013) *Master Plan for Tourism Development to 2020 with Vision to 2030*. VNAT, Hanoi.
VNAT (Vietnam National Administration of Tourism) (2018) *Master Plan for Việt Nam Tourism Development to 2030 with Vision to 2050*. VNAT, Hanoi.
VNAT (Vietnam National Administration of Tourism) (2020) An unforgettable year of Vietnam tourism. Available at: https://vietnamtourism.gov.vn/index.php/items/35618 (accessed 30 May 2021).
World Bank (2019) *Taking Stock: Recent Economic Development of Việt Nam (Special Focus: Việt Nam's Tourism Developments)*. The World Bank Group, Hanoi.

4 Tourism Planning in the Central Highlands: Policies and Practices

Thuy T. Duong[1], Tuan Q. Pham[1], Hai Q. Truong[1], Huong T. Bui[2]*, Thanh D. Tran[3] and Giang T. Phi[4]

[1]Vietnam National University, Hanoi, Vietnam; [2]Ritsumeikan Asia Pacific University (APU), Japan; [3]Saigon International University, Vietnam; [4]VinUniversity, Hanoi, Vietnam

Introduction

This chapter discusses the gaps between planning for sustainable tourism and the reality of resource allocation and their utilization for development. The study is based on the case of Đắk Lắk province located in South-central Vietnam. The area is not only the most populous province in the Central Highland Region, it also has the highest ratio of ethnic diversity in the country; 49 out of 54 ethnic minority groups find their homes there. Regional richness in cultural resources has derived from ethnic group lifestyles in harmony with the environment of mountains, highlands, powerful rivers and waterfalls.

The Đắk Lắk's Tourism Development Plan in 2016–2020 (Daklak's People's Committee, 2012) has major problems of tourism development in the province. First, it is the issues in relation to tourist markets. The short length of stay and very small number of international tourist arrivals are challenges facing tourism planners. The number of domestic visitors is also relatively low compared to other provinces in the South-Central area, and are concentrated only in Buôn Ma Thuột and its adjacent areas. Second, current tourism products are limited in range and repetitive in pattern. Third, there are problems of unequal distribution and limited multiplier effect of tourism incomes in local communities. Đắk Lắk's tourism development has faced numerous challenges. The social challenges are coming from the urgent needs for conservation of ethnic cultures and lifestyles that made up the unique tourism products of the region, such as lifestyles linked to elephant nurturing and riding. Environmental challenges driven by deforestation and climate change causing depletion of water and rainforest also require immediate attention. Owing to the diversity of tourism resources in the area, however, a one-size-fits-all approach to tourism development has not been effective, and there is a heightened need to develop specific tourism niches in specific localities.

Along with the national strategy emphasis on sustainable tourism development since the 2010s, there is a need to reorient tourism development and to introduce new forms of tourism that could better utilize the regional resource abundance. Based on the theories of sustainable tourism development and planning in the Vietnamese context, this chapter provides a comprehensive spatial analysis for tourism development in Đắk Lắk province. The study

*Corresponding author: huongbui@apu.ac.jp

DOI: 10.1079/9781789242782.0004

aims to draw practical implications for tourism planning and product development for the area.

Towards Sustainable Development of Tourism

For many developing countries, tourism often brings a comparative advantage in development terms. Tourism-led growth theory has placed tourism in the development strategy for half a century, along with the evolution of development paradigms of modernization, dependency, neoliberal, sustainable and alternative development (Bui and Dolezal, 2020). Modernization and neoliberal perspectives are implicit in the many debates over tourism's impacts in developing countries, being the default mode of thinking for policy makers (Telfer and Sharpley, 2015). Critiques of modernization can be noticed, building on the argument that tourism does not foster independence, but rather creates dependencies on foreign exchange, capital, visitors and assistance (Holden *et al.*, 2011). Those who are opposed to both capitalism and mass tourism are inclined to support sustainable and alternative development, paving the way for paralleling concepts of sustainable and alternative tourism to emerge.

Akin to the United Nations' (UN) Sustainable Development Goals, sustainable tourism rhetoric has embraced holistic and ambitious goals as a 'triple-bottom line' approach that purports to address economic, environmental and socio-cultural impacts concurrently (Elkington, 1998). Generally, sustainability should consider the wide range of tourism impacts by including economic and socio-cultural factors along with environmental ones (Sharpley and Telfer, 2002; Mowforth and Munt, 2015). Critically, Sharpley and Telfer (2002) contend 'the often-rhetorical commitment to the principles of environmental protection and conservation are subordinated by the primacy of economic growth, forming the ideological basis for mainstream sustainable development theory' (p. 253). Another critical viewpoint is that the 'triple bottom line' is very difficult to operate as a management strategy (Jenkins, 2015). For example, economic impact arising from tourism is immediate, when tourists arrive and begin spending. Socio-cultural and environmental changes take much longer to emerge and are difficult to control and manage. While the concept of sustainable development can enable the articulation of specific benchmarks and indicators, it is ambiguous, prone to misappropriation, manipulation and 'greenwashing' (Harrison, 2015).

Following the introduction of sustainable tourism is the concept of alternative tourism, that refers to anything which is not mass tourism, including 'nature based', 'backpacking' and 'pro-poor tourism' (Harrison, 2015). The accompanying ambiguity on matters of implementation and context, argued to confer advantageous flexibility (Hunter, 1997), is reflected in the evolving relationship between sustainable tourism and mass tourism. However, both large-scale mass tourism and small-scale (alternative) tourism can be sustainable or unsustainable depending on their destination and planning/management context (Weaver, 2014). This understanding has betted an emergent convergence argument, whereby mass and alternative tourism are seen to be situated merely as 'ideal type' polarities with reality characterized by hybridity. But alternative tourism is not mass tourism and will never replace it (Aramberri, 2010).

Policy and planning for sustainable tourism should reflect the opportunities, conditions and constraints of the planning zone at country or sub-country level. Four major issues in tourism planning can be outlined in the context of developing countries (Jenkins, 2015). First, governments primarily plan for tourism development. Second, the type, scale and location of tourism developments are crucial decisions in creating a sustainable sector. Careful evaluation of projects is required to optimize economic returns and to minimize negative socio-cultural and environmental impacts. This is an exceedingly difficult area of analysis and one which can be influenced by political considerations. Third, any development will depend on the availability of the necessary infrastructure. Fourth, the assurance of 'equitable distribution' of benefits is only an ideal outcome and is based on altruism rather than reality. Linking the theoretical debates on the policies for promoting sustainable tourism in Vietnam, the following section sets the contextual background for field investigation.

Sustainable Tourism Policy and Planning in Vietnam

The concern for sustainable tourism was addressed in the first *Master Plan for Tourism Development 1995–2010* (VNAT, 1994) under the technical assistance from the World Tourism Organization (UNWTO) and approved by the Prime Minister in 1995. The plan for 15 years of development set economic, social security, environmental and cultural objectives to be achieved through tourism development. According to this plan, the territory of Vietnam was divided in three tourism zones and 11 tourism sub-zones. Different zones develop distinct tourism products based on the resources in an area. Since then, the method of tourism zoning by resources has been established and has remained the main method of planning for tourism throughout the past 30 years.

The centralized mechanism for tourism and development in Vietnam aligned with Jenkin's (2015) contention of the ultimate role of government in policy making and planning. The mobilization of resources for development-led growth of the tourism sector is prioritized. The concern for sustainable utilization of resources was not yet stated in this resolution. The second *Master Plan for Tourism Development to 2020 with Vision to 2030* (VNAT, 2013) provided a roadmap for accelerating tourism development by economic objectives (e.g. contribution to GDP, market expansion, modernization) along with other objectives of social and environmental sustainability emphasizing the preservation of traditional values and environment protection. The development of tourism in Vietnam has achieved the goals of increasing visitor numbers and making a significant contribution to national GDP (see Chapter 3, this volume). The World Economic Forum Tourism Competitiveness Index in 2017 ranked Vietnam the 67th globally, on par with the average of the rest of its regional competitors (World Bank, 2019). However, Vietnam's weakness is in the area of environmental sustainability (129th globally), although as a whole the South-east Asia region also performs very poorly in this dimension. The World Bank Group report (World Bank, 2019) raises the concern that rapid visitor growth, declining visitor yields, and rising overcrowding pressures imposes important challenges that will need to be addressed to ensure that future tourism growth is more sustainable, inclusive, and geographically balanced.

Furthermore, Vietnam has emphasized a more structured planning approach to the tourism sector with master plans at both the national level and municipal levels (World Bank, 2019). However, adherence to these master plans has often proved problematic, potentially undermining tourism development objectives and sustainability (Daklak's People's Committee, 2012). Under this circumstance, the authors of the current study conducted empirical research to investigate the gaps between tourism planning ordered by the nation state, but designed and implemented at the municipal level in Đắk Lắk province, and the reality of tourism development under the sustainable direction in the plan.

Đắk Lắk Tourism Planning

Đắk Lắk is located in the centre of Tây Nguyên (the Central Highlands), accessible to and from Gia Lai and Lâm Đồng, to and from the South Central Coast and Cambodia, connected to Khánh Hòa province through Highway 26 and Lâm Đồng through Highway 27 and Highways 14, 14C (see Fig. 4.1).

The province has Buôn Ma Thuột airport, operating domestic flights. Đắk Lắk province is the most populous area in the Central Highlands. As of 2017, the population is recorded as 1.8 million, accounting for 32.9% of the whole region's population. The distribution of the population is uneven, more than 50% of the population is concentrated in Buôn Ma Thuột city, the municipal capital. Đắk Lắk has the highest ratio of ethnic diversity in the country; 49 minority groups (out of 54 in Vietnam) inhabit the area, where the Kinh accounts for 70% of the population and the remaining 30% belongs to the other 48 minority groups, including 12.9% Êđê, 3.3% M'Nông, Thái, Tày, etc. These groups also preserve a variety of unique ethnic cultural values, with one UNESCO Intangible Cultural Heritage list member – the *Cultural Space of Gong Culture* in the Vietnam Highlands, 32 historical cultural sites, and more than seven festivals from the 49 minority groups residing in the area (Duong, 2020). The outstanding natural resources of the region are Yok Đôn and Chư Yang Sin national parks, and five other

Fig. 4.1. The location of Dắk Lắk province and the Central Highlands of Vietnam. Map courtesy of Thuy T. Duong.

protected areas, preserving unique tropical flora and rare animals. Basaltic eruptions have also left traces in numerous volcanoes and caves. The resulting volcanic soil and mild plateau climate make it favourable for the region to develop a wide range of agricultural products, including high-value perennial crops. The mountainous and plateau terrain upstream of the Sêrêpốk, Ba, Ea H'leo rivers create 17 waterfalls, in addition to the system of volcanoes and caves. On the basis of these resources Đắk Lắk's Tourism Development Plan in 2016–2020 (Daklak's People's Committee, 2012) identifies four major types of niche sustainable tourism as discussed below.

Major Forms of Sustainable Tourism Considered for Planning

Ecotourism

Vietnam Tourism Law (GOV, 2017) defines ecotourism as a form of tourism that takes place in natural settings intertwined with local cultural identity, involves local communities, and engages educational activities for environment protection. In the national tourism development strategy, ecotourism is considered to be the most important approach to navigating the development of sustainable tourism (see Chapter 6, this volume on ecotourism in Vietnam's protected areas). On the one hand, ecotourism in many cases has contributed to the conservation of nature, where tourism activities take place, and to the conservation of local culture by creating sustainable livelihoods (i.e. financial incentives) for local communities in the area. Ecotourism can also help to raise tourist awareness of environmental and cultural sustainability via educational elements of the activities (Buckley, 2009; KC *et al.*, 2020). On the other hand, as ecotourism relies heavily on natural resources and opens up environmentally sensitive areas to tourists, poorly planned and managed ecotourism can create many negative impacts. These range from overcrowding, pollution of air/water, disturbing local wildlife, flora and fauna to the destruction of the whole delicate ecological system (e.g. Boracay island, the Philippines) (Chirenje, 2017). There may also be potential power conflicts between external developers and local communities, which result in raising the cost of living, commodification of cultures, and displacing indigenous people out of their ancestral lands (Valle-García, 2014). With its richness of natural resources, ecotourism is on the top of the range of tourism products being proposed for Đắk Lắk.

Wellness tourism

Wellness tourism engages travellers in various physical, psychological or spiritual activities (e.g. yoga, meditation, body massage, spa) in order to enhance their health and well-being. The global wellness tourism market has grown to US$639 billion in 2017 and has become more popular in Vietnam in the last decade (Vietnamnews, 2019). This research defines wellness tourism in its 'green' manifestation, intertwining wellness-seeking motivations in an ecotourism setting, the so-called eco-wellness tourism. Eco-wellness tourism is defined as tourism activities utilizing natural resources (e.g. forest and fresh/mineral water, fresh climate and clean environment) for health enhancement and stress relief. Hence similar issues to ecotourism can also be observed in eco-wellness tourism, especially in relation to impacts to the environment and the responsible behaviours of tourism suppliers and tourists. In the context of mountainous regions, land/spatial suitability and the proper selection of potential areas for wellness tourism are also important to ensure the success of this tourism niche (Pan *et al.*, 2019). The topography and diversity of natural resources and favourable climatic conditions make this option of development feasible for the province.

Cultural tourism

Cultural tourism is defined by UNWTO (2017) as 'a type of tourism activity in which the visitor's essential motivation is to learn, discover, experience and consume the tangible and intangible cultural attractions/products in a tourism destination'. In Vietnam, cultural tourism is seen as tourism which takes advantage of cultural diversity and uniqueness and contributes to the preservation and promotion of traditional culture as specified in *Vietnam Tourism Law* (GOV, 2017). These positive impacts can be achieved through the inter- and cross-cultural exchanges which can promote cultural understanding and increase community pride, and the provision of economic/social incentives for the revitalization of cultural traditions and practices (e.g. traditional handicrafts and performances) (Esman, 1984; Kumar, 2017). Cultural tourism is extremely relevant to Đắk Lắk, as diverse minority ethnic cultures (both tangible and intangible) are distinguished attractions of the Central Highlands of Vietnam. However, the utilization of culture as a tourism resource may lead to misrepresentation or commodification of local cultures, a loss of authenticity and reinforcing stereotypes if it is not planned and managed properly (Macleod, 2006; Tolkach and Pratt, 2019).

Agri-tourism

Agri-tourism utilizes the agricultural resources available for leisure activities via the involvement of and interaction with local farmers. This form of tourism brings economic benefits to local communities and raises awareness of agricultural activities among visitors. Mostly taking place in rural areas and contributing to diversify local income, agri-tourism in developing countries is often linked to a pro-poor tourism approach, which aims to increase 'net' tourism benefits to the economically poor population (Phi *et al.*, 2018). In principle, agri-tourism creates value-added agricultural products and can take place at small scales ranging from households to cooperatives, agricultural villages and fields that directly involve local farmers (peasants) in an area. Regardless of the scale, for agri-tourism to be successful, there needs to be good coordination among suppliers of agri-tourism products and services (beyond individual actions) in order to develop strong links between tourism activities and specific resources in local ecosystems and cultural identity (Contini *et al.*, 2009).

To what extent do these four proposed types of tourism match the geographical resources and current tourism activities in our study area? Without precise understanding of the resource-development matching in Đắk Lắk, spatial planning can be hindered in its effectiveness in guiding future tourism development.

Study Methods

The current study utilized geographical surveys using quantitative assessment tools and geographic information system (GIS) methods. Accordingly, the geographical survey method was employed in Study I and the quantitative survey method was employed in Study II. The combination of these methods is justified for its relevance to tackling the problems in managing resources and preserving natural and cultural values while allowing tourism to develop its full potential. In Study I, MapInfo 12.0 software was used to project a tourism resource map matching the resources and potential tourism activities. This map provides a visual zoning solution being used for the visitor and resident survey in the next step. Data input for this GIS analysis was collected from secondary data provided by local officers and the first author's fieldwork. Then the characteristics and conditions of each zone were matched to relevant types of sustainable tourism which the province has considered to develop. The outcomes of Study I were presented in maps of resource zones and resource-product matching tables. In Study II, drawing on the zoning solutions from Study I, the researchers developed a questionnaire to collect tourists' feedback on geographical conditions and resources for tourism in these geographical zones. Survey questions used a five-point Likert scale. Sources for questionnaire items are shown in Table 4.1. Visitors were surveyed at different destinations in Đắk Lắk province. In 2018, 468 survey responses from tourists were collected. In addition to the tourist survey, surveys of locals were also conducted that targeted groups who provide accommodation, elephant riding services, kayaking, small travel agencies and souvenir shops (197 respondents). At the end of 2018, a total of 665 survey questionnaires were collected from locals in Đắk Lắk. Survey data were analysed using SPSS 20 software. In the final stage of analysis, the findings from Study I and Study II were integrated when the current tourism products were projected on the backdrop of resource assessments to highlight possible outcomes towards maximization of resource potentials while accomplishing the goal of sustainable tourism development. Consequently, suggestions for tourism planning adjustments were then drawn from the analysis.

Findings

Study I. Geographical zones and tourism resources

Based on assessment of resource distribution and condition in the province using geographical zoning methods, possible resources for tourism development were stratified into five zones (see Fig. 4.2).

Zone 1: Semi-Plateau of Buôn Đôn – Ea Súp

The area spans over two municipalities of Buôn Đôn and Ea Súp. The Êđê group accounts for 70% of the ethnic minorities' population, in addition to a small number of H'Rê, M'Nông, H'Mông and Nùng inhabiting the area. The main livelihoods in the region are agriculture, forestry and

Table 4.1. Geographic conditions and resources visitors were asked about in the survey questions.* Compiled by the authors.

Items for consideration in the questionnaire	Source for items
Evaluation of tourism resources	
Diverse geographical terrain	Wang *et al.* (2015)
Fresh and cool climate	Yan *et al.* (2017)
Diverse biological resources	Philemon (2015)
Green, clean environment	Yan *et al.* (2017)
Effectiveness of biodiversity conservation	Generated from local resource survey
Attractiveness of elephant tourism activities	Philemon (2015)
Special cultural, historical and festival values	Philemon (2015); Yan *et al.* (2017)
Noise pollution in tourist season	Generated from local resource survey
Conservation of local culture and handicrafts	Generated from local resource survey
Crowding in festive seasons	Generated from local resource survey
Monotonous tourism products	Generated from local resource survey
Losing traditional values	Generated from local resource survey
Evaluation of tourism facilities	
Clean water supply	Tosun (1998)
Convenient and large-scale parking	Tosun (1998)
Quality of transportation	Tosun (1998)
Quality of restaurants	Tosun (1998)
Convenience and variety of souvenirs	Tosun (1998)
Quality of amusement and entertainment	Tosun (1998)
Quality of healthcare facilities	Yaakop (2013)
Priority for protection of resources and environment	
Elephant protection	Generated from local resource survey
Historical and cultural heritage conservation	Castellani and Sala (2010)
Conservation of ethnic culture in Êđê, M'Nông, Gia Rai villages	Generated from local resource survey
Biodiversity conservation in national parks and nature reserves	Castellani and Sala (2010)
Waste collection	Castellani and Sala (2010)
Environment education	Castellani and Sala (2010)
Periodical treatment for environment	Castellani and Sala (2010)

*Tourists who took part in the survey were asked to rank each item using a five-point Likert scale.

tourism. The largest national park in Vietnam, Yok Đôn is home to many globally endangered species and has the richest bird fauna in Indochina. The region's biological diversity and its natural features, and the Êđê group's cultural traditions and lifestyles are the background to some harmonious human–nature activities such as taming elephants, housing architecture and costumes.

Zone 2: Buôn Ma Thuột Plateau

This is the third largest volcanic plateau in the Central Highlands and has an area of

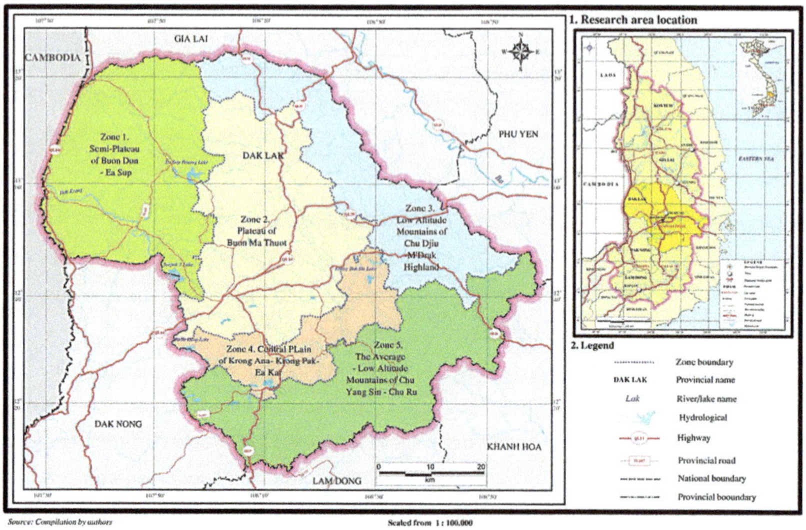

Fig. 4.2. Map of the five geographical zones in Dắk Lắk province. Map courtesy of Thuy T. Duong.

3000 km^2. From Buôn Ma Thuột, visitors can get around to other destinations. In terms of ethnic diversity, the dominant group, the Kinh inhabits the urban area of the plateau. The Êđê ethnic group originally settled there many generations ago, living in villages ringed around suburban areas. The traditional arts, performance and lifestyles of diverse ethnic minorities are major features for the annual Central Highland Gong Festival (see Fig. 4.3). Climate and soil conditions of the area are suitable for perennial crops such as coffee, pepper, avocado and cashews, the major livelihood for inhabitants of the area. In addition, the Ea Ral Habitat Conservation Area is another potential site for ecotourism activities.

Zone 3: Low Altitude Mountains of Chư Djiu – M'Đrắk Highland

The main ethnic groups living in the area are the Gia Rai, M'Nông, Kinh, Mường, Dao, Nùng and H'Mông peoples. Major economic activities are afforestation and agricultural production (planting perennial crops). There is Ea Sô Nature Reserve (27,800 ha in Ea Kar district), suitable for developing ecotourism, while coffee and pepper plantations have potential for agri-tourism.

Zone 4: Central Plain of Krông Ana – Krông Pắk – Ea Kar

This region is located in between Buôn Ma Thuột Plateau and the high-altitude mountains of Chư Yang Sin. Its topography is made up of accumulated deposits and eroded river valleys and hills. The distribution of ethnic groups is diverse throughout the area with the M'Nông in Krông Ana and Lak, the Kinh and Nùng in Krông Bông and the Êđê in Ea Kar. Scenic landscapes, festivals and elephant riding activities are among the resources for tourism development. In this lowland area, wet-rice agriculture and aquaculture are the major economic activities. The area has diverse terrain and attractions with more than ten water landscapes (waterfalls and lakes), such as the Đray Nur – Đray Sáp Thượng Waterfall and Lắk Lake. This region also has a pleasant and fresh climate.

Zone 5: The Average – Low Altitude Mountains of Chư Yang Sin – Chu Ru

The Chư Yang Sin Mountain (2442 m, the second roof of the Central Highlands), and many mountain peaks higher than 2000 m (Chư Mư, Vọng Phu, etc.) are found in this region. The cultural values link to the main ethnic groups living

Fig. 4.3. An ethnic festival. Photo courtesy of Thuy T. Duong.

along the mountainous parts of the area. The lowland area is the home of the Êđê, M'Nông, Kinh and Mường. Meanwhile, the less populous groups such as the Tày, Thái and Gia Rai live at altitudes below 1000 m, and the H'Mông people live in areas 1100–1500 m high. The main livelihood of locals depends on forestry (forest plantation and production), cultivation of industrial and food crops (sugarcane, cassava, maize, wet rice) and aquaculture. Forest conservation and reforestation are given priority. The evergreen forest in many mountainous areas offers a pleasant climate for ecotourism as well as wellness-ecotourism.

Matching geographical zones to tourism types

In analysing the concentration of tourism activities by the five geographical zones, the authors identified the following patterns:

1. Agricultural tourism takes place at coffee farms, fruit and bonsai gardens in Buôn Ma Thuột city and Cư M'Gar (Zone 2). Agritourism activities (Zone 1) add secondary activities in sightseeing of cultural landscapes, such as short visits to coffee farms, orchids and ornamental plant gardens.
2. Wellness tourism is being experienced mostly in the area of Lắk Lake (Zone 4) and in spa resorts located in Buôn Ma Thuột city (Zone 2). At a smaller scale, there are resorts in the low mountain area of Zone 3.
3. Ecotourism is the major type of tourism that utilizes the resources of various water-front landscapes, national parks and reserves. The sites for ecotourism are: Yok Đôn National Park and Buôn Đôn Ecotourism Area (Zone 1), trekking in Chư Yang Sin National Park and Nam Kar Nature Reserve (Zone 5), Lắk Lake Ecotourism Area, Đray Nur – Đray Sáp Thượng Waterfall (Zone 4), and Akô Dhông Ecotourism, Đầu Nguồn, Đồi Thông, Bến Nước Amarin, Suối Ông, M'Đrăk (Zones 2 and 3).
4. Cultural tourism has its typical products including the *Cultural Space of Gong Culture* in the Central Highlands designated as a UNESCO Intangible Cultural Heritage list member, historical relics, ethnic

Table 4.2. Matching geographical zones with types of tourism. Compiled by the authors.

Geographical zones	Types of tourism: Agri-tourism	Wellness tourism	Ecotourism	Cultural tourism
Zone 1: Semi-Plateau of Buôn Đôn – Ea Súp	O		O	O
Zone 2: Plateau of Buôn Ma Thuột	O	O		O
Zone 3: Low Altitude Mountains of Chư Djiu – M'Đrắk Highland		O	O	
Zone 4: Central Plain of Krông Ana – Krông Pắk – Ea Kar		O	O	O
Zone 5: The Average – Low Altitude Mountains of Chư Yang Sin – Chu Ru			O	

festivals, and leisure activities engaging with elephants, and unique cultures of the Êđê, M'Nông and Gia Rai's ethnics across all zones. Currently, cultural tourism has only been developed in Zones 1, 2 and 4. The matching of geographical zone and types of tourism is shown in Table 4.2.

Current tourism activities are highly concentrated in the urban area (Zone 2), where all four forms of tourism are present in the city of Buôn Ma Thuột (cultural tourism) and its suburban area (ecotourism, wellness and agri-tourism). Concentration of tourism activities is found in Zone 1, where Buôn Đôn offers cultural tourism and the surrounding areas have developed ecotourism and agri-tourism. In Zone 4, the Central Plain, tourism development has taken place around the Lắk Lake. Currently, limited ecotourism activities have been initiated in the Ea Sô Nature Reserve (Zone 3) and Chư Yang Sin National Park (Zone 5). In other words, potential resources in Zone 3 and Zone 5 have not yet been utilized for tourism development. In the meantime, tourism activities in Zones 1, 2 and 4 are rather simple and repetitive.

Study II. The reality of tourism development across the five zones

Tourist evaluation of resources

Domestic tourists accounted for more than 95% of the total number of visitors in 2018. International tourists increased at a slower pace compared to domestic tourists. Between 2000 and 2018, international tourists increased from 7825 to 41,752, but accounted for only 4.8% of the total number of visitors to the area (see Fig. 4.4). Across the five zones, tourists

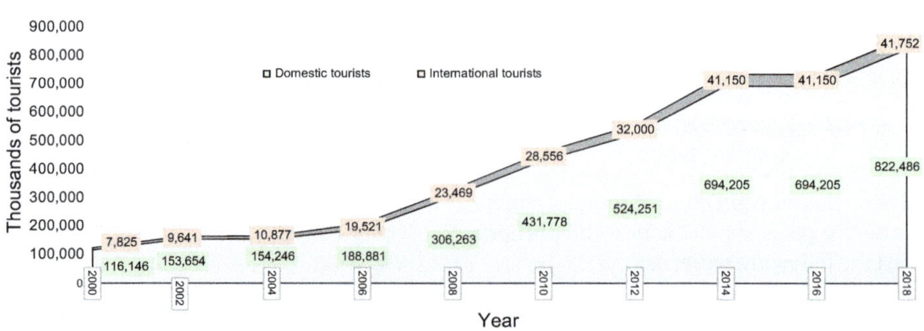

Fig. 4.4. Number of visitors to Đắk Lắk (2000–2018). Pink values are the number of international tourists, green values are the number of domestic tourists. From Daklak's People's Committee, 2019.

Table 4.3. Tourist evaluation of resources. Compiled by the authors from fieldwork data in 2018 in Đắk Lắk.

Evaluation criteria*	Zone 1	Zone 2	Zone 3	Zone 4	Zone 5
Diverse geographical terrain	3.48	4.39	4.05	4.53	4.68
Fresh and cool climate	2.65	4.08	4.35	4.05	4.69
Diverse biological resources	4.39	2.04	4.13	4.19	4.72
Green, clean environment	4.02	4.01	4.06	4.06	4.32
Effectiveness of biodiversity conservation	4.59	3.08	3.93	3.97	4.58
Attractiveness of elephant tourism activities	4.76	2.56	2.75	4.67	4.06
Special cultural, historical and festival values	3.27	4.86	3.97	4.39	3.98
Noise pollution in the tourist season	3.06	4.39	2.69	3.78	3.80
Effectiveness of the conservation of local culture and craft villages	3.87	4.82	3.00	4.16	3.01
Crowding in festive seasons	4.88	4.75	2.98	4.35	2.95
Monotonous tourism products	2.04	2.30	4.38	3.86	4.29
Losing traditional values	4.62	2.46	2.80	3.09	2.38

*Rating on a five-point Likert scale.

concentrate in Zone 2, in particular Buôn Ma Thuột city welcomes 30% of the total visitor arrivals. Zone 1 (mainly in Buôn Đôn) received around 9% of the overall visitor volume. The Lắk Lake and Đray Nur Waterfall in Zone 4 shared only 5% of the total visitor arrivals in 2017. In the period from 2005 to 2019, even though the length of stay of domestic tourists incrementally increased only by 4.83%, the daily expenditure increased by 300% from Vietnamese dong (VND) 350,000/day in 2005 to VND 1,050,000/day in 2019. International tourist expenditure doubled from US$30 (approximately VND 600,000/day) in 2005 to US$66 (VND 1,320,000/day) in 2019.

Visitor evaluation of tourism resources in the five zones was conducted in 2018 using a set of questions about the geographic conditions and resources shown in Table 4.1. The results of the visitor survey are shown in Table 4.3.

- Zone 1: Tourists are interested in tourism activities linked to elephants (4.76), biological diversity (4.39) and the environment (4.02). Visitors rate the effectiveness of biodiversity conservation at a relatively high level (4.59). However, visitors are highly concerned at the deterioration of traditional values (4.62) and overcrowding in the festive seasons (4.88).
- Zone 2: The value of ethnic culture, traditional craft villages, unique festivals are the main attractions (4.86) in addition to the basalt plateau (4.392). However, noise and overcrowding during the tourist season (4.75) are major concerns for tourists.
- Zone 3: The region attracts tourists by its cool, fresh climate (4.35), the Ea Sô Nature Reserve (4.13), the waterfall system (4.05) and the pristine environment (4.06). However, tourism has yet to be developed in the region due to a shortage of investment and promotion.
- Zone 4: Visitors are attracted by its diverse terrain with water landscapes, pleasant climate and the green environment in addition to the unique culture of the M'Nông ethnic (4.39) group, and elephant-related leisure activities (4.67).
- Zone 5: The place is endowed with mountains that are attractive because of a typical mountain ecosystem (4.72), cool and fresh climate (4.69), and biodiversity

Table 4.4. Tourist evaluation of facilities. Compiled by the authors from fieldwork data in 2018 in Đắk Lắk.

Evaluation criteria*	Zone 1	Zone 2	Zone 3	Zone 4	Zone 5
Clean water supply	3.06	4.07	3.97	4.59	4.01
Convenient and large-scale parking	4.54	4.03	3.59	3.87	3.01
Quality of shopping	4.02	4.67	3.88	4.15	3.02
Quality of restaurants	4.28	4.75	3.04	4.39	4.11
Convenient shopping and variety of souvenirs	4.14	4.46	3.76	4.06	3.23
Good quality of amusement and entertainment facilities	3.58	4.36	2.97	3.40	2.95
Quality of healthcare facilities	2.79	4.25	3.48	4.17	2.98

*Rating on a five-point Likert scale.

conservation and environmental protection (4.58), and the mountains are ideal locations for developing ecotourism and resorts. Two protected areas of Chư Yang Sin National Park and Nam Kar Nature Reserve are well reserved, however, these are rather difficult for tourists to gain access to, thus tourism products remain relatively simple (4.29).

The survey results show that Zones 1, 2 and 4 generally received higher scores on the quality of tourism facilities, while in Zones 3 and 5 the facilities are underdeveloped (see Table 4.4). Except for the urban area (Zone 2) that has established entertainment and healthcare facilities, other areas received relatively low scores on these items. The areas with great potential (Zone 3 and 5) thanks to the mountainous terrain, nevertheless still suffer from accessibility problems, hence the investment in tourism facilities has yet to be encouraged.

Local evaluation of economic and socio-cultural impacts of tourism

Commercial services. Households participating in the survey are those who sell food (63.3%), beverages (30.3%), and souvenirs and clothing (13.3%). The shopkeepers live in Buôn Ma Thuột city (23.3%), Lắk district (20%), Ea Kar district (10%), Buôn Hồ town (10%), and a smaller number residing outside the province. The tourist services offer opportunities for local employment especially for women. Opening a shop requires micro-finance, where 73.3% of the respondents state that they only need VND 5–10 million, while 37.02% need less than VND 5 million (see Fig. 4.5). Local income, however, varies. These micro-businesses attract the participation of ethnic minority groups. Of the respondents 46.7% state that income from tourism accounts for less than 40% of the household income, 33.3% of the respondents answer that this proportion is 40–70%, and 16.7% of respondents admit that 80% of the total household income is from commercial tourism services. During the festive and tourist season in summer, shops selling food and beverages in major tourist areas (e.g. Buôn Ma Thuột city, Buôn Đôn district, Krông Ana district, Lắk district) generate an income of VND 15–25 million/month in 1–3 months/year, but souvenir shops bring in VND 5–10 million/month. However, shops in remote areas, such as Zone 4 of Cư M'gar district, and M'Đrắk district, the shopkeepers benefit less than VND 2 million/month.

Accommodation services. The price of accommodation in Đắk Lắk ranges from VND 180,000 to VND 2,000,000/room/night and the price is increased by 1.5/2 times during festival seasons and in the first 3 months of each year. The number of employees working for accommodation services varies from two to three employees at a motel to 20–40 employees at high-end hotels and resorts. Accommodation services at Đắk Lắk are gradually integrating events (68%), healthcare services (25.3%), entertainment (26.5%) and shopping (14.28%) onto their portfolios. However, only 20% of all accommodation services cooperate with tourism agencies, restaurants and destination management boards. Supporting products provided by the accommodation sector are cuisines using local specialties

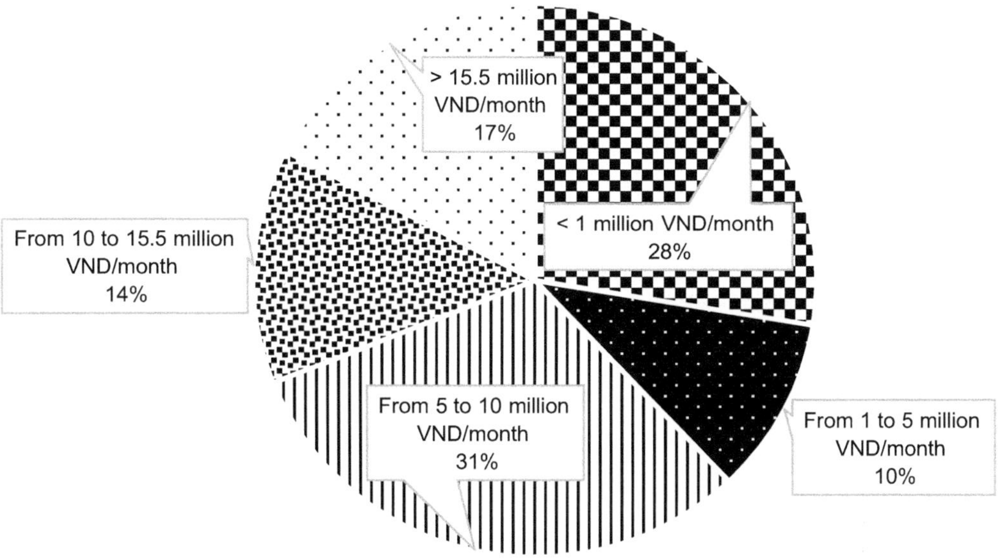

Fig. 4.5. The distribution of income from tourism. Authors' compilation from fieldwork data in 2018 in Đắk Lắk.

(33%), liquor (21.4%) and ethnic cultural items. The challenges that accommodation providers are facing include lack of visitors after the first 3 months of the year (100%), and lack of high-quality labour in the tourist season (92%). Local accommodation providers would like to have support in tourism promotion (100%), training (75%) and more incentives to expand their businesses (67%).

Homestay owners who participated in the survey are those who live in the ethnic villages of Êđê, M'Nông, Gia Rai, Dao, Nùng and Thái, found in Buôn Ma Thuột city (28.5%), and in remote areas of Lắk, Buôn Đôn and Cư M'Gar districts (14.28%). The price of homestay ranges from VND 150,000/night to VND 500–600,000/night and varies in accordance with room space, location and target customers. Homestays in ethnic villages offer an authentic cultural experience to foreign visitors, but a lack of foreign language skills, and certification issued by the Ministry of Culture, Sports and Tourism are major barriers for expansion. Therefore, 86.9% of all homestay owners expect to receive further training on foreign language and tourism-related skills (75.8%) in addition to financial support (59.2%).

Elephant riding services. This service has attracted many tourists in Buôn Đôn, Lắk Lake and Nam Kar Nature Reserve. The elephants have fairly long-life expectancy (45.5% of all the elephants are 50–85 years old, while 10% are about 10–12 years old). Their main food is elephant grass, bananas, sugarcane, maize (corn) bran and the price for one serving is around VND 200,000–250,000 per elephant. Most of the food (90%) is grown by the family, while the remaining 10% is bought from other people in the same village. The fee for riding elephants is about VND 150,000–250,000 per person.

Survey results show that 75% of the elephants serve between 20 and 30 tourists/day. The elephants are overworked, easily get tired and suffer from diseases in their digestive and visual organs. A village revealed that for many years, there have been no baby elephants born naturally in that village. Local households recognize the issues of animal abuse, when they suggest that 'elephants should not overwork', and emphasize 'the need to have financial subsidies for local people to pay for elephant food', and demand 'a particular policy that protects elephants'. In the long-run, elephant ageing and raising concerns on animal ethical issues

Table 4.5. Local evaluation of the priority for conservation and protection of resources (expressed as a percentage of respondents who took part in the survey). Compiled by the authors from fieldwork data in 2018 in Đắk Lắk.

Criteria	Very few priorities	Less priority	Medium priority	Quite a priority	Very high priority
Elephant protection	–	–	2.74	18.68	78.58
Historical and cultural heritage conservation	–	10.43	24.17	57.14	8.26
Conservation of ethnic culture in Ede, M'Nông, Gia Rai villages	0	12.63	22.52	28.02	36.83
Biodiversity conservation in national parks and nature reserves	0.54	2.74	13.18	42.85	40.69
Waste collection	–	4.9	25.27	45.6	24.23
Environment education	–	8.2	38.46	36.81	16.53
Periodical treatment for environment	1.64	15.38	28.02	39.01	15.95

have meant that the elephant riding service has declined significantly and might disappear in the near future. Lately, the elephant riding service, however, has been suspended.

LOCAL AWARENESS OF THE NEED FOR RESOURCE CONSERVATION AND PROTECTION. Table 4.5 shows the local awareness of the need for environment protection. The results show that protecting elephants is highly prioritized (97.26%), followed by protecting the ecosystem in national parks and protected areas (83.54%), preserving cultural values (64.85%), and conserving cultural and historical heritage (65.4%). Collecting waste is also prioritized (69.83%) but periodical environment treatment is still limited (17.02% answered that this is not a priority). Meanwhile, the effectiveness of campaigns on environmental conservation remains low.

Discussion

Đắk Lắk is considered to be a relatively remote and not-easy-to-access area for international tourists when the only airport in the province operates domestic routes. International tourism accounts for less than 5% of the total visitor arrivals to the area, therefore, Đắk Lắk remains primarily a domestic destination. Moving in the direction of sustainability, four forms of tourism are planned for the area. Spatial analysis was carried out to classify possible resources for tourism development into five zones and match these to appropriate forms of alternative tourism. The tourist opinions on evaluation of resources and local survey of tourism product offerings that were conducted in the five zone areas provide a testimony for the planning of tourism development.

Although Đắk Lắk's resources for tourism are abundant and varied, with a great potential to accommodate all four alternative forms of tourism, the current state of tourism development shows imbalance in distribution and inefficiencies in utilizing the resources. Tourism activities are mainly concentrated in Zone 2, around Buôn Ma Thuột city, and expand only to the nearby area of Zone 1 and to some limited extent in Zone 4. While visitor spending has increased, perhaps because of the relatively high rate of inflation in the country, the length of stay has not increased over the years, showing that tourism activities remain unattractive and do not contribute to extending tourist stay in the area.

The unique *cultural tourism* experience by interactions with elephants, however, has been at the edge of extinction. First, the elephants' natural reproductive capacity is gradually declining, while the number of domesticated elephants is decreasing. Second, Yok Đôn National Park has signed an agreement with the Animals Asia Foundation not to promote the elephant riding service; visitors can only 'watch' the elephants. Đắk Lắk has one UNESCO Intangible Cultural Heritage item – the *Cultural Space of Gong Culture*, but this has yet to be utilized effectively for branding regional cultural tourism. While all five zones have rich resources derived from the 49 ethnic groups living there, cultural tourism is still in its rudimentary development. There are limited homestay places where visitors can integrate and experience ethnic cultures. There are no cultural centres or regional museums where tourists can learn about the ethnic groups living in the area, to appreciate and consequently to experience them. Much work remains to restore, embellish and build cultural institutions in the central villages of the ethnic minorities Êđê, M'Nông and Gia Rai.

The option of *ecotourism* seems to be dominant; all five zones show diverse resources along with their unique terrain and topography, from plains to semi-highlands, and highlands that have low-to-average mountain ranges with different climate zones. Currently ecotourism is concentrated in Zone 1 near the Sêrêpốk river and several waterfalls. The utilization of such mountain resources for tourism remains limited though. The development of sustainable tourism that utilizes natural resources, however, is facing other challenges. Among these are the hydropower dam construction projects altering water flow in rivers, deforestation in the upstream areas causing droughts and lowering the water level in many lakes, making the region suffer shortages of water. Furthermore, illegal deforestation has long been a threat to biodiversity. Also, the tourism industry's competitiveness is relatively low compared to other industries such as cultivating and processing agricultural products, therefore local tourism attracts little attention in terms of both capital and policies.

Resources are under-valued for *agri-tourism*. In all five zones, there are significant agricultural production sites for perennial crops and other unique, high-value agricultural products, such as coffee (see Fig. 4.6). However, to date the limited agri-tourism is being offered mainly in Zones 1 and 2. These activities have the potential to be established in other zones, where local communities can benefit from value-adding to agricultural products from this type of tourism. Effective branding for agri-tourism can utilize Buôn Ma Thuột's self-claimed status as the coffee capital of Vietnam. The Buôn Ma Thuột Coffee Festival has been organized for 15

Fig. 4.6. A coffee farm. Photo courtesy of Thuy T. Duong.

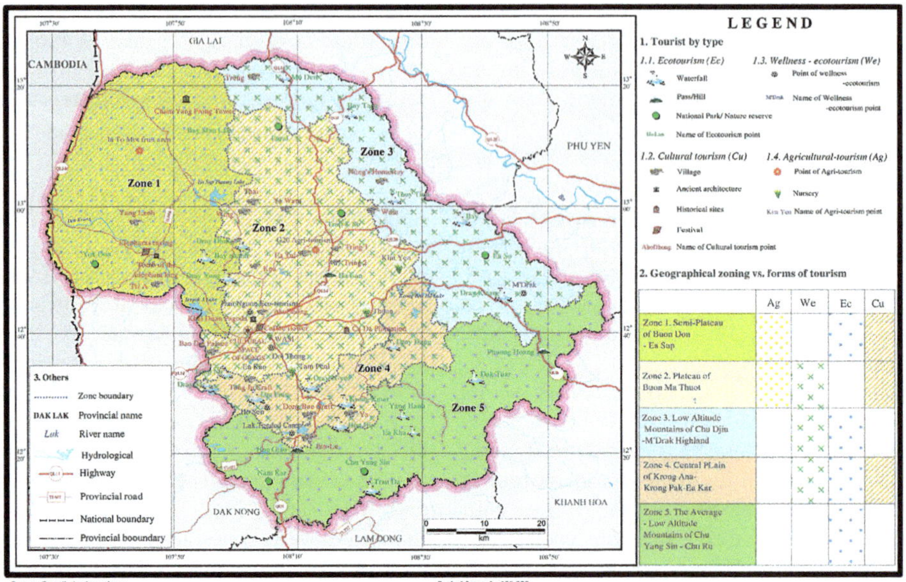

Source: Compilation by authors

Scaled from 1 : 100.000

Fig. 4.7. Geographical zones and tourism resources. Map courtesy of Thuy T. Duong.

years and could be used for agri-tourism branding focusing on coffee culture and art.

The Central Highlands, with a favourable climate and mountain environment for *wellness tourism*, have yet to be utilized for development. Wellness tourism could utilize various waterfront resources, such as waterfalls, lakes and forests. The mountain areas, with a cool climate and diverse ethnic cultures, have the potential to be developed into mountain resorts for wellness and cultural tourism. Not too far from Hồ Chí Minh city, Đắk Lắk could be an ideal location for wellness tourism if resources are properly utilized. These opportunities have been highlighted in research on mountain development in South-east Asia (Jones *et al.*, 2021). Mapping geographical resources and potential forms of tourism is shown in Fig. 4.7.

Conclusions

Overall, Đắk Lắk wants to shift tourism away from urban centres by planning for alternative forms of tourism utilizing natural and cultural resources. However, the current state of tourism development is still very much dependent on mass tourism for the domestic market. This trend is obvious when tourists stay for relatively short periods, concentrate in very few areas, and participate in a limited range of activities. Tourism products are similar in types and forms. While niche tourism has received attention, the demand for niche tourism products such as agri-tourism, wellness and ecotourism has not yet been researched properly. The current stage of alternative tourism, however, links to mass domestic tourists coming in the 3-month festive seasons that echo Harrison's (2015) contention that alternative tourism is dependent on mass tourism.

The tourism planning and policies in Đắk Lắk reflect all four points mentioned by Jenkins (2015). First, only central and municipal governments plan for tourism development, often in coordination with Vietnam National Administration of Tourism (VNAT). This centralized role and excessive planning approach for tourism was identified also in the World Bank report (2019). Second, because spatial planning is researched by scientists under government command, the type, scale and location of tourism development are decided by them to ensure the sustainability of tourism development. Ideally for the government, development should optimize economic returns and minimize negative socio-cultural and environmental impacts. The current situation of tourism in Đắk Lắk shows that some areas are excessively developed, while other areas with resource abundance are relatively underdeveloped. However, regional development has deeply been influenced by political considerations, thus, the decision to develop is not in the hands of local administrators. Third, the development of tourism is subject to the availability of the necessary infrastructure. The mountain areas are relatively poor in facilities, making it difficult to attract and utilize resources there. Fourth, the question of who should benefit from tourism remains unanswered. Although ethnic minorities are encouraged and given opportunities to participate in the tourism supply chain, the major barriers of training and language proficiency may hinder their accessibility and participation in tourism services. Moreover, tourism only contributes 1.5% of the GDP for Đắk Lắk, much less than agriculture. Thus, shifting from agriculture and forestry to tourism still has a long way to go.

References

Aramberri, J. (2010) *Modern Mass Tourism*. Emerald, London.

Buckley, R. (2009) Evaluating the net effects of ecotourism on the environment: a framework, first assessment and future research. *Journal of Sustainable Tourism* 17(6), 643–672.

Bui, H. and Dolezal, C. (2020) The tourism-development nexus in Southeast Asia: history and current issues. In: Dolezal, C., Trupp, A. and Bui, H. (eds) *Tourism and Development in Southeast Asia*. Routledge, London, pp. 23–40.

Castellani, V. and Sala, S. (2010) Sustainable performance index for tourism policy development. *Tourism Management* 31, 871–880.

Chirenje, L. I. (2017) Contribution of ecotourism to poverty alleviation in Nyanga, Zimbabwe. *Chinese Journal of Population Resources and Environment* 15(2), 87–92.

Contini, C., Scarpellini, P. and Polidori, R. (2009) Agri-tourism and rural development: the Low-Valdelsa case, Italy. *Tourism Review* 64(4), 27–36. DOI: 10.1108/16605370911004557.

Daklak's People's Committee (2012) *Master Plan for Daklak Tourism Development until 2020 with the Vision of 2030*. Daklak's People's Committee, Buon Ma Thuot, Vietnam.

Daklak's People's Committee (2019) *Report on Social and Economic Development of Daklak Provice – 2019*. Daklak's People's Committee, Buon Ma Thuot, Vietnam.

Duong, T.T. (2020) Assessement of geographial conditions and resources for sustainable tourism development in Daklak Province. PhD thesis, University of Natural Sciences, Vietnam National University, Hanoi.

Elkington, J. (1998) Partnerships from cannibals with forks: the triple bottom line of 21st-century business. *Environmental Quality Management* 8(1), 37–51.

Esman, M.R. (1984) Tourism as ethnic preservation: the Cajuns of Louisiana. *Annals of Tourism Research* 11(3), 451–467.

GOV (Government of Vietnam) (2017) *Vietnam Tourism Law*. Government of Vietnam, Hanoi.

Harrison, D. (2015) Development theory and tourism in developing countries: what has theory ever done for us? *International Journal of Asia Pacific Studies* 11(S1), 53–82.

Holden, A., Sonne, J. and Novelli, M. (2011) Tourism and poverty reduction: an interpretation by the poor of Elmina, Ghana. *Tourism Planning & Development* 8(3), 317–334.

Hunter, C. (1997) Sustainable tourism as an adaptive paradigm. *Annals of Tourism Research* 24(4), 850–867.

Jenkins, C.L. (2015) Tourism policy and planning for developing countries: some critical issues. *Tourism Recreation Research* 40(2), 144–156.

Jones, T., Bui, H. and Appollo, M. (2021) *Nature-based Tourism in Asia's Mountain Protected Areas – a Transregional Review of Peaks and Parks*. Springer, Cham, Switzerland.

KC, A., Ghimire, S. and Dhakal, A. (2020) Ecotourism and its impact on indigenous people and their local environment: case of Ghalegaun and Golaghat of Nepal. *GeoJournal*, 1–19. Available at: https://doi.org/10.1007/s10708-020-10222-3

Kumar, A. (2017) Cultural and heritage tourism: a tool for sustainable development. *Global Journal of Commerce & Management Perspective* 6(6), 56–59.
Macleod, N. (2006) Cultural tourism: aspects of authenticity and commodification. In: Smith, M. and Robinson, M. (eds) *Cultural Tourism in a Changing World. Politics, Participation and (Re)presentation*. Channel View Publications, Clevedon, UK, pp. 177–190.
Mowforth, M. and Munt, I. (2015) *Tourism and Sustainability: Development, Globalization and New Tourism in the Third World*. Routledge, London.
Pan, X., Yang, Z., Han, F., Lu, Y. and Liu, Q. (2019) Evaluating potential areas for mountain wellness tourism: a case study of Ili, Xinjiang Province. *Sustainability* 11(20), 5668.
Phi, G.T., Whitford, M. and Reid, S. (2018) What's in the black box? Evaluating anti-poverty tourism interventions utilizing theory of change. *Current Issues in Tourism* 21(17), 1930–1945.
Philemon, J. (2015) Assessment of tourist perception and satisfaction of Tanzania destination. *European Scientific Journal* 11(13), 107–119.
Sharpley, R. and Telfer, D. (eds) (2002) *Tourism and Development*. Channel View Publications, Clevedon, UK.
Telfer, D.J. and Sharpley, R. (2015) *Tourism and Development in the Developing World*. Routledge, London.
Tolkach, D. and Pratt, S. (2019) Globalization and cultural change in Pacific Island countries: the role of tourism. *Tourism Geographies* 23(3).
Tosun, C. (1998) Roots of sustainable tourism development at the local level: the case of Urgup in Turkey. *Tourism Management* 19(6), 595–610.
UNWTO (World Tourism Organization) (2017) Tourism and Culture. Available at: https://www.unwto.org/tourism-and-culture (accessed 10 May 2021).
Valle-García, S.E. (2014) Ecotourism: sustainable indigenous policies and its effects in Mayan communities, southern Mexico. In: Brebbia, C., Favro, S. and Pineda, F. (eds) *Sustainable Tourism VI. WIT Transactions on Ecology and the Environment*. vol. 187. WIT Press, Southampton, UK, pp. 239–250.
VNAT (Vietnam National Administration of Tourism) (1994) *Master Plan for Tourism Development 1995–2010*. VNAT, Hanoi.
VNAT (Vietnam National Administration of Tourism) (2013) *Master Plan for Tourism Development to 2020 with Vision to 2030*. VNAT, Hanoi.
Vietnamnews (2019) Wellness tourism on the rise in Việt Nam. Available at: https://vietnamnews.vn/economy/519170/wellness-tourism-on-the-rise-in-viet-nam.html (accessed 14 May 2021).
Wang, D., Niu, Y., Lu, L. and Qian, J. (2015) Tourism spatial organization of historical streets – a postmodern perspective: the examples of Pingjiang Road and Shantang Street, Suzhou, China. *Tourism Management* 48, 370–385.
Weaver, D.B. (2014) Asymmetrical dialectics of sustainable tourism: toward enlightened mass tourism. *Journal of Travel Research* 53(2), 131–140.
World Bank (2019) *Taking Stock: Recent Economic Development of Vietnam (Special Focus: Vietnam's Tourism Developments)*. The World Bank Group, Hanoi.
Yaakop, A. (2013) Tourist satisfaction in Malaysia. *International Journal of Business and Social Science* 4(3), 221–226.
Yan, L., Gao, W. and Zhang, M. (2017) A mathematical model for tourism potential assessment. *Tourism Management* 23(5), 12–24.

5 Marine and Island Tourism: Stakeholder Involvement in Policy Formulation and Implementation

Huong H. Do[1]* and Giang T. Phi[2]
[1]Thai Binh Duong University, Nha Trang, Vietnam; [2]VinUniversity, Hanoi, Vietnam

Introduction

Coastal marine areas and islands have long been valuable resources associated with mass tourism. Having a coastline that stretches over 3000 km plus thousands of islands, Vietnam records that over 70% of all tourists visit coastal areas and islands (Pham and Pham, 2017). The marine environment offers great opportunities for recreational activity, but at the same time is highly fragile. It is also a source of livelihood for millions of residents living and working in it. Therefore, tourist activities in the coastal areas and the islands must be carefully planned and controlled to ensure not only positive tourism outcomes but also to mitigate negative impacts on these fragile ecosystems and local communities.

Marine tourism 'includes recreational activities that involve travel away from one's place of residence and which have as their host or focus the marine environment (where the marine environment is defined as those waters which are saline and tide-affected)' (Orams, 1999, p. 9). The term coastal tourism has often been used interchangeably with that of marine tourism. Coastal tourism has close links with marine tourism as the coastal zone is the departure and return point for most marine tourism activities (Lück and Orams, 2014). Coastal tourism can be considered as one subsection of marine tourism according to the spectrum of marine recreation opportunities (SMARO) classifying tourism activities in terms of ease of accessibility (see Table 5.1). Finally, island tourism takes place in a unique setting that is perceived to be a distant and different destination from the urbanized locales that provide the source of island tourists (Carlsen and Butler, 2011). The SMARO framework, under marine tourism Classes 4 and 5 considers island tourism to be a subset of marine tourism.

In this chapter we adopt the definition of marine tourism proposed by Orams (1999) that includes any human leisure activity happening along the coastline, out to the reef line (if any) and under the sea. However, under the classification system used in Vietnam, marine and island tourism are two distinctive forms of tourism and are different from the notion of the coastal region. The Vietnam National Administration of Tourism (VNAT) defines 'marine and island tourism as a form of tourism that relies on the ocean potentials, takes place in potential ocean regions that satisfies the people's needs in leisure, entertainment, relaxation, visitation,

*Corresponding author: dhhuongtourism@gmail.com

DOI: 10.1079/9781789242782.0005

Table 5.1. The spectrum of marine recreation opportunities (SMARO). Adapted from Orams and Lück (2012).

Characteristics	Class 1 Easily accessible	Class 2 Accessible	Class 3 Less accessible	Class 4 Semi-remote	Class 5 Remote
Experiences	Much social interaction High level of services Usually crowded Noisy Lots of activity	Frequent contact with others Some spaces/times to escape from others	Some contact with others Locations/times when no other people present Quieter	Infrequent contact with others Peace and quiet Close to nature Limited evidence of human activity/structures High quality environment	Virtually no contact with others Solitude Tranquillity Self-sufficiency
Environment	Many human influences Highly modified Lower quality natural environment	Human structures and influences visible and close by Environment quality variable	Few human structures close by but some may be visible Higher environment quality	Limited evidence of human activity/structures High quality environment	Isolated Little to no evidence of human activity Pristine environment
Locations	Close to or in urban areas Beaches, docks, piers, urban coastal parks and walkways/cycleways Close to road, parking and mass-transport options	Intertidal zone and areas up to 100 m offshore Often requires walk of several hundred metres from car parking, mass transport	Between 100 m and 1 km offshore More than 20 km from major urban area More remote beaches, islands and coastal areas Often requires boat for access or walk in > 500 m	Between 1 km and 50 km offshore Isolated coasts/islands/reefs difficult to access Boats or overnight hiking needed to access	> 50 km offshore Coastal areas > 100 km from any significant human habitation
Examples of activities	Sunbathing People-watching Swimming Playing games Eating Social gatherings Sightseeing Special events (e.g. concerts)	Swimming Fishing Boating Surfing Windsurfing/kite-boarding	Boating Fishing Snorkelling/scuba diving Nature study Surfing Sailing Sea-kayaking	Coastal sailing Remote coast hiking and camping Live-aboard vessels (fishing, diving, surfing, etc.)	Offshore blue-water sailing Live-aboard offshore vessels (for fishing, diving, etc.) Remote coast/reef sea-kayaking and surfing

sea bathing and researching' (ITDR, 2020, p. 5). In a similar vein, Nguyen and Nguyen highlight the cultural aspect of marine tourism by introducing the following definition:

> marine and island tourism comprises human activities that involve travel away from one's place of residence to the sea and islands in a certain period of time for the main purpose of enjoying ecological, cultural and spiritual values of local fishermen and local community.
> (Nguyen and Nguyen, 2017, p. 464)

The concept of 'coastal region' defines the geographical territories adjacent to the coastline from the north (Quảng Ninh province) to the south end (Cà Mau province) that includes 28 provinces/cities (Dao, 2017) (see Table 5.2). Therefore, visits to the coastal regions and islands can be considered under the category of marine and island tourism.

This chapter starts with an overall evaluation of resources for marine tourism, then applies stakeholder theory and other methodological approaches to the analysis of marine tourism policy formulation and implementation. To highlight the current problems associated with marine tourism policies, two case studies on Phú Quốc and Côn Đảo islands are used to emphasize the local authority role in tourism development policy as well as the differences in the degree of openness for marine and coastal tourism development.

Table 5.2. List of provinces/cities adjacent to the ocean and islands. From ITDR, 2013.

Provinces/cities adjacent to:		
The ocean (28)		Islands (province) (12)
Quảng Ninh	Hải Phong city	Cô Tô (Quảng Ninh)
Thái Bình	Nam Định	Vân Đồn (Quảng Ninh)
Ninh Bình	Thanh Hóa	Bạch Long Vĩ (Hải Phòng)
Nghệ An	Hà Tĩnh	Cát Bà (Hải Phòng)
Quảng Bình	Quảng Trị	Cồn Cỏ (Quảng Trị)
Thừa Thiên-Huế	Đà Nẵng city	Hoàng Sa (Đà Nẵng)
Quảng Nam	Quảng Ngãi	Cù Lao Chàm (Quảng Nam)
Bình Định	Phú Yên	Lý Sơn (Quảng Ngãi)
Khánh Hòa	Ninh Thuận	Trường Sa (Khánh Hòa)
Bình Thuận	Hồ Chí Minh city	Phú Qui (Bình Thuận)
Bà Rịa-Vũng Tàu	Tiền Giang	Côn Đảo (Bà Rịa-Vũng Tàu)
Bến Tre	Trà Vinh	Phú Quốc (Kiên Giang)
Sóc Trăng	Kiên Giang	
Bạc Liêu	Cà Mau	

Resources and Infrastructure

Endowed with a 3260 km long coastline, Vietnam's sea territory covers 3000 islands, including the Paracel (Hoàng Sa) and Spratly (Trường Sa) archipelagos, three large islands of more than 100 km^2 (Phú Quốc, Cái Bàu and Cát Bà) (see Fig. 5.1), 24 islands of more than 10 km^2 and 84 islands larger than 1 km^2. Among the islands, 2773 islands with a total area of 1720 km^2 are close to the coastline and are found mostly in the east of the Gulf of Tonkin and in the south-west of the country. Quảng Ninh province possesses 2078 islands, which account for 75% of the total coastal islands, followed by Hải Phòng city (243 islands), Kiên Giang (157 islands) and Khánh Hòa (103 islands). Notably, Phú Quốc, Cát Bà, Côn Đảo, Cù Lao Chàm and Hạ Long Bay island systems are among the most valuable islands for tourism (Marine and Island Institute, 2020).

In addition, Vietnam has 48 bays with a total area of 4000 km^2, mostly in the South Central Coast region (64.6%) (Marine and Island Institute, 2020). Hạ Long Bay (UNESCO World Heritage listed in 1994; UNESCO, n.d.), Nha Trang, Lăng Cô and Xuân Đài Bay are the most well-known bays for international and domestic tourists thanks to their unique geologic structure and picturesque landscape (Nguyen and Nguyen, 2017). Ten marine protected areas (MPAs) are established in the islands and bay areas of Bạch Long Vĩ (Hải Phòng), Cát Bà (Hải Phòng), Cồn Cỏ (Quảng Trị), Cù Lao Chàm (Quảng Nam), Lý Sơn (Quảng Ngãi), Nha Trang

Fig. 5.1. Map of Vietnam coastline. Map courtesy of Thuy T. Duong.

bay (Khánh Hòa), Núi Chúa (Ninh Thuận), Hòn Cau (Bình Thuận), Phú Quốc (Kiên Giang) and Côn Đảo (Bà Rịa-Vũng Tàu) (Hoang, 2020). Four national parks (NPs) are established on the islands of Bái Tử Long National Park, Cát Bà National Park, Côn Đảo National Park and Phú Quốc National Park, and another 12 NPs in the coastal region. The country is one of the top ten marine biodiversity centres in the world and is one of the 20 ocean areas having abundant seafood. This is an additional benefit for marine and island tourism development compared with other territories and countries (ITDR, 2013; Hoang, 2020).

The 125 beaches spreading from the north to the south include some of the best beaches in the world, including Mỹ Khê (Đà Nẵng), voted by Forbes 2005 (Da Nang Today, 2019); and CNN endorsed Bãi Dài (Phú Quốc) and An Bàng (Hội An) (CNN, 2017). Meanwhile, Sầm Sơn, Vũng Tàu and Nha Trang beaches have been popular attractions from the early establishment of modern tourism dating back to the colonial period (see Chapter 2, this volume). The nature of the coastline and length of the beach are favourable conditions that attract visitors (Bigano *et al.*, 2006).

With reference to climate conditions, the northern and north central regions have a four seasons climate with hot summers (28°C) and cold winters influenced by north-east monsoon winds (below 20°C). Due to this seasonal pattern of climate, the beaches in the northern region are only attractive to tourists during the summer. The South Central and southern regional climates are less affected by the monsoon and distinguished by two seasons, dry and wet with stable warm weather (23–29°C), which makes beaches and islands in the southern areas favourable throughout the year (ITDR, 2013). From May to October, the coastline and islands from Quảng Ninh to Khánh Hòa are affected by five to eight tropical typhoons a year.

Air and water transportation to the islands and coastal destinations, and the connections between road and train systems to ports remain weak (Pham, 2020). Major sea ports are used for freight and there is a lack of basic services for international tourists such as telephones, currency exchange counters, toilets and souvenir shops (Hong-Trang, 2016b). Côn Đảo, Vân Đồn and Phú Quốc are the only islands with airports but the latter two have international connections to China, Thailand, Malaysia, Singapore, Cambodia, Russia and Korea (Airports Corporation of Vietnam, 2018). Transportation on the islands has been moved towards eco-friendly vehicles. For example Cô Tô island was among the pioneers in replacing fossil-fuel-engine cars with electric cars. This island has 300 electric cars with capacity of 7–14 seats, 500 motorbikes and 200 bicycles to accommodate 10,000 visitors/day (Lan-Anh, 2021).

Electricity and fresh water supply are crucial for tourism development, however they remain constraint factors for many of the islands. Except for Cát Bà and Phú Quốc, which have self-contained fresh water from underground wells, most of the islands have only limited fresh water supplies. In addition, diesel generators are used in almost all the islands. Phú Quốc is the only island that has power from the national grid supplied by a 56 km submarine cable from the mainland. Côn Đảo is a typical example of the impact of shortages in the supply of fresh water and electricity on visitor numbers, notwithstanding its great potential from its natural and historical values. Ultimately, natural disasters as well as electricity and fresh water shortage are unfavourable factors for marine and island tourism development in Vietnam, however, these remain less important than the favourable factors mentioned earlier.

Tourism Development in Coastal Regions: State of the Art

Fifty per cent of domestic tourists visited coastal regions and islands during 2004–2014 with an average growth rate of 17.4% (ITDR, 2013). The number of domestic visitors to marine and island destinations increased by four times, from 9.9 million trips in 2000 to 43.9 million trips in 2010. Notably, there has been an increasing percentage of domestic visitors to the northern central coasts lately. The increasing popularity of the North Central Coast region in marine and island tourism is attributed to the improvement of tourism investment and service quality at major beaches such as Sầm Sơn, Cửa Lò, Thiên Cầm, Nhật Lệ and Lăng Cô. The South Central and South Central Coast regions are home to the most popular marine destinations. Among the centres of coastal tourism, Bà Rịa-Vũng Tàu is the most attractive destination with an average of 4.78 million trips by domestic tourists that are mainly from Hồ Chí Minh city and other southern provinces. The international tourist market grew 16% during 2000–2010, an increase from 3.05 million in 2000 to 10.86 million in 2010. International tourists prefer the South Central region and southern coastal areas, centring on major hubs such as Nha Trang – Ninh Chữ and Hồ Chí Minh city – Long Hải – Bà Rịa. However, international tourists are gradually travelling outside these traditional routes thanks to the improvement of tourism services and infrastructure in other regions (ITDR, 2013).

International tourist visitors to the coastal islands also increased by 23% during 2015–2019. In 2019, more than 1.5 million international tourists visited the coastal islands, almost tripling the number of 663, 000 trips in 2015. The northern islands attract more Chinese, Japanese, Korean, South-east Asian, German and Dutch tourists; the central islands are preferred by other European visitors, Laotians, Japanese and Koreans; while European tourists, in particular UK, French, and Italian tourists tend to visit the southern islands. Meanwhile, their domestic counterparts displayed a similar growth rate of 23.49% during the same period, from 4.1 million in 2015 to 9.5 million in 2019. Phú Quốc and Cát Bà are two of the most attractive island destinations, accommodating the majority (80–90%) of domestic tourists. The revenue from island tourism has increased from US$129 million in 2015 to US$920 million in 2019 (ITDR, 2020). The acceleration of marine and island tourism in Vietnam has been largely attributed to the overall development of the marine economy and relevant policies. Table 5.3 shows that tourism ranked number three in economic contribution to the marine economy compared with other sectors in 2015.

Stakeholders and Policies

Stakeholders are defined as 'any group or individual that can affect a business or be affected by the realization of an organization's purpose' (Freeman, 2010, p. 53). Stakeholders are categorized into two groups: primary and secondary stakeholders. Primary stakeholders of a firm might comprise its employees, suppliers, financiers, customers and communities, while secondary stakeholders are identified as government, competitors, consumer advocate groups, special interest groups and media (Freeman *et al.*, 2007). The stakeholder concept has gained widespread acceptance not only by scholars researching business in general (Jawahar and McLaughlin, 2001) but also in tourism development policy to explore perceptions of tourism policy (Stoffelen *et al.*, 2020) and participation in formulating and implementing tourism policy (Kimbu and Ngoasong, 2013).

Table 5.3. Marine economy in Vietnam (2015). Adapted from Vu, 2017.

Economy sector	Total GDP (thousand billion VND)*	Proportions (%) in the whole country GDP†
Aquaculture and fisheries	275.14	6.5
Seafood processing	71.72	1.7
Off-shore oil and gas production	310.92	7.4
Ship building and maintenance	45.66	1.1
Ports and marine transportation	40.80	1.0
Tourism	122.57	2.9
Rescue service	4.00	0.1
Total	870.81	20.7

*GDP, gross domestic product; VND, Vietnamese dong.
†GDP of the whole country was 4192.9 thousand billion VND.

Policy is defined as 'a set of interrelated decisions (or non-decisions) and actions (or inactions) formulated and implemented by government and public authorities to deal with public problems, concerns and opportunities' (Airey and Chong, 2011, p. 42). The formulation of tourism policy requires the involvement and satisfaction of all stakeholder interests and needs (Liasidou, 2019). From the stakeholder perspective, national tourism agencies pursuing a tourism-led growth strategy turn tourism into an 'extractive' industry (Petridou *et al.*, 2019). For the improvement of the local economy, government stakeholders in remote destinations often advocate growth-driven and market-oriented tourism development strategies (Schmallegger and Carson, 2010). Meanwhile, the private sector contributes to the society and economy by pursuing its monetary interests and motives (Liasidou, 2019). Using a holistic approach, governments can balance the development by involving the private sector in the strategic planning process and creating a more supportive and cooperative policy for tourism (WTTC, 2015). Having prioritized tourist needs in the tourism-led growth strategy however, the local community has often been eliminated from tourism governance and planning in many emerging economies (Moscardo, 2011). Stakeholder involvement in tourism policy has thus revealed major problems, including: (i) the lack of a unified vision towards the future; (ii) a lack of trust from the tourism entrepreneurs towards the national government; (iii) a lack of time for regular and face-to-face interactions; and (iv) a lack of institutional forms of knowledge, resource and information exchange (Petridou *et al.*, 2019).

Governments at different levels have taken the major role in tourism destination planning and development (Ruhanen, 2013). In many countries, the hierarchical administration of tourism governance generally splits into three levels: (i) national tourism administration at the top; (ii) regional governments; and (iii) local-level administrations (local authorities) (Marafa *et al.*, 2019). Eventually, how the development plan and policy from the central government agencies is understood and implemented in the local destinations is a critical issue. By offering services and infrastructure, local authorities play a crucial role in the development of tourism (Andriotis, 2002). Local authorities are also the most suitably placed to implement and control tourism policies because:

- They have democratic legitimacy.
- They have relative permanence and the ability to take a long-term view.
- They are responsible for a range of functions that can influence tourism development, including spatial planning, development control, environmental management and community service (UNEP and ICLEI, 2003, p. 8).

Being a communist country with a highly centralized policy-making process, Vietnam is an interesting case to examine the issues of tourism stakeholder involvement in making tourism development policy. This issue has proved to be important and significant within the context of growing interest in recreational marine resources.

Study Methods

To clarify the research themes, the authors employed the thematic analysis of secondary data gathered from different government legal documents, strategic plans and reports, as well as published conference papers, journal articles and mass-media online articles. The findings from secondary data analysis were validated by interviews with key government officers. Fieldwork in the South Central coastal area was undertaken in January 2020, the first author visited Nha Trang Institute of Oceanography library and conducted an interview with the Vice-Director of this institute. A telephone interview with an officer of VNAT was conducted in June 2021 with the aim of clarifying the actual governance mechanism and issues relating to marine and island tourism. The author(s) had been involved in several tourism development projects in southern coastal and islands areas in the past. Drawing on both officers' opinions, secondary data, and their auto-ethnographic experience of being a local consultant, the authors structured the findings around the roles and functions of stakeholders such as government, local authorities, private sector business and local communities involved in tourism policy making and implementation. The stakeholder analysis is exemplified by two case studies, Phú Quốc and Côn Đảo islands. Case analysis highlights the role of the local authorities in supporting tourism development policy and reveals the different views in opening islands for tourism development.

Stakeholders Roles in Formulation and Implementation of Tourism Policies

Centralized policy making

Tourism policy making in Vietnam is hierarchical, highly centralized and uses a top-down approach (see Chapters 3 and 4, this volume). The process of policy making is mainly carried out at the highest level of the central government and the cabinet of the ruling party. The development strategies and objectives are officially stated in the Resolution of the Central Committee of the Vietnamese Communist Party, being issued and renewed every 5 years. The socio-economic objectives set in the Resolution provide guidelines for 5-year planning by state government and relevant ministries, and their implementation by local governments and respective departments. The latest objectives for sustainable development of the marine economy (including tourism) have been set and decided by the political bureau and are as stated in the latest document, Resolution No. 36/NQ-TW 2018 issued by the Central Committee of the Vietnamese Communist Party (The Socialist Republic of Vietnam Online Newspaper of the Government, 2019). In this government document, tourism and the marine service sector have been identified as a top priority for the sustainable development of the marine economy towards 2030 and as a vision to 2045. Section B.1.2 of the Resolution states:

> Investment for the development of tourism infrastructure in key coastal areas and islands; encourage and create conditions for all economic sectors to participate in developing tourism resorts, large scale service projects, high quality shopping and entertainment centres, eco-tourism, scientific exploration, community-based tourism...
>
> (The Socialist Republic of Vietnam Online Newspaper of the Government, 2019)

Actions towards accomplishment of these goals are:

- Building favourable mechanism and policy for coastal inhabitants to change their career, directly participate in tourism business, and receive benefits from tourism.
- Building favourable mechanism and policy for tourism businesses developing marine and island ecotourism products, tourism routes to offshore islands.
- Building mechanism and policy to strengthen capacity of search and rescue; promote roles of social groups and local community in development of tourism and marine service.

(The Socialist Republic of Vietnam Online Newspaper of the Government, 2019)

The implementation of this strategy for sustainable development of the marine economy stated in Resolution No. 36/NQ-TW in 2018, is guided by the government's Resolution No. 26/NQ-CP issued in 2020 (Vietnam Communist Party Central Committee XII, 2018) that formulates the objectives of 5-year development plans. Notably, there is the plan for the 'development of high-quality resorts in Quảng Ninh, Khánh Hòa, Kiên Giang and pilot development of ecotourism in some offshore islands' (Section II.2.a). According to this 5-year plan, VNAT has shifted its planning focus from resource analysis for development (ITDR, 2013) in the Proposal for Development of Marine and Island Tourism to 2020, to the affirmation of this form of tourism as one of the four strategic products from 2020. The Vietnam Strategic Tourism Development Plan toward 2030 states:

> Priority to develop marine and island leisure tourism and marine sport and leisure tourism products which align to sustainable development strategy for Vietnam marine economy. Focus resources to invest and develop certain tourism major destinations, strong branded and luxury marine leisure centres in the international tourist market.
>
> (Prime Minister, 2020, p. 5)

A system of legal documents regulating tourism activities in coastal regions and islands over the last 20 years, including the Tourism Law (2017), Ordinance No. 92/2007NĐ-CP detailed rules for implementation of Tourism, and the Environmental Protection Regulations in Tourism published with Decision No. 02/2003/QĐ-BTNMT dated 29 July 2003 by the Minister of Environment and Natural Resources. In addition, to facilitate cruise tourism, VNAT coordinated with the Ministry of Finance to enact Decision No. 61/2003 QĐ-BTC, dated 25 April 2003, that published fee schedules and marine and port services fees. Other related laws and regulations are also in effect on tourism activities in this region, such as the Environment Protection Act (2005), the Cultural Heritage Act (2001), the Forest Protection and Development Act (2004), the Aquaculture Act (2003), the Education Act (2005), the Water Resources Act (1998), the Land Law (2005) and the Biodiversity Law (2008).

The shortage of regulations is in the areas of governing tourist health and safety, using waterway transportation, limiting tourism activities in fragile marine areas, and regulating the environment, building and construction on the seashore and islands (ITDR, 2013). In their study of coastal and marine cities in Vietnam (Tsai *et al.*, 2021) found that there is an inadequate policy and legal framework covering sustainable solid waste management, a lack of intervention pertaining to prohibitions and compulsory regulations, and the fact that the tourism infrastructure of these cities has been developed without careful examination of their impact on marine resources.

Local authorities: policies implementation and monitoring

Provincial People's Committees (PPCs) are responsible for local zoning plans, tourism master plans, and the approval of investment projects. Accordingly, local tourism master plans follow the directions and guidelines of the national master plan for tourism. While the restructure of VNAT under the jurisdiction of the Ministry of Culture, Sport and Tourism (MCST) in 2007, merged the provincial departments of tourism with the Department of Culture, Sport and Tourism, provinces and cities with significant tourism resources and activities (e.g. Hà Nội, Hồ Chí Minh city, Đà Nẵng, Ninh Bình, Bình Định, Quảng Bình, Thừa Thiên-Huế and Khánh Hòa) set up independent departments of tourism. These were made responsible for local tourism master planning, annual plans and investment projects. These authorities include many provinces which have significant marine tourism activities. Hạ Long Bay in Quảng Ninh province, for example, benefits from its special status of being a natural UNESCO World Heritage Site and is an important tourist destination. To match this special status the Hạ Long Bay Management Board was established in 1995 to coordinate site management between provincial government, local businesses, local inhabitants and other stakeholders (Hampton *et al.*, 2018). Currently, the Management Board is under the MCST, Quảng Ninh PPC and the UNESCO Committee of Vietnam. The Hạ Long Bay Management Board is mainly responsible for planning and implementing the protection and management of the World Heritage Site, controlling activities on the bay, implementing investment projects in maintenance of the heritage site, and marketing, interpretation and ticketing. Investment project approval and master plan establishment related to Hạ Long Bay are under the jurisdiction of Quảng Ninh PPC (Ban Quan Ly Vinh Ha Long, 2019).

The major issues that the local authorities have originate from their position in between the central government and local communities. The line of communication and coordination is getting more complicated and ineffective, which results in two major obstacles (outlined below): (i) getting stuck in the middle; and (ii) ambiguity by multiple responsibility.

Getting stuck in the middle

Although all coastal provinces have established their local tourism master plans (ITDR, 2020), the pace at which tourism development has taken place, including massive construction on various islands, has made these plans obsolete. Many of the approvals for construction in the fragile marine and island environments were obtained prior to or during the period of preparation for local master planning, therefore local authorities have limited power to deal with the complaints from local communities and the public about ongoing construction (Tu-Truc, 2015; Hieu-Cong, 2019). The district authorities located on the islands have little power over the investment and construction activities happening in their localities, because the approval of construction of resorts and facilities for recreation are made by provincial authorities residing in the mainland area of the respective provinces.

Moreover, coordination between the central governmental agencies, such as MCST, VNAT and the local Department of Culture, Sport and Tourism/Department of Tourism is also a challenging issue. An interview with a VNAT officer revealed a lack of consistent guidelines and communication between the central and the provincial governments resulted in incomplete statistical data on tourism. However, there is no mechanism to enforce the reporting system owing to the transferring and rotating of officials between departments. This might explain why the country's statistical data, forecasting and other follow-ups using these databases remain inefficient.

Ambiguity by multiple responsibility

The management system of MPA is highly ambiguous under a different mechanism of jurisdiction. The four MPAs managed as NPs are Bái Tử Long, Côn Đảo and Núi Chúa under the PPCs, while Cát Bà is managed by Hải Phòng city's Department of Agriculture and Rural Development (DARD). Other MPAs are managed at provincial levels, by either a provincial DARD (Hòn Cau, Phú Quốc, Cồn Cỏ) or a provincial/ city People's Committee (Cù Lao Chàm, Nha Trang, Hạ Long Bay) (Walton *et al.*, 2015). Although government and local authorities have made efforts in managing MPAs, violations such as introducing new destructive fishing methods and equipment, and the rapid increase of tourists have worsened environmental pollution and threatened biodiversity (Hoang-Nam, 2020). In fact, the MPA management boards are non-administrative technical units and have no authority to impose any penalty for violations (Walton *et al.*, 2015).

Overcoming this ambiguity, Cù Lao Chàm (Hội An – Quảng Nam province) has set an example for the active involvement of local authorities in managing MPAs. Fox *et al.* (2011) reported that 10% of the MPA is dedicated to becoming a core zone (e.g. absolutely prohibiting any fishing activity) (see Fig. 5.2) resulting from successful negotiation with the local people. The significantly increasing number of shellfish and invertebrates observed in local markets show a positive outcome from effective MPA regulation in just a short period of time. Another example is the case of the hotel-managed marine reserve in the Whale Island resort, Vân Phong, which proved to increase fish stocks rapidly while giving them effective protection. However, this method creates some tension among fishermen because their catchment area has been narrowed (Svensson *et al.*, 2009).

Private sectors: mass tourism's growth engines

Seaside resorts were first established in the early 20th century during the French colonial period in Đồ Sơn, Sầm Sơn and Vũng Tàu (Tran, 2010) (also see Chapter 2, this volume). Hotels in coastal regions, Nha Trang and Đà Nẵng in particular, continued this development in the 1960s and 1970s. After the *Đổi Mới* era, the private sector has taken a leading role in establishing accommodation and tourism services along the coastline of Vietnam.

Established in 1993, the Coco Beach Resort was built in the small fishing village of Mũi Né (Bình Thuận) and was the first tourist resort, unleashing a wave of investment holiday accommodation making Mũi Né a 'resort capital' with 387 resorts and hotels by July 2020 (Ha-Truong, 2020). Another popular beach destination is Đà Nẵng, the city now has 943 hotels and resorts with 40,000 rooms, including 20 luxury five-star accommodation facilities developed since the first five-star Furama hotel in the Central coast was established in 1997 (Hoang-Son, 2019). The number of accommodation facilities on the coastline and islands by 2010 was 6646 with 148,972 rooms, three times the number in 2000 (2229 accommodation facilities and 48,750 rooms). Since 2004, the accommodation sector has thus become well developed, and the improvement observed is not only in quantity but is also in quality (ITDR, 2013).

Fig. 5.2. Poster of prohibition of land crab collection in Cù Lao Chàm. Photo courtesy of Huong H. Do.

The private sector has also played a pioneer role in tourism development, in some cases before the government policy and regulations, by diversifying tourism attractions, products and marketing for coastal and island destinations. In 2006, the VinGroup marked a new model for integrated resort destination in Vietnam by introducing the Vinpearland Nha Trang Complex, which includes a five-star resort, an amusement park, an aquarium, a shopping street, food courts and restaurants on Hòn Tre Island. The company also invested in a cable car system which was advertised as the longest cable car over the ocean in the world at that time (Hai-Dang, 2016). The project has transformed Nha Trang from a 'stagnation' beach city into an international resort destination. After this success, Nha Trang was selected to host various international beauty contests that have promoted the image of Nha Trang and Vietnam tourism to international audiences by live broadcasting during the events (Tien-Minh, 2018). Pursuing a different strategy, the central coastal city of Đà Nẵng focuses on MICE (meetings, incentives, conferences, exhibitions) and sports events by hosting world-class meetings such as APEC 2017, an international marathon, a triathlon, and Asia Golf tourism convention in 2017 (Pham and Truong-Dinh, 2018). In the north, a road connecting Tuần Châu Island to the mainland in 1997 built by the Tuần Châu Corporation was the starting point in creating success for Tuần Châu Island as the first tourism complex with a port for all tourist cruise ships visiting Hạ Long Bay (Dinh-Son, 2019; Tuan Chau Corporation, 2021).

However, when tourism businesses solely pursue profit and ignore other responsibilities, this can hinder the overall sustainability of tourism development (ESRT, 2015). Evidence of this can be found easily in any beach destination, such as in the case of the hotels in Đà Nẵng's illegal discharge of wastewater to the ocean (Canh sat moi truong, 2019) (see Fig. 5.3). A study of the water quality in Hạ Long Bay revealed that while captains of tourist boats agree that oil separator equipment must effectively function before discharging wastewater into the sea, they only turn on the equipment when being watched by Ministry of Transportation inspectors (Le, 2015).

Fig. 5.3. Direct discharge of wastewater to the beach. Photo courtesy of Long H. Pham.

Local communities: the rights of the grassroots in question

Coastal regions have higher population densities and lower educational levels compared to other regions of the country, accommodating more than 37.2 million inhabitants in 2010 (ITDR, 2013). In remote small islands and coastal areas, local communities face more difficulties as their living heavily relies on fishing and other seafood harvesting. Poor infrastructure, harsh weather and a lack of fresh water and electricity are unfavourable conditions for the livelihood of the residents in island areas. In this situation community-based tourism has been introduced and has proved successful in Hòn Mun MPA, Cù Lao Chàm Island and Xuân Thủy National Park. In Cù Lao Chàm, no commercial hotels are permitted on the island, but residents can offer a home-stay service for tourists and earn direct income from tourism. Local people are trained in housekeeping, reception, cooking, food safety, and accommodation-business financial calculations. This is a rare case in the context of a developing country like Vietnam to prioritize long-term benefits for residents (Fox *et al.*, 2011).

These initiatives are exceptional and have proved to be successful on islands with limited resources. In many coastal regions though, the role of local communities in tourism development policy formulation and implementation is rather insignificant. While residents are the first to experience the direct impact from tourism activities, they are usually the last to be informed of the development plans for their residential space. In the famous beach area of Sầm Sơn – Thanh Hóa province, 3.5 km of the coast with beaches have been privatized for the FLC Corporation to establish a new leisure complex with playgrounds, botanic garden, water music square, and so on. This development plan directly impacts on the livelihood of the local fishermen as 500 m of the beach used as a dock for small fishing boats must be removed for construction (Nguyen, 2016). Prior to the construction of the resort complex, the development and management plan were not publicized to collect opinions of the locals; that decision caused much resentment among residents in early 2016. In the central southern coastal areas, the development of seaside resorts has limited residents and fishermen's rights to access the beaches (Tam *et al.*, 2020). Mass tourism in the seaside resorts also causes other social problems. Prior to the pandemic, in late 2019, the beach area of Nha Trang was overloaded by Chinese tourists, and increased waste from tourism activities resulted (Cong and Chi, 2020). The fragile environment, with limited water and electricity supplies, and the infrastructure of many islands has only slowly adapted to the rapidly increased number of tourists. The difficulty is that tourism exceeds the carrying capacity of the islands, a situation well understood by the local people in Côn Đảo (An Nhien and Manh Cuong, 2021) and Phú Quốc island (Nhat-Huy, 2018).

In a nutshell, the findings from stakeholder analysis show the role of the Vietnam central government in leading and providing legal frameworks for the development of marine and island tourism. However, local authorities at provincial/city level have more power in their local tourism planning and approval for particular investment projects. When rapid tourism development pulls massive investment from the private sector, the approval process can be accelerated without the participation of the destination local authority and community who are directly impacted by the development policies. Two cases presented below further discuss the issues of stakeholders' involvement in island tourism embracing different approaches.

Case Studies of Marine and Island Tourism

Phú Quốc: a resort destination for mass tourism?

Phú Quốc, also known as the 'Pearl Island' is an island district of Kiên Giang province, Southern Vietnam. Located in the Gulf of Thailand not far from the mainland (45 km from Hà Tiên and 120 km from Rạch Giá), Phú Quốc has a population of around 100,000 people. It comprises 22 large and small islands, spanning close to 600 km^2, a long coastline of 150 km, 99 hills and 63.74% of the area is covered by primeval forests (Hoan, 2015). In 2006, Phú Quốc National Park was recognized by UNESCO as a global biosphere reserve (Hong-Trang, 2016a). Prior to 2004, the island

Fig. 5.4. Bãi Sao – Phú Quốc in 2011. Photo courtesy of Huong H. Do.

was full of pristine beaches with very few tourists (see Fig. 5.4). In 2004, the Prime Minister issued Decision No. 178, approving a strategic project to develop Phú Quốc into a large and modern centre of tourism, resorts and international trade for the whole Mekong Delta region. The decision paved the way for the opening of Phú Quốc International Airport in 2012, with air connections to countries such as Singapore, Malaysia, Thailand, South Korea, the UK and Russia. Concurrently, infrastructure projects, such as the upgrade of Dương Đông water plant and connection to the national electrical grid, contributed to substantially reduce operational costs for businesses on the island. More importantly, a visa waiver policy that allows foreigners a 30-day visa-free stay was adopted in 2014, and the government also offered special incentives for the Phú Quốc area in terms of land rental contracts, deductions in corporation tax (10% instead of 28%) and personal income tax (50% reduction) (Hong-Trang, 2016a).

These policies were soon supported by private investors, starting a trend in mass tourism resort development in the island. Vingroup was a pioneer investor with the 304 ha Vinpearl Phú Quốc Complex, which includes five-star hotels, a safari tour and and a chain of commercial and entertainment facilities. Shortly after, a plethora of other large corporations, both domestic and international, continuously added large hotels and resorts to serve the high-end tourist segment. By 2019, Phú Quốc had become a big construction site with around 300 approved development projects, totalling VND 370.000 billion (Hieu-Cong, 2019). In 2020, there were 15,000 hotel rooms rated three-star and above, including 6000 rooms rated five-star (Tan-Thai and Hoang-Dung, 2021). This rapid tourism development has meant that Phú Quốc has become one of the most popular destinations in Vietnam, attracting millions of tourists annually who are interested in coastal, wellness, MICE and nature-based tourism, thanks to convenient domestic flight routes. Fig. 5.5 shows Bãi Dài which was full of resorts and beach umbrellas in 2019.

Having turned Phú Quốc into a 'resort paradise', the policy in favour of mass tourism has created several challenges for this island district. First, while tourism has been a main

Fig. 5.5. Bãi Dài – Phú Quốc in 2019. Photo courtesy of Long H. Pham.

driver for improvement of accessibility, the establishment of infrastructure is concentrated mainly in the tourist areas. In many peripheral areas of the island, basic infrastructure is still poor, such as narrow roads in bad condition and lack of electricity (Bui, 2020). Second, touristic transportation is a priority, but the island has attracted taxi companies because there is no plan for the development of public transportation (Nhat-Huy, 2018). Third, while locals' income in general has increased thanks to tourism, there is a lack of training for locals to further increase their capacity. Given the small number of residents on the island, tourism and hotels in the island rely heavily on external workers from the mainland (Phi-Nhat, 2021).

The master plan for Phú Quốc devoted a significant portion of public coastal land for private use of resorts, leading to 'concretization' along the major beaches. The high density of resorts without proper water drainage systems at the coastline has been identified as one of the major factors for Phú Quốc's historical flood in 2019 (Hieu-Cong, 2019). Also, this 'privatization' of the beachfront not only creates visual pollution but also blocks access to the beaches for locals (especially fishermen) and tourists. Phú Quốc's Communist Party secretary, Mai Văn Huỳnh, also acknowledged the ongoing infrastructure overload. He gave the example that Dương Đông town's population had quickly increased from 10,000 to over 50,000 people in recent years due to rapid urbanization, which constantly overloads the technical and social infrastructure of the town.

Phú Quốc's many issues are partly due to the weak local authority. This links to our analysis of the ambiguity of responsibility and capacity of the local authorities in tourism development administration. It is shown that new development projects are approved at the provincial level by the Kiên Giang government. Despite being the implementing agency operated in the island, Phú Quốc local administration does not have sufficient resources and capacity such as specialized departments to monitor the implementation of existing and new projects (Hieu-Cong, 2019). This limitation of local administrators has been stated by an officer of Phú Quốc PPC:

> The district's management and planning of construction and urban areas are still weak. According to the regulations of the Provincial People's Committee, the 50m area from the sea edge upwards is public beach; but in reality, the Phú Quốc development plan does

> not have public roads to access the beach areas. According to the current planning and construction, after 2030 – the year Phú Quốc expects to welcome about 12 million visitors, there will be no public beaches for tourists.
> (Thiennhien, 2015, p. 1)

In January 2021, Phú Quốc district has officially become 'Phú Quốc city' (Tan-Thai and Hoang-Dung, 2021). Phú Quốc has been nominated as a destination to trial COVID-vaccine passports for the early reopening of international tourism in Vietnam. At the same time, there is a plan to upgrade and widen domestic flight routes to serve the domestic market (Viet-Tuan, 2021). There is no doubt that the 'Pearl Island' will continue to remain a tourist hotspot in the post-pandemic period, yet this continuous and rapid growth of tourism may lead to irreversible environmental damage to the fragile island's ecosystem and a decrease in the local quality of life if not managed properly.

Côn Đảo islands: the right path towards sustainable tourism?

Côn Đảo, also known as Côn Lôn or Côn Sơn, is an island district of Bà Rịa – Vũng Tàu province, Vietnam. Côn Đảo is an archipelago consisting of 16 small islands with a total area of 76 km^2, located approximately 117 km south-east of Vietnam's mainland (VOV World, 2011). Its remote and isolated location has contributed to create Côn Đảo's unique characteristics as a tourist destination. In terms of environment, Côn Đảo possesses pristine nature with one of the most diverse and abundant marine and island ecological systems in Vietnam. Côn Đảo National Park is among Vietnam's largest natural reserve, spanning nearly 20,000 ha with many rare species of corals, turtles, dolphins and endangered dugongs. Côn Đảo also has 200 km of coastline with many untouched beaches such as Đầm Trầu. An Hải and Côn Đảo waters are included in the list of 'Highest priority areas for protection' in the system of MPAs (Vneconomy, 2021). In terms of heritage, Côn Đảo's history is associated with the two periods of Vietnamese resistance against the French and the Americans to protect the country's independence. Côn Đảo was turned into prison islands for political prisoners during the French colonial era, and from 1968 to 1975, the South Vietnamese regime backed by the USA imprisoned Vietnamese revolutionary soldiers in the infamous cells known as the 'tiger cages' (see Fig. 5.6). In 2012, the relics of Côn Đảo Prison was recognized by the Vietnamese government as a special national war heritage site with significant historical and educational values (Vietnamtourism, 2021).

Tourism development has taken off in the last decade. Initially, Côn Đảo was to be developed with the ambition of becoming 'a modern tourist economic zone of regional and international stature', as stated in Decision No. 1518/QD-TTG signed by the Prime Minister of Vietnam on 5 September 2011. Similar to Phú Quốc, since 2013, the central government has approved a number of preferential mechanisms and policies for Côn Đảo, including more efficient investment procedures, offering corporate income-tax incentives (the tax rate is reduced to 10% and the incentive period can be up to 30 years) and a 50% reduction of personal income tax, exempting import tax and applying a value-added tax rate of 0% for capital construction investment expenses (Bariavungtau tourism, 2015).

However, in acknowledging Côn Đảo's historical significance and ecological fragility, the policy directions have changed in the following years, adopting a more careful and sustainable approach with impacts of the development being closely observed and mitigated. This cautious approach to tourism development is stated in Decision No. 870/QD-TTg (2015), approving the 'Master plan for development of Côn Đảo National Tourist Area, Bà Rịa – Vũng Tàu province to 2030':

> Constructing Côn Đảo to ensure a harmonious relationship between economic development and conservation of historical and cultural relics, protection of forest-sea ecosystems, ensuring national security and defence towards modern and sustainable development.
> (Bariavungtau tourism, 2015)

The alternative policy has directed Côn Đảo's tourism development towards sustainability. On the one hand, the pro-tourism policy with selective development has attracted investors to high-end accommodation projects

Fig. 5.6. Côn Đảo Prison – the 'hell on earth'. Photo courtesy of Huong H. Do.

such as the Six Senses Resort (Fig. 5.7), Côn Đảo Resort, Poulo Condor Boutique Resort and Spa. In addition, accessibility to Côn Đảo has improved, including new sea connections via boats from Vũng Tàu or Cần Thơ and the speedboat Superdong, connecting Sóc Trăng and Côn Đảo within 2.5 hours with a capacity of 806 passengers. By air, Côn Sơn airport now has direct connections with Hồ Chí Minh city, Cần Thơ and Sóc Trăng, or tourists can opt to use private helicopters from Vũng Tàu. Telecommunications infrastructure on the island has also been improved with more reliable Internet, television and phone networks. In 2021, a project worth VND 4.8 trillion (US$207 million) was approved to connect Côn Đảo island to the national electricity grid (Truong-Ha, 2021). Until now, Côn Đảo's electricity has been produced by diesel generator, which has failed to meet the demand of both locals and the growing numbers of tourists. The current diesel generators supply 11.82 MW while the island's demand is 18.62 MW, therefore resorts like Six Senses Côn Đảo and Sài Gòn-Côn Đảo hotel had to install self-supply electricity using their own

Fig. 5.7. A view from Six Senses Resort in Côn Đảo. Photo courtesy of Huong H. Do.

diesel generators. Solar power and wind power are considered as a supportive source, but are not a reliable source of power in unfavourable weather conditions (Dong-Hieu, 2021). Côn Sơn is the only one among 16 islands in the Côn Đảo archipelago with limited underground fresh water and five freshwater lakes. Water shortages are critical during April and May, which is the dry season and also the tourism high season (Nguyen and The, 2020). On the other hand, efforts to develop alternative and small-scale tourism have kept the growth of tourist numbers to Côn Đảo at a somewhat stable level. In the period between 2012 and 2017, Côn Đảo tourism revenue increased by an average of 35%/year, leading to an increase of more than 14% in commercial services annually (Vietnamtourism, 2019).

However, along with the continuous improvement of accessibility and tourism products/services and the promotional efforts of the destination authority, Côn Đảo is in danger of heading towards the so-called 'mass-alternative tourism', a convergence of mass tourism and alternative tourism. In 2019, the number of tourists reached close to 400,000, a rapid increase of 37% compared to 2018. Within the same year, Vietnam Airlines Corporation proposed a plan to upgrade and expand Cỏ Ống airport on Côn Đảo, with the aim of increasing the capacity to 4–5 million passengers a year, ten times more than the current capacity (Hoang-Nam, 2020). In addition, it is inevitable that new investment projects will continue to flock to the island district to take advantage of its great tourism potential. As a small, isolated island district, Côn Đảo will face significant challenges in the coming decades in terms of limited land for development, freshwater scarcity and difficulty in waste treatment as tourism growth takes place. Concurrently, in a similar way to Phú Quốc, local authorities at the district level have limited capacity and decision-making power to evaluate, control and monitor investment activities.

In recognizing these challenges, in December 2020 the PPCs of Bà Rịa – Vũng Tàu province submitted an adjustment to the general construction plan for Côn Đảo to the Prime Minister (Vneconomy, 2021). The new plan seeks the containment of most new development projects within the central town of Côn Sơn. Any new construction outside this area must meet the strict criteria of minimizing impacts on the environment, conserving the forest-sea ecosystem and controlling population growth. In addition, the adjustment plan focuses on turning the Côn

Đảo district into a famous island ecotourism area at both regional and international levels. Importantly, tourist numbers are to be kept at the lowest level in this plan, with a focus on balancing between domestic and international tourists. In particular, the expected number of tourists by 2030 is about 350,000 arrivals/year, of which international tourists will account for about 40%. To reduce the negative impacts on nature and the ecosystem, Côn Đảo tourism in the coming time will also extend beyond ecotourism to exploring the depth of cultural/historical values, monuments, legends and festivals. These timely actions hopefully will contribute to ensure long-term sustainable development in Côn Đảo.

Conclusions

This chapter has contributed up-to-date information on the current issues of Vietnam marine and island tourism from the perspective of stakeholders' involvement in policy making and implementation. Through critically examining relevant tourism development policies from resource availability analysis, policies and stakeholder involvement in policy making and tourism governance, the chapter reveals the high level of complexity that drives tourism development in the marine and island context. The literature suggested that in most countries, tourism governance generally splits into three levels: (i) national; (ii) regional; and (iii) local-level administrations. While this same governance structure was also observed in Vietnam, the country's communist political system means that the central government has the highest power in shaping the direction of tourism development in the destinations. Policy making by central government is directed by the objectives of socio-economic development set by the Central Communist Party secretary. This is the legacy of the central-planning economy under the jurisdiction of a single ruling party as outlined in Chapter 3, this volume. At regional and local levels, the findings of this chapter show that the provincial governments' roles are mainly in monitoring and implementing top-down, centralized policies. However, they also possess the decision-making power to approve development projects, despite being located on the mainland away from the actual island destinations in many cases. There is also the strong influence of the private sector on both central and provincial governments, to push forward a pro-growth agenda that is geared towards mass tourism development and to limit strict regulations on development activities. While the local authorities and local community are directly impacted by these policies, they often have little role in the policy planning process and are left to reactively deal with many of the consequences of tourism development. The current policy-making process is thus partly responsible for many sustainability issues in marine and island destinations, for example overcrowding, lack of basic resources such as water and electricity, and environmental pollution and destruction.

The critical issues concerning tourism development policies and stakeholders' involvement in marine and island tourism are exemplified and discussed in two case studies. In the Phú Quốc case, hundreds of private development projects have been approved in a short period of time, almost all focusing on serving tourists to maximize profit, and most without consideration for the natural environment or the local community. Although the Côn Đảo district government also has relatively little influence on tourism policies, thanks to its historical significance, Côn Đảo has been more carefully directed and monitored by the central and provincial governments towards the direction of sustainable development from the early stages. This has allowed Côn Đảo to develop at a slower pace and to better manage impacts on local resources, albeit also thanks to it having smaller land size and being less accessible compared to Phú Quốc.

Throughout the COVID-19 pandemic, tourist consciousness and awareness about social responsibility in general and responsible travel in particular have been constantly raised (O'Connor and Assaker, 2021). Vietnamese governments in recent years have also increasingly directed tourism policies towards sustainable tourism development, and hopefully these new directions will soon be reflected in the policies post-pandemic. However, as the central government has limited capacity to monitor the implementation of every policy on the ground, for sustainable tourism development to take place (especially in the context of isolated islands), local authorities and local communities must be given more active roles and decision-making power in the planning and implementation process of the tourism master plan that directly exploits local resources and impacts on their (and their future generations') lives.

References

Airey, D. and Chong, K. (2011) *Tourism in China: Policy and Development Since 1949*. Routledge, London.

Airports Corporation of Vietnam (2018) Phu Quoc International Airport. Available at: https://vietnamairport.vn/phuquocairport/duong-bay-2 (accessed 22 September 2021).

Andriotis, K. (2002) Local authorities in Crete and the development of tourism. *Journal of Tourism Studies* 13(2), 53–62.

An Nhien and Manh Cuong (2021) Con Dao is facing clean water pressure. Ba Ria-Vung Tau. Available at: http://www.baobariavungtau.com.vn/xa-hoi/202104/con-dao-truoc-ap-luc-thieu-nuoc-sinh-hoat-923828/ (accessed 18 July 2021).

Ban Quan Ly Vinh Ha Long (2019) Available at: https://halongbay.com.vn/ban-quan-ly-vinh-ha-long-cc2353.html (accessed 15 July 2021).

Bariavungtau tourism (2015) Con Dao looks forward to high end tourism destination. Available at: http://sodl.baria-vungtau.gov.vn/quy-hoach-ke-hoach-phat-trien/-/view_content/content/12758/con-%C4%91ao-huong-toi-thanh-pho-du-lich-chat-luong-cao (accessed 15 July 2021).

Bigano, A., Hamilton, J.M. and Tol, R.S.J. (2006) The impact of climate on holiday destination choice. *Climatic Change* 76, 389–406.

Bui, L.H. (2020) A case study on Phu Quoc Island, Vietnam: is tourism a useful tool towards sustainability? Master's thesis. Aalborg University, Aalborg, Denmark.

Canh sat moi truong (2019) Series of hotel violating waste water regulations. *Tai Nguyen and Moi Truong*, 01/06. Available at: https://baotainguyenmoitruong.vn/hang-loat-khach-san-vi-pham-quy-dinh-ve-xu-ly-nuoc-thai-260981.html (accessed 15 July 2021).

Carlsen, J. and Butler, R. (eds) (2011) *Island Tourism: Towards a Sustainable Perspective*, Vol. 8. CAB International, Wallingford, UK.

CNN (Cable News Network) (2017) 100 Best Beaches Around the World. Available at: https://edition.cnn.com/travel/article/100-best-beaches/index.html (accessed 4 October 2021).

Cong, L.C. and Chi, T.T.V. (2020) The sustainability of marine tourism development in the South Central Coast, Vietnam. *Tourism Planning & Development* 2020, 1–19. Available at: https://doi.org/10.1080/21568316.2020.1837226

Da Nang Today (2019) The city wins numerous international tourism awards. Available at: https://baodanang.vn/english/infographics/201903/the-city-wins-numerous-international-tourism-awards-3170066/ (accessed 28 May 2021).

Dao, T.B.T. (2017) Status of marine tourism in Vietnam tourism. In: *Conference proceedings of Sustainable Development of Marine Economy: From Strategy to Reality*. VNU University of Economics and Business, Vietnam National University, Hanoi, pp. 83–92.

Dinh-Son (2019) Tuan Chau island lord threw money into the ocean. *Thanh Nien*. Available at: https://thanhnien.vn/tai-chinh-kinh-doanh/chua-dao-tuan-chau-mang-tien-do-xuong-bien-1142090.html (accessed 15 July 2021).

Dong-Hieu (2021) Power cable crossing ocean is the best option. Ba Ria-Vung Tau. Available at: http://www.baobariavungtau.com.vn/kinh-te/202103/keo-dien-vuot-bien-ra-con-dao-la-toi-uu-nhat-922434/ (accessed 28 May 2021).

ESRT (Environmentally and Socially Responsible Tourism) (2015) Establishing Vietnam Responsible Tourism Policy Framework. Available at: https://vietnamtourism.gov.vn/esrt/default.aspx-portalid=1&tabid=352&itemid=712.htm (accessed 15 July 2021).

Fox, E., Hien, B.T. and Brunner, J. (2011) Marine Protected Areas and Sea Turtles Conservation in Vietnam: Challenges and Opportunities. International Union for Conservation of Nature (IUCN). Available at: https://www.iucn.org/downloads/mpa_paper_june_2011_final__3____vn.pdf (accessed 15 July 2021).

Freeman, E.R. (2010) *Stakeholder Management: Framework and Philosophy*. Cambridge University Press, Cambridge, UK.

Freeman, E.R., Harrison, J.S. and Wicks, A.C. (2007) *Managing for Stakeholders: Survival, Reputation, and Success*. Yale University Press, New Haven, Connecticut.

Hai-Dang (2016) Vinpearland – 10 years as a pioneer. *Dau Tu Online*, 7 September. Available at: https://baodautu.vn/vinpearl-land-10-nam-vi-the-tien-phong-d51209.html (accessed 18 July 2021).

Hampton, M.P., Jeyacheya, J. and Pham, H.L. (2018) Can tourism promote inclusive growth? Supply chains, ownership and employment in Ha Long Bay, Vietnam. *The Journal of Development Studies* 54(2), 359–376.

Ha-Truong (2020) Infrastructure revives once 'resort capital' of Binh Thuan. *VnExpress*, 30 July. Available at: https://vnexpress.net/ha-tang-hoi-sinh-thu-phu-resort-mot-thoi-cua-binh-thuan-4136695.html (accessed 31 May 2021).

Hieu-Cong (2019) Phu Quoc on serious flood – consequence of booming property projects besieging beaches. *Zing News*. Available at: https://zingnews.vn/phu-quoc-ngap-sau-cai-gia-cua-bat-dong-san-o-at-vay-bien-post977731.html (accessed 15 July 2021).

Hoan, N. (2015) Tourism destination products management – case studies of Phu Quoc Island Kien Giang Province, Vietnam by 2020. *Universal Journal of Management* 3(12), 509–513.

Hoang, D.C. (2020) New approach in Vietnam marine conservation: relying on ecologically or biologically significant marine area. In: *Conference Proceedings of the 4th National Conference on Environment and Sustainable Development*. Science and Technology Publishing House, Hanoi, pp. 443–453.

Hoang-Nam (2020) Sea turtle conservation in Vietnam: realities and urgent issues. *Vietnam Plus*. Available at: https://www.vietnamplus.vn/bao-ton-bien-viet-nam-thuc-trang-va-nhung-van-de-cap-bach-hien-nay/668924.vnp on 15/07/2002 (accessed 15 July 2021).

Hoang-Son (2019) Da Nang is facing overproduction in hotel sector? *Thanh Nien*. Available at: https://thanhnien.vn/tai-chinh-kinh-doanh/da-nang-khung-hoang-thua-khach-san-1162335.html on 31/5/2021 (accessed 31 May 2021).

Hong-Trang (2016a) Phú Quốc and its journey to become a tourism paradise. Nhan Dan. 12 January. Available at: https://nhandan.vn/du-lich/phu-quoc-va-hanh-trinh-tro-thanh-thien-duong-du-lich-252784 (accessed 30 May 2021).

Hong-Trang (2016b) Cruise is ready 'to havest'. Nhan Dan. Available at: https://nhandan.vn/dien-dan/du-lich-tau-bien-vao-mua-282011 (accessed 15 July 2021).

ITDR (Institute for Tourism Development Research) (2013) *Tourism Development Plan in Marine, Islands and Coastal Areas*. Ministry of Culture, Sport and Tourism, Hanoi, p. 127.

ITDR (Institute for Tourism Development Research (2020) Introduction. In: *Proceedings of Conference, Current Situation and Recommendations for Tourism Development in Vietnam Coastal Islands, 11 December, Hà Nội*, pp. 3–32.

Jawahar, I.M. and McLaughlin, G.L. (2001) Toward a descriptive stakeholder theory: an organizational life cycle. *The Academy of Management Review* 26(3), 397–414.

Kimbu, A.N. and Ngoasong, M.Z. (2013) Centralised decentralisation of tourism development: a network perspective. *Annals of Tourism Research* 40, 235–259.

Lan-Anh (2021) Cô Tô to develop high quality marine and island ecotourism. Du Lich. Available at: http://www.vtr.org.vn/co-to-phat-trien-du-lich-bien-dao-sinh-thai-chat-luong-cao.html (accessed 15 July 2021).

Le, T.A. (2015) *Analysing Water Condition in Ha Long Bay, Quang Ninh Province, Vietnam: a Social Study from Tourism Businesses Perspective*. International Union for Conservation of Nature (IUCN), Gland, Switzerland.

Liasidou, S. (2019) Understanding tourism policy development: a documentary analysis. *Journal of Policy Research in Tourism, Leisure and Events* 11(1), 70–93.

Lück, M. and Orams, M. (2014) Coastal tourism. In: Jafari, J. and Xiao, H. (eds) *Encyclopedia of Tourism*. Springer, Cham, Switzerland.

Marafa, L.M., Qi, H. and Chan, C.S. (2019) The roles of hierarchical administrations of tourism governance in China: a content analysis. *Journal of Policy Research in Tourism, Leisure and Events* 11(2), 354–369.

Marine and Island Institute (2020) Strengthen status and role of coastal islands in Vietnam tourism development: issues and solutions. In: *Proceedings of Conference, Current Situation and Recommendations for Tourism Development in Vietnam Coastal Islands, 11 December, Hà Nội*. Institute for Tourism Development Research (ITDR), Hanoi, pp. 46–56.

Moscardo, G. (2011) Exploring social representations of tourism planning: issues for governance. *Journal of Sustainable Tourism* 19(4–5), 423–436.

Nguyen, D. (2016) Local people continue to strike, Thanh Hoa reinforce security. *Zing News*. Available at: https://zingnews.vn/dan-tiep-tuc-doi-bai-bien-thanh-hoa-tang-cuong-an-ninh-post631451.html (accessed 15 July 2021).

Nguyen, N. and The, H. (2020) Lack of fresh water threat in Con Dao. *Nhan Dan*. Available at: https://nhandan.vn/tin-tuc-xa-hoi/nguy-co-thieu-nuoc-sach-tai-con-dao-624549/ (accessed 31 May 2021).

Nguyen, T.D.P. and Nguyen, H.N. (2017) Position of marine and island tourism in Vietnam. In: *Researching, Teaching Vietnam Studies and Vietnamese: Theories and Practices*. Vietnam National University, Hanoi, pp. 464–472.

Nhat-Huy (2018) Taxis invade Phú Quốc. *Tien Phong*. Available at: https://tienphong.vn/taxi-tran-ra-dao-phu-quoc-post1030257.tpo (accessed 18 July 2021).

O'Connor, P. and Assaker, G. (2021) COVID-19's effects on future pro-environmental traveller behavior: an empirical examination using norm activation, economic sacrifices, and risk perception theories. *Journal of Sustainable Tourism*, 1–19. Available at: https://www.tandfonline.com/doi/full/10.1080/09669582.2021.1879821

Orams, M. (1999) *Marine Tourism: Development, Impacts and Management*. Routledge, London.

Orams, M. and Lück, M. (2012) Marine system and tourism. In: Holden, A. and Fennell, D.A. (eds) *Handbook of Tourism and the Environment*. Routledge, Abingdon, UK, pp. 170–182.

Petridou, E., Olausson, P.M. and Ioannides, D. (2019) Nascent island tourism policy development in Greenland: a network perspective. *Island Studies Journal* 14(2), 227–244.

Pham, H.C. (2020) Examining current situation of transportation infrastructure for tourism in Vietnam coastal islands: issues and recommendations. In: *Proceedings of Conference, Current Situation and Recommendations for Tourism Development in Vietnam Coastal Islands, 11 December, Hà Nội*. Institute for Tourism Development Research (ITDR), Hanoi, pp. 40–45.

Pham, H.L. and Pham, T.A. (2017) Marine tourism and the reality of resorts booming in coastal resorts. *Bản Tin Chính Sách Tài Nguyên, Môi Trường, Phát Triển Bền Vững* 25(Quarter I/2017), 18–21.

Pham, T.L. and Truong-Dinh, B.Q. (2018) An exploratory study of the perception of co-creation experience in the tourism industry: a case study in Đà Nẵng city, Vietnam. *Academy of Marketing Studies Journal* 22(3), 1–14.

Phi-Nhat (2021) Phú Quốc turns into Southeast Asia's must-go destination of tourists and investors. *Hanoitimes*. Available at: http://hanoitimes.vn/phu-quoc-turns-into-vietnams-must-go-destination-of-tourists-and-investors-317112.html (accessed 15 July 2021).

Prime Minister (2020) Decision on Approval of Sustainable Development Strategic Plan toward 2030. Available at: http://vtr.org.vn/FileManager/anh_web_2020/thang1/0110/QD147.PDF (accessed 3 July 2021).

Ruhanen, L. (2013) Local government: facilitator or inhibitor of sustainable tourism development? *Journal of Sustainable Tourism* 21(1), 80–98.

Schmallegger, D. and Carson, D. (2010) Is tourism just another staple? A new perspective on tourism in remote regions. *Current Issues in Tourism* 13(3), 201–221.

Stoffelen, A., Adiyia, B., Vanneste, D. and Kotze, N. (2020) Post-apartheid local sustainable development through tourism: an analysis of policy perceptions among 'responsible' tourism stakeholders around Pilanesberg National Park, South Africa. *Journal of Sustainable Tourism* 28(3), 414–432.

Svensson, P., Rodwell, L.D. and Attrill, M.J. (2009) Privately managed marine reserves as a mechanism for the conservation of coral reef ecosystems: a case study from Vietnam. *Ambio* 38(2), 72–78. DOI: 10.1579/0044-7447-38.2.72.

Tam, N.T., Truong, T. and Ha, Q. (2020) Must ask for when wishing to have sea bathing. *Thanh Nien*. Available at: https://thanhnien.vn/tai-chinh-kinh-doanh/muon-tam-bien-phai-xin-197765.html (accessed 15 July 2021).

Tan-Thai and Hoang-Dung (2021) Expectation from Phu Quoc island city. *Saigondautu*. Available at: https://saigondautu.com.vn/kinh-te/ky-vong-tu-thanh-pho-dao-phu-quoc-87249.html (accessed 15 July 2021).

The Socialist Republic of Vietnam Online Newspaper of the Government (2019) Party's Resolution on Strategy for Sustainable Development of Marine Economy. Available at: http://vanban.chinhphu.vn/portal/page/portal/chinhphu/hethongvanban?class_id=509&_page=1&mode=detail&document_id=199353 (accessed 2 July 2021).

Thiennhien (2015) Mass 'concretisation' Phu Quoc beaches. Available at: https://www.thiennhien.net/2015/04/02/o-at-be-tong-hoa-cac-bai-bien-phu-quoc/ (accessed 15 July 2021).

Tien-Minh (2018) Nha Trang – a beauty contest place. *Tin Tuc*. Available at: https://baotintuc.vn/du-lich/nha-trang-noi-hoi-tu-cac-cuoc-thi-nhan-sac-20180117080914994.htm (accessed 8 July 2021).

Tran, V.N. (2010) Vietnam tourism in early XX century. *Ha Noi National University Journal of Social Sciences and Humanities* 26, 164–173.

Truong-Ha (2021) Proposal for 4,800 billion project to link Con Dao with national grid. *VnExpress*. Available at: https://vnexpress.net/de-xuat-4-800-ty-dong-keo-dien-ra-con-dao-4252319.html (accessed 18 July 2021).

Tsai, F.M., Bui, T.D., Tseng, M.L., Lim, M.K. and Tan, R.R. (2021) Sustainable solid-waste management in coastal and marine tourism cities in Vietnam: a hierarchical-level approach. *Resources, Conservation and Recycling* 168, 1–15.

Tuan Chau Corporation (2021) Available at: https://tuanchau-halong.com.vn/ (accessed 12 July 2021).

Tu-Truc (2015) Ly Son is being concretised. *Nguoi Lao Dong*. Available at: https://nld.com.vn/thoi-su-trong-nuoc/ly-son-dang-bi-be-tong-hoa-20151123214059069.htm (accessed 15 July 2021).

UNEP (United Nations Environment Programme) and ICLEI (Local Governments for Sustainability) (2003) *Tourism and Local Agenda 21: the Role of Local Authorities in Sustainable Tourism*. UNEP, Geneva, Switzerland.

UNESCO (United Nations Educational, Scientific and Cultural Organization) (n.d.) Ha Long Bay. Available at: https://whc.unesco.org/en/list/672/ (accessed 30 May 2021).

Vietnam Communist Party Central Committee XII (2018) Resolution no.36-NQ-TW dated 22/10/2018 by on sustainable development strategy of marine economy towards 2030 with outlook to 2045. Dan Van. Available at: http://www.danvan.vn/Home/Bien-dao-Viet-Nam/8568/Nghi-quyet-so-36-NQTW-ngay-22102018-cua-Ban-Chap-hanh-Trung-uong-khoa-XII-ve-Chien-luoc-phat-trien-ben-vung-kinh-te-bien-Viet-Nam-den-nam-2030-tam (accessed 18 July 2021).

Vietnamtourism (2019) In 2019, nearly 400 thousand visitation to Côn Đảo. Available at: https://vietnamtourism.gov.vn/index.php/items/30759 (accessed 15 July 2021).

Vietnamtourism (2021) Pivotal solutions for developing high end and competitive tourism in Côn Đảo. Available at: https://vetaucondao.vn/nhung-giai-phap-trong-tam-co-ban-de-tap-trung-phat-trien-du-lich-chat-luong-cao-co-tinh-dac-thu-rieng-va-khac-biet-cua-du-lich-con-dao/ (accessed 15 July 2021).

Viet-Tuan (2021) Pilot vaccine passport program for Phu Quoc nears completion. VnExpress. Available at: https://e.vnexpress.net/news/travel/places/pilot-vaccine-passport-program-for-phu-quoc-nears-completion-4307694.html (accessed 15 July 2021).

Vneconomy (2021) Côn Đảo: Quy hoạch lại theo hướng khu du lịch quốc gia thu hút 350.000 lượt khách/năm. Available at: https://vneconomy.vn/con-dao-quy-hoach-lai-theo-huong-khu-du-lich-quoc-gia-thu-hut-350-000-luot-khach-nam.htm (accessed 18 July 2021).

VOV World (2011) Con Dao island, an impressive historical and ecological site. Available at: https://vovworld.vn/en-US/discovery-vietnam/con-dao-island-an-impressive-historical-and-ecological-site-60482.vov (accessed 18 July 2021).

Vu, T.C. (2017) Sustainable development of Vietnam marine economy: current situation, potentials, challenges and solutions. In: *conference proceedings of Sustainable Development of Marine Economy: From Strategy to Reality*. Hanoi, VNU University of Economics and Business, Vietnam National University, pp. 42–49.

Walton, A., Hang, T.M., Hien, B.T.T., Duong, K.T. and Hien, N.B. (2015) *Vietnam Marine Protected Area Management Effectiveness Evaluation*. International Union for Conservation of Nature (IUCN), Gland, Switzerland.

WTTC (World Travel and Tourism Council) (2015) Governing National Tourism Policy. Available at: https://fac.ksu.edu.sa/sites/default/files/governing_national_tourism_policy_final.pdf (accessed 15 July 2021).

6 Ecotourism in Forests and Protected Areas

Long H. Pham[1*], Huong T. Bui[2], Huong H. Do[3] and Thomas E. Jones[2]
[1]Vietnam National University, Hanoi, Vietnam; [2]Ritsumeikan Asia Pacific University (APU), Japan; [3]Thai Binh Duong University, Nha Trang, Vietnam

Introduction

Vietnam has much potential for developing ecotourism reflected in its high biodiversity with plenty of wild endemic species, diverse ecosystems and beautiful scenery. It ranks as the 16th most biodiversity-rich country globally, hosting 110 Key Biodiversity Areas (Mittermeier *et al.*, 2004). UNESCO Vietnam recognizes 'Vietnam is one of the most important biodiversity ecosystems in the world, both in the marine and terrestrial ecosystems, especially in forest ecology and mangrove forests' (Van, 2019). Besides the global importance of its naturally occurring biodiversity, Vietnam is one of the richest countries in agro-biodiversity (USAID, 2013).

Vietnam's territory lies within South-east Asia's two climate zones; the zone around the equator, the equatorial zone, covers roughly the South Central and southern regions, and the intermediate tropical zone covers northern and North Central Vietnam (Boomgaard, 2007). The equatorial zone is characterized by the absence of prolonged dry periods and rainfall throughout the year. The type of vegetation corresponding to this rainfall pattern is that of evergreen and tropical forest that is always wet, also called tropical rainforest. However, in the intermediate tropical zone, the rainfall patterns are governed by the monsoon, which means that a rainy season and a dry season alternate. The existence of a real dry season corresponds with the presence of monsoon forests that include a number of deciduous trees. Such forests are more open than the rainforest. In the past two centuries, the physical environment has been significantly degraded, first by the establishment of plantation and extraction economies of the colonial times, followed by military conflicts and wars in the 20th century. Second, widespread deforestation, agricultural expansion, resettlement and urbanization have severely impacted local vegetation and fauna (Gupta, 2005).

This chapter offers an overview of ecotourism in forest-protected areas in Vietnam from theoretical concepts and historical evolution of forest conservation to the current structure of governance and management of forest resources and the economic and socio-cultural perspectives of forest resources ecotourism and ecotourists. Differentiated from conventional approaches using cases or empirical research focus on particular sites, the current study systematically analyses national policies and practices of ecotourism in national parks and protected areas (PAs). The review synthesizes

*Corresponding author: phamhonglong@gmail.com

DOI: 10.1079/9781789242782.0006

existing studies published on various channels in and outside the country and numerous government reports and research conducted by NGOs on ecotourism, PAs and forest conservation. The analysis pertinent in this chapter, however, excludes marine-based ecotourism, which is covered in a separate chapter (see Chapter 5, this volume).

Conceptual Perspectives

The definition of ecotourism refers to 'travel to fragile, pristine and usually PAs that strive to be low impact and (usually) small scale,' and that are 'primarily concerned with the direct enjoyment of some relatively undisturbed phenomenon of nature' (Valentine, 1992, p. 108). Ecotourism provides 'funds for conservation; directly benefits the economic development and political empowerment of local communities; and fosters respect for different cultures and human rights' (Honey, 1999, p. 25), highlighting its economic and socio-cultural outcomes. Blamey (1997) introduces a concise set of three ecotourism criteria that capture the intent and content of most of the previous definitions. First, ecotourism places primary emphasis on nature-based attractions that occur in or entail relatively undisturbed settings. Second, ecotourism involves the provision of opportunities for environmental learning and education. Third, the management of ecotourism sites and associated visitations should be compatible with the principles of environmental, socio-cultural and economic sustainability, or the so-called 'triple bottom line'. None the less, there is potential for ecotourism to cross over to the mainstream market, especially mass tourism. The current trend of ecotourism in Asia is inclined towards this trend, referred to as 'enlightened mass tourism' (Weaver, 2014).

The concept of ecotourism has been introduced to Vietnam since the mid-1990s, being defined as 'a type of nature-based and cultural indigenous tourism, including environmental education, contributing to conservation and sustainable development efforts with the active participation of local communities' (VNPPA, 2011, p. 16). Vietnam *Tourism Law* 2017 specifies 'ecotourism is a type of nature-based tourism, linking to local culture and participation of local communities, combining the purpose of education and conservation of environment' (MARD, 2020). In the strategy for Vietnam's tourism development released in 2020, looking forward to 2030, 'ecotourism' is used in parallel with the concept of 'green tourism' with three major objectives: (i) the development of 'green tourism' is inseparable from the utilization of valuable resources and protection of the environment; (ii) the environment is the prime factor that attracts tourists, therefore it ensures quality and value of consumption, reinforcing the tourism brand; and (iii) that tourism development projects adhere to environmental laws and regulations (MARD, 2020).

Historical Perspectives

The present-day territories of Laos, Cambodia and Vietnam came to constitute French Indochina in the late 19th century. The Pacific War (1941–1945) and World War II spelled the end of colonial rule in the region, and Vietnam became independent from 1945 and re-established its control over the territory in 1954. The various wars of independence and other uprisings and conflicts characterized the years between 1945 and the late 1970s in the Indochinese Peninsula, followed by post-war economic sanctions and stagnation until the early 1990s (Boomgaard, 2007). The conservation of forests and reforestation have only become focal policies since the mid-1990s, together with the emergence of ecotourism in forest-based settings.

Under French colonial rule, the rate of deforestation could be mainly explained by the expansion of agriculture and the need for land for livestock to graze (Boomgaard, 2007). French foresters set up the forest service in Indochina in the later 19th century by introducing scientific measures to facilitate sustainable forest exploitation. Their knowledge and expertise, however, was based on forests in the temperate zone, thus the foresters had little experience of monsoons or wet, tropical evergreen forests. Nature reserves, wildlife reserves and national parks with institutionalized by-laws were established later, from 1920 onwards (Boomgaard, 2007). Dated back then, 'forest politics' was one of the primary mechanisms by which the colonial government would enclose vast tracts of land and resources in the process of dispossessing forest farmers and other forest-dependent people. It was through these enclosed forests that primitive accumulation by state agencies, corporate interests, and influential individuals in strategic positions became possible in both the colonial and later periods (Nevins and Peluso, 2008). Paradoxically, forest services were established to save the forests from destruction, but they were the institutions that facilitated the exploitation of these forests.

Marking the early days of introducing nature-based tourism to mountainous areas were European patterns of seasonal trips to upland areas to the 'hill stations', where the climate is much cooler. In the past, emphasis was placed on the accessibility of hill stations (e.g. Bà Nà Hills, Sa Pa, Đà Lạt), and the majority of these settlements could be reached by railway within a range of fewer than 100 km. The development of the hill stations focused on Europeans. However, it required the permanent presence of Vietnamese and ethnic minority labourers; and also attracted Vietnamese tourists seeking to participate in the new 'mountain cure' (DeWald, 2008). For example, the establishment of Sa Pa, which went from being a French military sanatorium in 1912 to a full-scale European summer enclave by the end of World War II, implies that colonial legacies behind the emergence of the modern concept of ecotourism intertwined splendid nature and exotic culture. The same search for temperate respite from the tropical climate that once propelled the development of nature-based tourism thus remains relevant today (DeWald, 2008).

The forest coverage of Vietnam has changed dramatically under the pressures of prolonged wars and post-war economic development. In 1943, the forested area of Vietnam was 14.3 million ha, covering 43% of the country's territory. That was reduced to 11.2 million ha with a territorial coverage of 34% between 1943 and 1975, as much of the central and northern regions of the country were heavily defoliated by aerial application of herbicides during the Vietnam War (Nguyen and Vu, 2007). From 1976 to 1990, forests were excessively exploited for post-war reconstruction, leading to further reduction of the forest area by 2 million ha, reducing the remaining coverage ratio to 27.8%. Since 1990, with a policy of reforestation and central government incentives, forest coverage has increased. Statistics from 2005 show that the entire country's forest area has rebounded to 12.6 million ha, covering 37% of the territory (Nguyen and Vu, 2007), with the growth attributed to the efforts of reforestation programmes. However, the reforestation programmes focused almost entirely on plantations with exotic species, which have very low biodiversity value, and the remaining forests are often fragmented and degraded. The Forest Investigation and Planning Institution (FIPI) of Vietnam estimates that only 6–8% of the remaining cover consists of primary forest (i.e. old-growth forest untouched by human activity) (USAID, 2013). Due to the continued reforestation programme, forest cover reached 14.45 million ha in 2018, representing some 41.65% of Vietnam's landmass and comprising 10.3 million ha of natural forests and 4.15 million ha of plantation forests, respectively (VN FOREST, 2020).

Structural Perspectives

Since the late 1990s, the emphasis of the national forest policy has shifted from production to protection, including the management of forests for conservation along with local livelihoods and economic development. There has been significant reform of state-owned enterprises, a loosening of government controls in agriculture, and, from 1998, a decentralization of authority to the lowest appropriate level. This decentralization process has important implications for PA management. Responsibilities for all nature reserves and most national parks have been devolved to Provincial People's Committees (PPCs) (ICEM, 2003). With decentralization, provincial and local-level government institutions have assumed an important role in the conservation of PAs. Functions and responsibilities of the different institutions at national and provincial levels hint at the lack of clarity in mandates, the fragmentation of decision making, and overlap in responsibilities (USAID, 2013).

The governance structure delegates PAs' conservation responsibilities to two ministries. At the national level, the Ministry of Natural

Table 6.1. National parks and protected areas (PAs) in Vietnam. (Mountainous national parks are in italics.) From VN FOREST, 2020.

Region	Name	Year established	Area (ha)	Province
Northwest	*Hoàng Liên*	*2002*	*29,845*	*Lào Cai, Lai Châu*
	Ba Bể	*1992*	*7,610*	*Bắc Kạn*
	Bái Tử Long	2001	15,783	Quảng Ninh
	Xuân Sơn	*2002*	*15,048*	*Phú Thọ*
	Tam Đảo	*1996*	*36,883*	*Vĩnh Phúc, Thái Nguyên, Tuyên Quang*
	Du Già	*2015*	*15,006*	*Hà Giang*
	Phia Oắc - Phia Đén	*2018*	*10,593*	*Cao Bằng*
Red River Delta	*Ba Vì*	*1991*	*10,815*	*Hà Nội, Hòa Bình*
	Cát Bà	1986	15,200	Hải Phòng
	Cúc Phương	*1962*	*22,200*	*Ninh Bình, Thanh Hóa, Hòa Bình*
	Xuân Thủy	2003	7,100	Nam Định
North Central Coast	*Bến En*	*1992*	*14,735*	*Thanh Hóa*
	Pù Mát	*2001*	*91,113*	*Nghệ An*
	Vũ Quang	*2002*	*55,029*	*Hà Tĩnh*
	Phong Nha-Kẻ Bàng	*2001*	*85,754*	*Quảng Bình*
	Bạch Mã	*1991*	*22,030*	*Thừa Thiên–Huế, Quảng Nam*
South Central Coast	*Phước Bình*	*2006*	*19,814*	*Ninh Thuận*
	Núi Chúa	2003	29,865	*Ninh Thuận*
Central Highlands	*Chư Mom Ray*	*2002*	*56,621*	*Kon Tum*
	Kon Ka Kinh	*2002*	*41,780*	*Gia Lai*
	Yok Đôn	*1991*	*115,545*	*Đắk Nông, Đắk Lắk*
	Chư Yang Sin	*2002*	*58,947*	*Đắk Lắk*
	Bidoup Núi Bà	*2004*	*64,800*	*Lâm Đồng*
	Tà Đùng	*2018*	*20,937*	*Đắc Nông*
Southeast	*Cát Tiên*	*1992*	*73,878*	*Đồng Nai, Lâm Đồng, Bình Phước*
	Bù Gia Mập	*2002*	*26,032*	*Bình Phước*
	Côn Đảo	1993	15,043	Bà Rịa–Vũng Tàu
	Lò Gò-Xa Mát	2002	18,765	Tây Ninh
Mekong Delta	Tràm Chim	1994	7,588	Đồng Tháp
	U Minh Thượng	2002	8,053	Kiên Giang
	Mũi Cà Mau	2003	41,862	Cà Mau
	U Minh Hạ	2006	8,286	Cà Mau
	Phú Quốc	2001	31,422	Kiên Giang

Fig. 6.1. Distribution of protected areas (PAs) in Vietnam. Map courtesy of Thuy T. Duong.

Resources and Environment (MoNRE) takes the lead for biodiversity. However, the Ministry of Agriculture and Rural Development (MARD) and its provincial departments are responsible for all special-use forests (SUFs). While MoNRE is legally responsible for biodiversity conservation, MARD is responsible for the management of SUFs and the provision of technical support services to PPCs and lower-level administrative units, such as Community People's Committees (CPCs). The agencies responsible for managing these areas in the provinces are the Department of Agriculture and Rural Development, the Department of Science, Technology and Environment, and the Forest Protection Department (USAID, 2013).

There are 167 PAs nationwide, including 33 national parks, 57 nature reserves, 14 species and habitat conservation areas, 54 landscaped PAs, and nine experimental and scientific research areas (VN FOREST, 2020) (see Table 6.1). The distribution of these PAs can be found in Fig. 6.1.

Among the 33 national parks, six are under the Vietnam Administration of Forestry and 20 under a PPC, and seven under the Department of Agricultural and Rural Development. Responsibility for managing a PA in Vietnam depends on whether it falls entirely within a single province or straddles several provinces (Jones *et al.*, 2016). MARD has management responsibility for those PAs that are in more than one province while municipal governments are responsible for PAs contained entirely within a single province. This arrangement has resulted in a highly decentralized governance system, with only six national parks being directly managed by MARD. The structure of forest resource governance delegates responsibilities for managing resources to central and local governments, where ecotourism activities could be carried out and managed. Further responsibility lies with forest owners and investors, while regulations on types and forms of ecotourism have been furthered strengthened by the *Law of Forestry*, the *Tourism Law*, and the *Law of Biodiversity* (Vietnam National Congress, 2017a, b; 2018), and related regulations. These laws and regulations lay a foundation for achieving desirable economic and socio-cultural outcomes of ecotourism as analysed in the sections below.

Economic Perspectives

In 2015, 61 national parks and SUFs welcomed 1.15 million visitors, generating VND 77.3 billion. The year 2016 witnessed over 1.76 million visitors, up 53.3% compared to 2015, with total revenue reaching VND 130 billion, a 68% increase compared to 2015. The visitor number dropped to 1.67 million in 2017, down 5.2% from the previous year, but the revenue reached VND 145 billion, up 11.1% compared to 2016. The year 2018 saw 2.39 million visitors generate VND 180 billion, an increase of 18.7% from the previous year. There were 2.5 million visitors in 2019, an increase of 5% compared to 2018, and the revenue was about VND 185 billion, up 12% compared to 2018. Since removing the fixed price for entrance fees to national parks in 2016, the revenue per visitor has dramatically increased. In 2020, due to the impact of climate change and the COVID-19 pandemic, the total number of tourists decreased by 66% (to 900,000), and total revenue decreased by 69% (to VND 44 billion) compared to 2019 (MARD, 2020).

There are four primary ecotourism-related sources of revenue. The breakdown of the sources of revenue of VND 180 billion in 2018 is revenue from entrance tickets at VND 115 billion (64%), with the remainder divided among the revenue from three forms of ecotourism services: self-organized, joint venture, and leasing forest resources, with revenues of VND 34.3 billion (> 19%), VND 16.6 billion (> 9.2%) and VND 13.56 billion (> 8%), respectively. Among 167 SUFs, 61 sites (25 national parks and 36 nature reserve areas) have provided ecotourism. The distribution of sites by the three forms of operations are: (i) self-organized (37 sites); (ii) joint venture (eight sites); and (iii) leased forest resources (13 sites) (MARD, 2020). Among these, 15 SUFs apply two forms of ecotourism services and two SUFs apply all three forms.

Entrance fee

Under regulations, visitors to national parks should purchase an entrance ticket. Revenue from entrance fees to national parks and natural reserve areas contributes the largest share of the revenue. The current fee system reveals several issues.

First, the fee system has been based on a fixed price and is not contingent on the resources and conditions of national parks. Entrance fees for national parks have remained unchanged since the promulgation of the Decree on Fees and Payments in 2001, despite variation in market demand and product quality. Since 2016, the Ministry of Finance has allowed the entrance fees to be changed. The fixed price that applies to all national parks managed by MARD is VND 60,000 for adults; discounted fees for students and pupils at VND 20,000; and a price of VND 10,000 for children; as well as a 50% discount (a fee of VND 30,000) for persons with disabilities and seniors (Pham and Bui, 2020).

Second, the entrance fee does not change according to the location and level of facilities established in various national parks across Vietnam. In reality, visitor arrivals to national parks vary significantly due to the national

Table 6.2. Revenue from entrance fees of six national parks, 2013–2017 (in VND million). From General Administration of Forestry, adapted from Pham, 2018.

	Year				
National park	2013	2014	2015	2016	2017
Ba Vì	4,934	5,247	7,559	13,009	19,593
Tam Đảo	0	204	348	274	143
Cúc Phương	2,624	2,176	2,574	3,393	5,561
Bạch Mã	369	580	552	620	860
Yok Don	0	0	0	36	29
Cát Tiên	652	802	917	1,135	1,556
Total	8,579	9,009	11,950	18,467	27,742

park's competitive advantage based on biological conditions, location, accessibility, landscape, infrastructure, and product quality. The national parks with locations that are closer to large cities are easy to access. Good infrastructure is also in place at these national parks, including such elements as the condition of roads, accommodation, the establishment of visitor centres, and qualified human resources. They also have well-established images and brands via advertisement and promotion. These national parks have competitive advantages over other parks, resulting in higher revenue from entrance fees (see Table 6.2).

Third, from the viewpoint of a market-based fee payment system, the resources for ecotourism in national parks are undervalued. Findings from a study of visitors' willingness to pay show that 88.5% of visitors to national parks were willing to pay a higher fee for entering national parks and 74% of visitors at travel agents showed their willingness to do so (Ngo and Pham, 2016). The average rate that tourists were willing to pay was 2.5 times higher than the current prices for entrance. In addition, interviews with park rangers also indicate the possibility of increasing entrance fees. A higher fee was confirmed among rangers of Cát Tiên National Park with visitors willing to pay VND 100,000 each. The rangers suggested that the park implement a two-price system that would apply to foreign and domestic visitors, respectively (Pham and Bui, 2020).

Finally, problems also arise from organizations involved in managing ecotourism activities in national parks. Ecotourism is a leisure activity. Therefore revenue and investment for ecotourism should be under the direction of the Ministry of Culture, Tourism and Sport, which is different from the main provider of state funding to national parks – the MARD. Consequently, the entrance fee falls under the mandate of the Ministry of Culture, Tourism and Sport, while national park management and investment are the responsibilities of MARD. Therefore, ecotourism development in national parks and the related fee-payment activities have not yet received enough attention from park rangers and mandated organizations. Responsibility for ecotourism activities is unclearly regulated due to the two ministries' overlapping mandates (Pham and Bui, 2020).

Self-organized ecotourism by park administrators

The operation of ecotourism entirely depends on the budget provided by the state to the national parks. Therefore, it directly affects the level of conservation, policy and capacity of the national parks. Park administrators have encountered numerous difficulties in trying to utilize limited resources for investment in ecotourism services. The majority of PAs are located in unfavourable, remote locations, with low socio-economic development levels, poor transportation infrastructure and a lack of qualified human resources. Thus, the opportunities are limited for the parks to generate additional revenue from this form of operation (VN FOREST, 2020). The ineffectiveness of ecotourism services self-organized by park administrators can be explained from the aspect that Vietnamese national park management lacks experience in managing tourism within the parks, especially after the decentralization of power to lower levels of government (Suntikul *et al.*, 2010). Despite the MARD and/or PPC taking on the responsibility to manage national parks directly, they do not participate in the daily operational management of the parks. This task belongs to the national park management board, a state-owned organization, which performs the functions and tasks of forest owners and implements the state-driven conditions for managing, protecting and developing PAs. This unit conducts and manages tourism business in parks via a subunit called the Tourism Management Unit. Even though a number of national parks and nature reserves have ecotourism development plans (e.g. Côn Đảo, Phong Nha-Kẻ Bàng, Yok Đôn, Cúc Phương, Cát Tiên and Ba Vì), these plans cannot be fully implemented due to shortages of human resources, such as staff who specialize in ecotourism, to meet the requirements of operating professional tourism activities. In addition, there are no quality standards in terms of infrastructure, technical facilities, and services for tourism to facilitate meeting the standard requirements of ecotourists. Shortages of investment in information gathering and displays result in a lack of a unified and synchronous information system linking all national parks and nature reserves to tourist centres. International travel companies have limited access to information on the ecotourism potential of national parks and nature reserves for large numbers of tourists, both domestic and international (Pham and Bui, 2020).

Joint ventures and public–private partnerships

In this partnership model, the national parks and nature reserves partially contribute capital and jointly operate ecotourism activities with private partners. A joint venture to develop ecotourism balances the interests of the national parks and nature reserves, on the one hand, and travel companies, on the other hand, to offset the weaknesses of each partner, as national parks/nature reserves often lack the necessary capital, while travel companies lack experience and knowledge of ecotourism and nature reserves. There are eight national parks and nature reserves applying these kinds of operations. Ba Bể National Park and Đồng Nai Culture-Nature Protected Area are preparing proposals for the establishment of joint ventures. Revenue generated from six sites operating under the partnership model accounts for 9.2% of the total revenue from ecotourism (MARD, 2020). As shown in Table 6.3 Phong Nha-Kẻ Bàng National

Table 6.3. Revenue from joint ventures and partnerships for ecotourism in 2016–2019. Adapted from MARD, 2020.

	Revenue (in a million VND)			
Name of the site	2016	2017	2018	2019
Cát Bà	1,700	2,355	2,631	(No update)
Hoàng Liên	1,814	1,246	1,176	(No update)
Phong Nha-Kẻ Bàng	5,210	17,363	11,750	4,800
Tràm Chim	216	359	277	(No update)
Ba Vì	3,036	No longer in operation in this category	No longer in operation in this category	No longer in operation in this category
Cát Tiên	0	0	180	1,302
Ba Bể	0	0	0	In preparation
Đồng Nai Cultural and Natural Reserve Area	0	0	0	In preparation
Total	11,976	21,323	16,014	(Incomplete)

Fig. 6.2. Phong Nha-Kẻ Bàng National Park. Photo courtesy of Long H. Pham.

Park has been the main player successfully operating under the joint venture and partnership model. Being at the popular tourist site of Phong Nha-Kẻ Bàng (see Fig. 6.2), major stakeholders including the national park management board, the Tourism Management Unit, and a private company make decisions on park tourism and recreation businesses (Ly and Xiao, 2016). However, this co-management model might not apply to all national parks, particularly the provincially managed parks. Generally, the operation of joint ventures remains small in scale, and they are not yet very effective. Due to the shortage of specific instructions, the national parks/nature reserves and other travel companies are facing numerous difficulties in setting up joint ventures, discussing agreements, and determining the powers and responsibilities of each party in management and operations (Bui, 2012).

Leasing forest resources

Although policies for national parks/nature reserves to lease forest resources to the private sector to establish tourism services are both available and encouraged by the state government, the ambiguity of the guidelines, policies and mechanisms has made it challenging to implement. Thus, forest resource leasing remains in its test phase. Among 13 areas leasing forest resources, only eight have reported revenue as of 2018, while one has yet to finalize revenue, and four are preparing proposals for opening ecotourism services (MARD, 2020). Revenue from leasing forest resources accounts for a small fraction of total revenue, at 8%. As shown in Table 6.4, Phong Nha-Kẻ Bàng National Park and Hoàng Liên National Park are major players that have managed to gain significant revenue from leasing existing resources for ecotourism, recreation and entertainment services.

Leasing forest resources has significant positive aspects such as attracting investment to potential national parks and nature reserves, thereby contributing revenue to conservation; the forest areas being contracted to companies are usually better protected. The drawbacks of leasing forest resources to the private sector lie in the ability of administrators to control and monitor the progress of projects and their construction in accordance with what has been approved. In reality, some companies do not execute the plan as proposed: (i) companies propose a project to lease a very large area of a forest, but only small or no deployments are

Table 6.4. Revenue from leasing forest resources for ecotourism in 2016–2019. Adapted from MARD, 2020.

	Revenue (in a million VND)			
Name of the site	2016	2017	2018	2019
Ba Vì	400	778	778	636
Hoàng Liên	1,814	1,246	1,176	1,100
Phong Nha-Kẻ Bàng	6,352	8,180	8,014	7,500
Phú Quốc	260	260	260	250
Tràm Chim	216	359	277	135
Bình Châu Phước Bửu	534	534	534	534
Núi Ông	0	0	1,693	–
Bạch Mã	0	0	0	32
Total	9,576	11,357	12,732	10,187

carried out due to failing to call for investment; or (ii) companies use the leased forest areas to develop tourism facilities without attention to conservation and environment protection (Bui, 2012).

Having assessed three operations of ecotourism services in forest-protected areas, several insights into resource management have emerged. It was found that active involvement in ecotourism services exists at those sites that adopt multiple forms of operation, such as Hoàng Liên, Phong Nha-Kẻ Bàng and Ba Vì National Park. In Hoàng Liên National Park, self-organized ecotourism services are provided by the Hoàng Liên Centre for Ecotourism and Environmental Education, including joint ventures with different partners since 2014, and the forest resources were leased for the project 'The Complex of cultural tourism, cable cars, entertainments and hotels Fansipan-Sapa' covering an area of 53.3 ha. Similarly, Phong Nha-Kẻ Bàng operates with all three forms of operations (i.e. self-organized, joint ventures and leasing). The leasing of resources is done in two ways: (i) long-term service supply for 50 years; and (ii) collecting expenses from each trip for cave exploration. Somewhat differently, Ba Vì National Park suspended the joint venture model and shifted to leasing forest resources. Meanwhile Phú Quốc National Park leases the resources for the project 'Vinpeal Phú Quốc Safari' over 498.9 ha, and Bến En National Park leases the forest to Sun Group comprising an area of 678 ha (MARD (Ministry of Agriculture and Rural Development, 2020). The investors cover expenses for reforestation and conservation of biodiversity, while park administrators collect fees for reinvesting in forest protection and reforestation and provide benefits for workers and ecosystem restoration.

Socio-cultural Perspectives

The unique relationship between humans and nature is deeply rooted in Asian cultural and religious characteristics. As Frost *et al.* (2014) contend 'there appears to be widespread respect for and interest in nature among Asia-Pacific societies, which has strong cultural and religious roots, contrary to the popular view that this is mainly a Western concern' (p. 721), Confucian and Buddhist thinking tends to 'regard all elements of the universe as one entity' (Lee *et al.*, 2013, p. 523). Thus human elements such as cultural attractions in natural surroundings are not seen as problematic, as nature and human beings are interdependent and one can shape or even improve the other (Weaver *et al.*, 2020). In this view, the socio-cultural notion of ecotourism in Vietnam in particular centres around two major competing concepts of ecotourism and/or ethno-tourism in natural forest settings (Jones *et al.*, 2021).

As the development of ecotourism in Vietnam cannot be separated from the socio-cultural environment of the local communities inhabiting the forests, there are major issues in relation to the status of interdependence between indigenous communities and natural surroundings in remote areas. First, local communities have a long history of engagement with the localities, well before the nature reserves were founded; most of the local people living in the nature reserves are ethnic minorities or indigenous people with their own socio-cultural characteristics. Therefore, their traditional culture is very rich and diverse, and it needs to be preserved in a complete system of natural and human ecology (Nguyen, 2008). Second, ethnic minorities and indigenous people are generally concentrated in remote and inaccessible areas that have been affected by pressures from forest closures and exploitation. These include land disputes involving tribal minorities, and the state and private logging interests which have become frequent and intense. Third, local people are usually poor, and dependent on the natural resources available in nature reserves. Their agricultural production is often ineffective due to the continued use of outdated farming methods such as shifting and slash-and-burn cultivation and lack of access to modern farming and husbandry techniques (Nguyen, 2008). Such activities pose a direct threat to the biodiversity of nature reserves. In Vietnam, sedentarization has long been a central feature of government policy towards minority groups in the northern mountainous areas and Central Highlands. Under the policy, indigenous peoples are encouraged to live in permanent villages and cultivate cash crops. During the 1980s, the policy was crudely enforced and the results were mixed. From 1989, however, new measures were introduced to return the land to indigenous peoples, make fixed cultivation more compatible with forest protection, and strengthen ethnic minority representation in national institutions (Clarke, 2001).

Hoàng Liên National Park in Vietnam illustrates the notion of the socio-cultural aspect of ecotourism development in remote areas. The national park is located south-east of the Himalayas in the north-west of Vietnam, within the two provinces of Lai Châu and Lào Cai. Ethnic minorities that speak local languages and follow traditional lifestyles based on subsistence farming predominantly populate the forest areas of the national park. The livelihood of these local people strongly depends on the available natural resources, while the scarcity of arable land coupled with population growth has led to increasing pressure on the forests. The creation of non-agricultural jobs through the development of ecotourism activities in rural areas has been suggested as a viable means to offset pressures on forests that can generate additional income for households and encourage the abandonment of less-productive farmland and the spontaneous establishment of secondary forest on former agricultural plots (Gaughan *et al.*, 2009). The diversity and distribution of ethnic groups along the Hoàng Liên Mountain range at different altitudes impacts their roles in the supply chain of ecotourism in several ways. For example, the H'Mông group provides porter and guide services to mountaineers. The Tày group, with larger houses located in middle-altitude areas, use these advantages to establish homestay services for soft-adventure visitors. The Dao and Giáy ethnic groups often participate in tourism souvenir production, while the Kinh group dominates tourism infrastructure and services in Sapa town lowland areas. Maintaining their traditional mountainous lifestyles, ethnic minority groups inhabit the core and buffer zones of the national park, and earn their living from forest products and resources. The main challenges are to find a balance between the rapid development of tourism activities and the preservation of the authentic socio-cultural elements of the ethnic minorities that make the area attractive for tourists. In the past, small numbers of Vietnamese and foreign tourists could access Mount Fansipan via a 3-day hike. With the inauguration of the Fansipan Legend (Indochina Summit) cable car (see Fig. 6.3), the mountain's summit can now be reached within just 18 minutes. The cable car starts at the height of 1500 m on the outskirts of the town of Sapa, transporting tourists to the summit of Mount Fansipan, the highest peak in Vietnam, landing just 143 m from Fansipan's summit (Jones *et al.*, 2021). The engineering of modern Sapa can be interpreted as a statement of the Vietnamese state's intent to bring these remote highlands under central political control through exhaustive 'distance demolishing technologies' (Michaud and Turner, 2017). The operation of this cable-car system exemplifies the environmental resource leasing model. Within this partnership, tourism development policy is directed by the state government,

Fig. 6.3. Cable car in Hoàng Liên National Park. Photo courtesy of Long H. Pham.

and tourism infrastructure is invested in by private groups (i.e. Sun Group) and targets the en masse domestic market, exemplifying Weaver's (2014) 'enlightened' or 'pragmatic' mass tourism.

Visitor Perspectives

There are several approaches to understanding the evolving market for ecotourists. The most popular approach remains one that is supply-based, with ecotourists frequently considered as synonymous with those who visit nature-based destinations such as national parks (Cochrane, 2006), bio-reserves (Lindberg *et al.*, 1997), or other relatively wild and 'undisturbed' areas (Ballantine and Eagles, 1994). Another approach is to research the clients of a specialized ecotourism company or members of an organization who have previously taken trips to nature-based destinations for purposes that are more clearly aligned with ecotourism (Fennell, 2002). However, to distinguish ecotourists from other nature-based tourists, often visiting the same places, a number of demand-side approaches are also adopted, including place of residence, demographics, motivation (Ballantine and Eagles, 1994; Meric and Hunt, 1998; Galloway, 2002), behaviour (Weaver and Lawton, 2002), self-definition (Tao *et al.*, 2004), benefit segmentation (Palacio and McCool, 1997), social value (Blamey and Braithwaite, 1997) and environmental value (Zografos and Allcroft, 2007).

Visitor perspectives: a supply-based approach

From a supply-based perspective, ecotourists visiting national parks and nature reserves in Vietnam are also unevenly distributed spatially. National parks with easy access to major national roads, with good connections to provincial roads and adequate accommodation regularly receive large numbers of visitors. These include Phong Nha-Kẻ Bàng, Ba Vì, Cát Bà, Bạch Mã and Cát Tiên national parks (Fig. 6.4). Though nature-based-tourism activities are varied, the majority of the activities are concentrated in areas with better infrastructure and more convenient access. Despite great potential for more diversified activities, remote areas with biodiversity or unique landscapes still struggle to attract visitors and persuade them to extend their stay in the respective national parks. Domestic visitors account for more than 90% of the total number of visitors to national parks and

Fig. 6.4. Cát Tiên National Park. Photo courtesy of Long H. Pham.

Fig. 6.5. Cúc Phương National Park. Photo courtesy of Long H. Pham.

SUFs. National parks with established ecotourism services and infrastructure account for 97.5% of the total number of visitors and 99% of the total ecotourism-generated revenue. Only 2.5% of the visitors came to visit nature reserves, bringing in about 1% of the total revenue (MARD, 2020).

Group travel is the most popular form of travel arrangement. Group travel is often organized by schools and universities for educational purposes or by public or private-sector organizations for their workers and family as a part of annual holidays offered as fringe benefits (Bui and Jolliffe, 2011). A study in Cát Tiên National Park (Do, 2014) revealed that almost all visitors to the park (92%) travelled in groups. The group sizes were found to be 12 persons on average for Vietnamese, while foreign visitors usually formed a smaller group of five people on average (Do *et al.*, 2012). Fieldwork conducted for MARD in 2016 by consultant teams (Pham, 2018) reveals that the majority of the visits are day trips or within 2 days and a night in Fig. 6.5, Cát Tiên and Phong Nha-Kẻ Bàng. International tourists tend to stay in national parks and nature reserves longer than domestic visitors. For example, in 2015, international visitors stayed for an average of 3.1 days compared to just 1.5 days for their domestic counterparts. That has been confirmed in the later study where only 7% of visitors in Cát Tiên National Park were repeat visitors and 73% reported sleeping in a park guesthouse or ecolodge (Do *et al.*, 2012).

Visitor characteristics to Cát Tiên National Park are unveiled in research conducted by Do (2014) using a random stratified sampling method, whereby receptionists asked every second visitor whether they were willing to participate in the survey upon checking out from their accommodation. A total of 1532 paper-based questionnaires were distributed, and 1267 visitors returned the questionnaires, resulting in a response rate of 82.7%, of which 1082 questionnaires remained valid. Just under half of the respondents (46.2%) were Vietnamese, and 53.8% were foreigners from six dominant source countries: (i) the USA (16%); (ii) Germany (15%); (iii) Australia (15%); (iv) the Netherlands (12%); (v) the UK (9%); and (vi) France (7%). When asked if they considered themselves to be ecotourists, 70% of those who answered this question self-identified as 'ecotourists'. The most interesting finding is that a large majority of Vietnamese (83%) agreed with the question compared with 58% of foreign tourists. The Vietnamese visitors tended to be younger (mean (M) = 28.5 years of age, standard deviation (SD) = 10.5) while their foreign counterparts were older, but display more variability (M = 41.2 years of age, SD = 15.2). There was an overall equilibrium between females (50%) and males (50%). However, there were 133 fewer females (46%) in the international visitors' group, compared with the Vietnamese group, which had a slight majority of females (55%). Overall, 51% reported having an undergraduate university background, 23% listed postgraduate qualifications, 15% had a diploma and 11% had completed high school. In terms of visitors' origin, among 582 foreign visitors, 549 reported their country of residence. In terms of visitor satisfaction, foreign visitors were significantly more satisfied than Vietnamese visitors but were also more likely to be very dissatisfied. More specifically, 58% of visitors were satisfied with their park visit, while 16% were very satisfied. Only 3% of visitors were dissatisfied, and the same proportion was very dissatisfied.

As illustrated in Table 6.5, a short walk was reported as the most preferred activity in the park (79%), followed by wildlife viewing and long walks (65% and 64%, respectively). Ten per cent of respondents came to the park to study or undertake research. One-third visited the Bear/Primate Rescue Centre, 27% joined in bird watching, 56% took photos of the natural landscape in Cát Tiên during their visits, 24% went boating (mostly on Crocodile Lake) and 16% went cycling. The *t*-test confirmed there were significant differences between the Vietnamese and the Westerners across all activities except for wildlife viewing and visiting the Rescue Centre. Western visitors were more likely to take part in the two walking activities and bird watching. More Vietnamese conducted study/research and enjoyed camping and taking photographs.

Visitor perspectives: a demand-based approach

Segmentation analysis of visitors to Cát Tiên National Park (Do, 2014) found that there are six homogeneous groups of visitors who all conform to the core expectations of the 'ideal' ecotourist reflecting the preference of being close to wild nature, commitment to a certain level of learning about nature, awareness of environmental protection, and striving to be a responsible traveller in tandem at least with the neutral option of status quo sustainability. One group resembles 'harder' ecotourists (i.e. physically active people who take specialized and long trips, with emphasis on personal experience, have expectations of few services and a strong commitment to the environment and enhanced sustainability), and one group

Table 6.5. Tourism activities undertaken by visitors to Cát Tiên National Park. Adapted from Do *et al.*, 2012.

Activities	Frequency (%)	Vietnamese (%)	Western visitors (%)	*t*-test value	*P*
Short walking tour (<5 km)	79	72	85	27.6	0.000
Long walking tour (≥5 km)	64	58	70	15.8	0.000
Wildlife viewing	65	63	67	1.97	0.160
Studying/research	10	19	2	95.7	0.000
Visiting Bear/Primate Rescue Centre	34	34	35	0.011	0.949
Camping	17	32	4	147.6	0.000
Bird watching	27	14	39	85.9	0.000
Photography	56	72	43	90.2	0.000
Boating	24	21	26	4.8	0.028
Cycling	16	13	18	4.6	0.032

presents characteristics of 'softer' ecotourists (i.e. physically passive people who take multi-purpose and short trips, expect services from the destination, and learn about natural history and aboriginal culture mostly from interpretation; they display a tepid commitment to the environment and sustainability). Three groups appear to be 'structured' ecotourists (i.e. those who resemble both 'harder' ecotourists and 'softer' ecotourists in certain characteristics) (Weaver and Lawton, 2002). The centrality of the softer-to-harder spectrum, as well as the dualistic structured ecotourists, suggests the utility of the 95/5 spatial model as a core protected area management framework, with the vast majority of domestic visitors preferring to restrict their activities to the confined site-hardened areas near gateways and trailheads. Softer and structured foreign ecotourists are also likely to avail themselves of such areas for most of their visit, with the undeveloped bulk of the parks focused on biodiversity and hydrological maintenance as well as infrequent visits from the mainly foreign hard-core ecotourists.

Implications suggest suitable ecotourism management, products and segment-targeted marketing and public relations tools for domestic ecotourist markets as well as effective interpretation and experiential programmes for enhancing learning/education (Do *et al.*, 2014). A new regime of learning facilitation for domestic visitors should take into account the broader regional impulses of collectivism (i.e. cooperative group learning) as well as the proclivities to touch and otherwise interact closely with desirable flora and fauna. The complex learning context of Vietnamese visitors, moreover, should accommodate opportunities, perhaps through sensory botanical gardens and wildlife rehabilitation facilities, to satisfy those who seek sensory impressions (Ballantyne *et al.*, 2011). This could foster a model of what might be described as 'complementary' or 'green' anthropocentrism that seeks deliberately and simultaneously to satisfy and enrich visitors' expectations without compromising parallel mandates to protect and restore the environment. It is perhaps even possible that interactions that attract feelings of awe and delight may further stimulate the sensitivities that are apparent in Vietnamese visitors through 'enhanced sustainability', and inspire their participation in different kinds of site enhancement activities.

Recently, the number of international visitors to national parks and PAs has significantly dropped as an impact of border closures due to COVID-19. Domestic tourism has proved its resilience to recover faster and become the main source of growth for the tourism industry while international borders remain closed. The current crisis offers an opportunity for the country to re-evaluate its tourism policies to encourage domestic travel. With the large volume of local and domestic visitors flocking to national parks, the increasing health concerns and safety issues at densely populated destinations, the requirement to social distance during the pandemic, and international travel restrictions, ecotourism in remote regions has the potential to recover quickly and sustain growth momentum.

Conclusions

Vietnam has great potential for the development of ecotourism. The duality of tourism development and poverty eradication has made it possible for remote locations with forest resources to adopt ecotourism for development. Additionally, the evolution of ecotourism in Vietnam cannot be independent of the structure of governance and regulations dating back to the colonial period and carried on by the post-war socialist state of Vietnam. By providing an analysis of the factors comprising ecotourism characteristics of Vietnam from conceptual, historical, structural, economic, socio-cultural and behavioural aspects, this chapter offers insights into the external contexts and internal issues of ecotourism on a national level.

This work contributes to widening current knowledge on ecotourism in Vietnam through a systematic synthesis of literature and data on a vast number of national parks and PAs operating ecotourism services. The chapter, however, has some inherent limitations. First, it only covers ecotourism either in forest settings or related to forest conservation and protection policies. The marine PAs are covered in Chapter 5, this volume. Second, the analysis is carried out on a macro-level, interspersed with case-study insights to operations at the micro-level. These limitations, however, open opportunities for future study to further explore this growing segment of tourism that is yet to be fully understood.

References

Ballantine, J.L. and Eagles, P.F.L. (1994) Defining Canadian ecotourists. *Journal of Sustainable Tourism* 2(4), 210–214. DOI: 10.1080/09669589409510698.

Ballantyne, R., Packer, J. and Sutherland, L.A. (2011) Visitors' memories of wildlife tourism: implications for the design of powerful interpretive experiences. *Tourism Management* 32(4), 770–779. DOI: 10.1016/j.tourman.2010.06.012.

Blamey, R.K. (1997) Ecotourism: the search for an operational definition. *Journal of Sustainable Tourism* 5(2), 109–130. DOI: 10.1080/09669589708667280.

Blamey, R. and Braithwaite, V. (1997) A social values segmentation of the potential ecotourism market. *Journal of Sustainable Tourism* 5(1), 29–45. DOI: 10.1080/09669589708667274.

Boomgaard, P. (2007) *Southeast Asia: an Environmental History*. ABC Clio, Santa Barbara, California.

Bui, H.T. and Jolliffe, L. (2011) Vietnamese domestic tourism: an investigation of travel motivations. *Austrian Journal of South-East Asian Studies* 4(1), 10–29.

Bui, X.T. (2012) *Assessing the Potential of Ecotourism and the Need for Ecotourism Investment in Vietnam*. WWF (World Wide Fund for Nature), Hanoi.

Clarke, G. (2001) From ethnocide to ethnodevelopment? Ethnic minorities and indigenous peoples in Southeast Asia. *Third World Quarterly* 22(3), 413–436. DOI: 10.1080/01436590120061688.

Cochrane, J. (2006) Indonesian national parks: understanding leisure users. *Annals of Tourism Research* 33(4), 979–997.

DeWald, E. (2008) The development of tourism in French Colonial Vietnam, 1918–1940. In: Cochrane, J. (ed.) *Asian Tourism: Growth and Change*. Elsevier, Amsterdam, pp. 221–231.

Do, H.H. (2014) Differences between Vietnamese and Western protected area visitors in Viet Nam: evidence from Cat Tien National Park. PhD thesis, Griffith University, Brisbane, Queensland, Australia.

Do, H.H., Weaver, D. and Lawton, L.J. (2012) Differences between Western and Vietnamese park visitors: evidence from Cat Tien National Park, Viet Nam. Paper presented at the 6th World Conference for Graduate Research in Tourism, Hospitality and Leisure, 24–29 April 2012, Fethiye, Turkey.

Do, H.H., Weaver, D. and Lawton, L.J. (2014) Evidence of an emerging domestic ecotourist market from Cat Tien National Park, Vietnam. Paper presented at the 2nd Tourism and Hospitality International Conference, 5–6 November 2014,Langkawi, Malaysia.

Fennell, D. (2002) The Canadian ecotourists in Costa Rica: ten years down the road. *International Journal of Sustainable Development* 5(3), 282–297. DOI: 10.1504/IJSD.2002.003754.

Frost, W., Laing, J. and Beeton, S. (2014) The future of nature-based tourism in the Asia-Pacific region. *Journal of Travel Research* 53(6), 721–732. DOI: 10.1177/0047287513517421.

Galloway, G. (2002) Psychographic segmentation of park visitor markets: evidence for the utility of sensation seeking. *Tourism Management* 23(6), 581–596. DOI: 10.1016/S0261-5177(02)00025-0.

Gaughan, A.E., Binford, M.W. and Southworth, J. (2009) Tourism, forest conversion, and land transformations in the Angkor basin, Cambodia. *Applied Geography* 29(2), 212–223. DOI: 10.1016/j.apgeog.2008.09.007.

Gupta, A. (2005) The landforms of Southeast Asia. In: Gupta, A. (ed.) *The Physical Geography of Southeast Asia*. Oxford University Press, New York, pp. 38–64.

Honey, M. (1999) *Ecotourism and Sustainable Development: Who Owns Paradise?* Island Press, Washington, DC.

ICEM (International Centre for Environmental Management) (2003) *Vietnam National Report on Protected Areas and Development. Review of Protected Areas and Development in the Lower Mekong River Region*. ICEM, Indooroopilly, Queensland, Australia.

Jones, T., Dinh, H.N. and Taranov, I. (2016) Comparing approaches to environmental governance: a review of protected areas and nature-based tourism in Kyrgyzstan and Vietnam. *Meiji University Journal of Governance Studies* 3, 43–57.

Jones, T., Bui, H. and Apollo, M. (eds) (2021) *Nature-Based Tourism in Asia's Mountainous Protected Areas*. Springer Nature, Cham, Switzerland.

Lee, Y.-S., Lawton, L.J. and Weaver, D.B. (2013) Evidence for a South Korean model of ecotourism. *Journal of Travel Research* 52(4), 520–533. DOI: 10.1177/0047287512467703.

Lindberg, K., Goulding, C., Huang, Z., Mo, J., Wei, P. *et al*. (1997) Ecotourism in China: selected issues and challenges. In: Oppermann, M. (ed.) *Pacific Rim Tourism*. CAB International, Wallingford, UK, pp. 128–143.

Ly, T.P. and Xiao, H. (2016) An innovative model of park governance: evidence from Vietnam. *Journal of Ecotourism* 15(2), 99–121. DOI: 10.1080/14724049.2016.1145227.

MARD (Ministry of Agriculture and Rural Development) (2020) *Current State of Ecotourism Development in Spetial-Use and Protected Forests in Vietnam*. MARD – Department of Forestry, Hanoi.

Meric, H.J. and Hunt, J. (1998) Motivational and demographic characteristics: a case of North Carolina travelers. *Journal of Travel Research* 36(4), 57–61. DOI: 10.1177/004728759803600407.

Michaud, J. and Turner, S. (2017) Reaching new heights. State legibility in Sa Pa, a Vietnam hill station. *Annals of Tourism Research* 66(3), 37–48. DOI: 10.1016/j.annals.2017.05.014.

Mittermeier, R.A., Robles Gil, P., Hoffmann, M., Pilgrim, J.D. and Brooks, T.M. (2004) *Hotspots Revisited: Earth's Biologically Richest and Most Endangered Ecoregions*. CEMEX, Mexico City.

Nevins, J. and Peluso, N.L. (2008) Introduction. In: Nevins, J. and Peluso, N.L. (eds) *Taking Southeast Asia to Market: Commodities, Nature, and People in the Neoliberal Age*. Cornell University Press, Ithaca, New York, pp. 1–26.

Ngo, A.T. and Pham, L.H. (2016) *Research on Financial Mechanism for Entrance Ticket Pays for Forest Environmental Services in National Parks and Special Use Forest*. General Administration of Forestry and GIZ, Hanoi.

Nguyen, Đ.K. (2008) *Environmental Education for Local Communities in Nature Reserves*. Youth Publishing House, Hanoi.

Nguyen, H.D. and Vu, V.D. (2007) Bảotồnđadạngsinhhọc ở Việt Nam – mốiliênhệvớiPháttriểnbềnvững (SD) v biếnđổikhíhậu [Biodiversity conservation in Vietnam in relation to sustainable development and climate change]. Paper presented at the International Symposium on Biodiversity and Climate Change, 22–23 May 2007, Hanoi.

Palacio, V. and McCool, S.F. (1997) Identifying ecotourists in Belize through benefit segmentation: a preliminary analysis. *Journal of Sustainable Tourism* 5(3), 234–243. DOI: 10.1080/09669589708667288.

Pham, L.H. (2018) Payments for forest environmental services in Vietnam: situation and solution. *Tạp chí Khoa học v Công nghệ Lâm nghiệp [Journal of Forest Sciences and Technology]* 1, 198–202.

Pham, L.H. and Bui, H.T. (2020) Ecotourism and sustainable development in Vietnam's protected areas. In: Trupp, A., Dolezal, C. and Bui, H.T. (eds) *Tourism and Development in Southeast Asia*. Routledge, London, pp. 59–72.

Suntikul, W., Butler, R. and Airey, D. (2010) Implications of political change on national park operations: Doi Moi and tourism to Vietnam's national parks. *Journal of Ecotourism* 9(3), 201–218. DOI: 10.1080/14724040903144360.

Tao, C. (Teresa)., Eagles, P.F.J. and Smith, S.L.J. (2004) Profiling Taiwanese ecotourists using a self-definition approach. *Journal of Ecotourism* 12(2), 149–168.

USAID (United States Agency for International Development) (2013) *Vietnam Tropical Forest and Biodiversity Assessment: US Foreign Assistance Act, Section 118–119 Report, August 2013*. USAID, Quito.

Valentine, P. (1992) Nature-based tourism. In: Weiler, B. and Micheal, H. (eds) *Special Interest Tourism*. Belhaven Press, London, pp. 105–127.

Van, N.T. (2019) Biodiversity conservation in special use forests in Vietnam – the efforts of WWF and its recommendations. Paper presented at the National Symposium on Management of SUFs, PAs and the Solutions for Sustainable Development, December 2019, Hanoi.

Vietnam National Congress (2017a) *Law of Forestry*. Available at: https://thuvienphapluat.vn/van-ban/Linh-vuc-khac/Luat-lam-nghiep-367277.aspx (accessed 10 May 2021).

Vietnam National Congress (2017b) *Tourism Law*. Available at: https://thuvienphapluat.vn/van-ban/Van-hoa-Xa-hoi/Luat-du-lich-2017-322936.aspx (accessed 10 May 2021).

Vietnam National Congress (2018) *Law of Biodiversity*. Available at: https://thuvienphapluat.vn/van-ban/Tai-nguyen-Moi-truong/Van-ban-hop-nhat-32-VBHN-VPQH-2018-hop-nhat-Luat-Da-dang-sinh-hoc-407228.aspx (accessed 10 May 2021).

VN FOREST (2020) *Report on Evaluation of the Performance of the Special Use and Protection of Forests Nationwide in 2020 and the Deployment of the Key Tasks in 2021*. Ministry of Agriculture and Rural Development – Department of Forestry, Hanoi.

VNPPA (Vietnam National Park and Protected Area Association) (2011) Proceedings of the workshop 'Proposing Policy Mechanisms to Develop Ecotourism in National Parks and Conservation Areas in Vietnam'.Association of National Parks and Nature Reserves in Vietnam, March 2011, Hanoi.

Weaver, D.B. (2014) Asymmetrical dialectics of sustainable tourism: toward enlightened mass tourism. *Journal of Travel Research* 53(2), 131–140.

Weaver, D.B. and Lawton, L.J. (2002) Overnight ecotourist market segmentation in the Gold Coast hinterland of Australia. *Journal of Travel Research* 40(3), 270–280. DOI: 10.1177/004728750204000305.

Weaver, D., Tang, C. and Zhao, Y. (2020) Facilitating sustainable tourism by endogenization: China as exemplar. *Annals of Tourism Research* 81(4), 102890. DOI: 10.1016/j.annals.2020.102890.

Zografos, C. and Allcroft, D. (2007) The environmental values of potential ecotourists: a segmentation study. *Journal of Sustainable Tourism* 15(1), 44–66. DOI: 10.2167/jost572.0.

7 Community-based Tourism: Challenges and Opportunities for Sustainable Development

Giang T. Phi[1]* and Long H. Pham[2]
[1]VinUniversity, Hanoi, Vietnam; [2]Vietnam National University, Hanoi, Vietnam

Introduction

Community-based tourism (CBT) is a form of alternative tourism that seeks to engage the host community throughout the planning, development and management of tourism. Built around the principles of social equity and cultural and environmental protection, CBT is seen as having great potential to contribute to local communities' sustainable development (Scheyvens, 2012). In Vietnam, CBT emerged as early as the 1990s and continued to expand rapidly since 2000 as part of the national and global poverty alleviation strategies (Phi and Whitford, 2017). As of 2020, there exist hundreds of CBT sites throughout the three regions of Vietnam (Vietnam National Administration of Tourism, 2020). Despite this rapid growth, many Vietnamese CBT projects still face a wide range of challenges to achieving short- and long-term sustainability. Concurrently, some CBT projects have reported initial success, and opportunities exist in the broader environment to replicate these successes elsewhere.

To date, no state-of-the-art review currently exists that identifies and brings together the diverse challenges and opportunities of CBT in Vietnam. Thus, this chapter seeks to address the gap in both the literature and practice and to contribute to the future development of CBT in Vietnam. The chapter first provides background information on the evolution of CBT in Vietnam before exploring critical areas of the CBT literature. Next, the qualitative approaches utilized for the study, on which the findings and discussions are drawn upon, are presented. The findings are further discussed within the context of three CBT case studies in Vietnam. Finally, recommendations are suggested for the future sustainable development of CBT in Vietnam.

Community-based Tourism: Definitions, Key Models and Impacts

Definitions of CBT

CBT is rooted in the alternative development school of thought. The 1980s witnessed the rise of alternative development, which aimed to address the limitations of mainstream development's 'Eurocentric, meta-narrative, economic models' (Sharpley and Telfer, 2008, p. 47). Alternative development encompasses several

*Corresponding author: giang.ptl@vinuni.edu.vn

 Vietnam Tourism: Policies and Practices
(eds Huong T. Bui *et al.*)
DOI: 10.1079/9781789242782.0007

theoretical models, including the provision of basic human needs, grassroots, bottom-up development planning, gender empowerment and sustainable development. These models arose out of the analysis of traditional development biases (i.e. a top-down approach to planning and decision making; excluding indigenous peoples, the poor and women's needs and roles in development; with no consideration to the limited nature of environmental resources) (Korten, 1980; Phi, 2017). During this period, the top-down, external approach to tourism development was also frequently criticized for contributing little to improving local communities' conditions (Britton, 1982). Subsequently, CBT emerged in the 1980s, among many other tourism approaches that seek to combat the negative impacts of neoliberal tourism development and ensure that residents benefit from tourism activities (Blackstock, 2005). Many CBT definitions can be found in both academic and industry sources, for instance:

> CBT incorporates tourism projects that involve the host community in the planning, developing, and maintenance of those projects.
> (Bauer, 2008, p. 181)

> CBT can be defined as tourism owned and/ or managed by communities and intended to deliver wider community benefit.
> (Goodwin and Santilli, 2009, p. 12)

> CBT provides visitors with an experience of local life, in which local communities are directly involved in tourism activities and get socio-economic benefits from tourism activities, and take responsibility for protecting the natural resources, environment, and local culture.
> (The Environmentally and Socially Responsible Tourism (ESRT) Capacity Development Programme; ESRT and WWF-Vietnam, 2013, p. 5)

The central focus of CBT, based on the above definitions, is who initiates, owns and manages the tourism activities at a destination (Blackstock, 2005). Advocates of CBT believe that by engaging and empowering local communities while concurrently limiting external actors' power in tourism development, CBT will deliver social equity and fair distribution of tourism benefits (Saayman and Giampiccoli, 2016). ESRT and WWF-Vietnam's definition also highlights vital elements of sustainable development in CBT, with which all three economic, social-cultural and environmental pillars ideally should be in balance. From a market-based perspective, CBT activities' focus may centre on the notion of allowing tourists to learn about local cultures and/or experience authenticity (Giampiccoli and Mtapuri, 2017). CBT's definitions thus suggest a potential overlap with cultural tourism or ethnic tourism.

Impacts and key models of CBT

CBT has frequently been promoted as an essential strategy to diversify local economies, especially those in rural and remote areas (Dezsi and Benedek, 2011). With its characteristic of bringing tourists to the site of consumption, CBT's development can enable rural communities to integrate their traditional agricultural and artisanal production with the provision of other services and commodities. Also, CBT has low entry barriers relative to other sectors; it is generally labour intensive and can employ young people and less-skilled individuals (Phi, 2020a). CBT can therefore enable rural communities to better cope with challenges like crop failure and price volatility in the subsistence agricultural market. This could lead to more sustainable livelihoods, improved food security, decreased outmigration of young people, and increased living standards for rural households (Haggblade *et al.*, 2010).

Depending on the modes of distribution, part of CBT profit can also be used to fund social and community projects such as community events or educational initiatives or facilities and services that otherwise might not be available (e.g. clinics or schools). Successful CBT can attract government and private investment to upgrade local infrastructure (e.g. roads and electricity), improving overall access and linkages between the community and other areas (Zapata *et al.*, 2011). Aside from material benefits, CBT is designed to respect local people and showcase their identity and strengths for outsiders' appreciation, which can also increase community pride, self-esteem and confidence (Phi, 2017). Additionally, community cohesion and integrity can be strengthened as the whole community works together to beautify and clean the local environment and develop and share benefits from tourism ventures and activities (Ellis, 2011; Phi and Whitford, 2017).

Nevertheless, many empirical studies have shown that CBT development does not automatically translate into positive benefits for the local community. This is partly because various CBT models exist in practice, and the level of community ownership, participation and control vary under each model. For instance, Giampiccoli and Mtapuri (2017) identify the three most popular CBT models: (i) community-owned CBT; (ii) individually and locally owned CBT; and (iii) externally owned or managed CBT. In the community-owned CBT model, the whole community develops and manages tourism under a single structure, such as community lodges/community tours, etc. The second model refers to the different levels of CBT ownership, where CBT does not occur on a community-wide level but comprises multiple local micro and small enterprises that provide tourism products and services. Alternatively, the third model is often associated with the concept of luxury CBT, where external investors develop tourism in local areas by entering into specific ownership and management agreements that help to 'retain local control and CBT principles' (Giampiccoli and Mtapuri, 2017, p. 9). These agreements may be underlined by unequal power relations, where local communities only receive marginal benefits in exchange for their tourism resources. Under the third model, some local enterprises may also opt to rent or sell out their businesses to external investors, leading to a lesser degree of local control over tourism management and profit sharing over time.

Issues with the exclusion of residents can also be experienced in the first two models, although these supposedly should have allowed a higher level of local engagement. In reality, CBT's development may marginalize those who cannot cope with the principles of participating in CBT's value chains (van Beers, 2011). Extant studies (Ellis, 1999; Zapata *et al.*, 2011; Lucchetti and Font, 2013) suggest that the introduction of tourism into a community is likely to face many barriers (e.g. low human capital and lack of physical/social/political resources) that prevent the poorer members of the community from gaining meaningful participation in tourism activities. Empirical studies reveal that most benefits from CBT projects are captured by the non-poor or the more 'elite' members of the community (Blackstock, 2005; Mitchell and Muckosy, 2008; Zapata *et al.*, 2011). CBT development without an understanding and application of sustainability principles can lead to environmental problems (e.g. noise and waste pollution, and overuse of natural resources), the commodification of culture, as well as the reinforcement of 'othering' and cultural stereotypes (Scheyvens, 2012).

Research Methods

This chapter adopts a qualitative research approach, with both secondary and primary data having been collected and utilized for this study. Collectively, the authors have been involved in CBT field research in Vietnam for decades. Thus, the authors draw from earlier participant observation and informal conversations, and interviews in the field with diverse CBT stakeholders, such as homestay owners, coordinators, consultants, tourists, and residents, directly and indirectly, involved in tourism. These ethnographic methods are relevant to capturing diverse stakeholders' perceptions of CBT challenges and opportunities in Vietnam. The trust built over a long term of engagement with local communities also enables honest viewpoints and unique insights to be captured (Pelto, 2017).

Several secondary sources of data were also utilized, including relevant academic publications and industry and NGO reports. More importantly, we also analysed the proceedings of the 2018 Conference on Policies for CBT Development in Vietnam (Vietnam Tourism Research Institute, 2018). The conference was organized under the direction of the Ministry of Culture, Sport and Tourism and took place on 3 August 2018 at Mai Châu, Hòa Bình province. Its proceedings (a total of 209 pages) consist of three parts: (i) policies; (ii) capacity development; and (iii) cases and models of CBT in Vietnam. Contributed by attendance from private-sector practitioners, academics, government officials, NGO staff, consultants and local CBT entrepreneurs, these proceedings offer up-to-date data and viewpoints on CBT in Vietnam, especially concerning sustainable development. Thematic analysis was applied to analyse the collected data by identifying and reporting key patterns and themes that 'capture something important about the data with the research question' (Braun and Clarke, 2006, p. 82).

Finally, a case study approach is utilized to provide further understanding of the findings identified through thematic analysis. Arising out of 'the desire to understand complex social phenomena' (Yin, 2009, p. 4), a case study approach favours 'thick descriptions' that help to reveal various layers of the social world and enable a much more in-depth examination of a phenomenon (Flyvbjerg, 2006). In particular, two case studies are chosen to highlight the challenges and opportunities, as well as failure and success factors in the development of CBT in Vietnam, namely: (i) Lác village in Hòa Bình province; and (ii) Sìn Suối Hồ village in Lai Châu province. Through triangulation of secondary and primary sources, along with a case-study approach, the trustworthiness and credibility of the research are enhanced.

The Development of CBT in Vietnam

The policies on CBT have emerged since the year 2000, aligning first with the Millennium Development Goals (MDGs) and then the Sustainable Development Goals (SDGs), over the period of 2000–2015 and from 2015 onwards, respectively.

2000–2015: the MDGs and pro-poor tourism framework

In the wake of the New Poverty Agenda at the beginning of the 21st century, poverty alleviation became the focus of the global tourism agenda, mirroring the shift away from economic development towards poverty alleviation (UNWTO, 2002). This shift is perhaps best demonstrated by the MDGs, a document signed in the year 2000 by leaders and representatives of 189 nations that aimed to free people from extreme poverty by 2015. Supported by the MDGs, the pro-poor tourism (PPT) approach emerged and was endorsed by the World Tourism Organization (UNWTO), which sought to change 'the distribution of benefits from tourism favouring poor people' (Pro-poor Tourism Partnership, 2005, p. 1).

CBT principles align well with the PPT framework, and since the early 2000s, many development agencies and international organizations (e.g. AusAid, Japan International Cooperation Agency, SNV Netherlands Development Organisation, the Asian Development Bank, UNESCO, the International Labour Organization) started to utilize CBT as part of their poverty reduction strategies in Vietnam. With generous funding allocated by development agencies, international and local NGOs became the key stakeholders, leading the exponential growth of CBT projects in Vietnam during this period.

As poor populations are often concentrated in remote and rural areas, CBT in Vietnam has also predominantly developed in rural areas such as the northern mountainous region (e.g. Hòa Bình, Lào Cai, Bắc Kạn, Hà Giang), the central coast (e.g. Quảng Nam, Huế, Nha Trang) and Central Highlands (e.g. Kon Tum), and around the Mekong Delta in the south (e.g. Cần Thơ, Tiền Giang, Bến Tre, Vĩnh Long) (ESRT and WWF-Vietnam, 2013). CBT activities are an important option to help locals diversify their livelihoods and become less vulnerable to the risks associated with subsistence agriculture and fisheries. These rural sites also possess beautiful natural landscapes and are rich in cultural heritage (primarily ethnic minorities) that appeal to domestic and international tourist markets (Phi *et al.*, 2017).

As CBT activities grow in Vietnam, they may appear under different names such as: (i) community-based tourism; (ii) community development in tourism; (iii) community-based ecotourism; (iv) community participation in tourism; and (v) community-based mountain tourism. Nevertheless, CBT in Vietnam is often defined as tourism activities primarily developed and/or managed by local communities, in order to bring economic benefits to local people and protect the local environment (Doan, 2018).

A market-based approach reported by ESRT and WWF-Vietnam (2013) identified four key market segments for Vietnam CBT. The international market includes 'independent travellers' and 'package travellers'. While independent travellers have more time to stay and explore CBT destinations, this experience is only a small part of the package travellers' overall itinerary. The fundamental similarity between these segments is their interest in learning about Vietnamese history, culture and the natural environment. The domestic market includes 'city-worker weekend breaks' and 'Vietnamese backpackers' (i.e. '*Phượt thủ*' in Vietnamese). Both are attracted by the fresh air and rural scenery, although the city workers tend to spend more on food and beverages and wellness services, while the Vietnamese backpackers prefer purchasing local products as souvenirs and gifts (ESRT and WWF-Vietnam, 2013).

In acknowledging CBT's fast growth, this period also witnessed the emergence of various Vietnamese government policies aimed at providing legal frameworks for CBT development. Some key policies that affect CBT include Vietnam's *Tourism Law* (2005), Decree Number 109 (2003/NĐ-CP), Decision Number 104 (2007/QĐ-BNN), and the *Environmental Protection Law* (2005) (ESRT and WWF-Vietnam, 2013). However, in contrast to the non-profit sector's focus on utilizing CBT as a tool for poverty alleviation, these policies focused more on environmental conservation as part of sustainable development, via the links between ecotourism, tourism development in wetland areas, national parks and natural protected areas on the one hand, and CBT on the other.

2015 onwards: the SDGs and cross-sectoral collaboration

Towards the end of 2015, the MDGs were superseded by the SDGs, another global commitment that seeks to end poverty in all of its forms everywhere (United Nations, 2016). The SDGs provide an important framework for promoting partnerships and coordination of resources across sectors.

While the non-profit sector contributed enormously to the expansion of CBT, NGO-led CBT projects tend to lack connections with the private tourism sector and the tourism market, leading many such destinations to be less commercially competitive. There is also an over-reliance on donor aid, and many communities struggle to continue CBT development once the funded projects have ended (Phi *et al.*, 2017). Guided by the SDGs, there has also been a rise in cross-sectoral collaboration in Vietnam over the last few years, whereby CBT development is now also supported by different levels of government, social enterprises and the private sector.

These partnerships are further enhanced by the most recent national policies on CBT. The 2017 Vietnam *Tourism Law* specifies, in Clause 6, the 'Participation of local communities in tourism development', and Clause 19 highlights the role of provincial governments in supporting CBT development, fair profit distribution among community members, and protection of the environment (Vietnamtourism, 2017). This law also focuses on cultural aspects of CBT, which enables tourists to experience local cultures and ways of life. In addition, the law recognizes the importance of CBT value-added products and services (aside from homestays or guest houses) such as food and beverages, cultural performances and local handicrafts. The 'New Rural Development Policy 2016–2020' (*Chính Sách Nông Thôn Mới*) is another important policy that affects CBT. Many provincial governments in Vietnam have associated CBT development with One Village One Product (OVOP) as part of this policy, aiming to improve rural infrastructure, alleviate poverty, and develop rural communities' capacity for job and livelihood creation and preservation (Vietnam Tourism Research Institute, 2018).

Overall, the evolution of Vietnam's CBT is strongly linked to the broader direct and indirect policies that support tourism, as well as the global agenda reflected via the MDGs and, now, the SDGs. The findings on Vietnam CBT's challenges and opportunities will be presented and discussed in the next section.

Challenges for CBT

Vietnam CBT's key challenges can be identified in the areas of awareness, marketing, capacity development and regulation. These challenges will be elaborated on in this section.

Awareness

Early CBT projects were often characterized by limitations in terms of both understanding tourists' needs, and having the capacity to develop tourism products. Until now, CBT has not been thoroughly understood, which leads to the failure of many CBT projects in Vietnam. The starting of CBT often follows an inductive or deductive path. The inductive path is often applied when CBT is thought to be prevailing, and applying CBT brings more benefits for the whole community. The deductive path is when many communities believe that they have the capacity to engage in CBT. In fact, both ways have weaknesses. Therefore, it is important to follow the path that allows communities to carefully assess their resources for CBT application.

Marketing

Traditional marketing revolves around the four Ps of: (i) product design and development; (ii) pricing; (iii) place or distribution channels; and (iv) promotional strategies (Kotler, 2012). Many Vietnam CBT projects have struggled to develop sustainable tourism products and services that meet tourists' needs while maintaining local communities' culture and environment (Ngo *et al.*, 2020). Many CBT projects were either self-organized by local people or led by NGOs as a poverty alleviation strategy, which led in many cases to a lack of understanding of tourism market needs, resulting in unattractive, low quality and/or over-commercialized products and services. For instance, many homestays in CBT projects lack clean and flushing toilets, hot water for showers, or comfortable mattresses for guests. There is also too much focus on accommodation services while neglecting other potential tourism products and services that could promote local values and distribute tourism benefits more equally to the whole community (e.g. cultural performances, local activity experiences) (Phi *et al.*, 2017).

The lack of expertise in tourism also affects power relations regarding pricing and distribution strategies. A study of Vietnam rural tourism found that local CBT entrepreneurs were often in inferior relationships with external tourism intermediaries and were subsequently forced to lower prices and receive only a small share of visitors' expenses (Chau, 2014). In some cases, local communities also entered into an exclusive contractual agreement with a sole tourism operator in the early stages of CBT development, later preventing them from reaching out to other intermediaries (Ngo and Doan, 2014). In terms of promotion, intermediaries such as tour operators or government tourism organizations tend to promote the concept of CBT as a product, while many tourists have very little understanding of CBT. Thus, Vietnam CBT needs updated promotional strategies that focus on niche markets and multiple transformative experiences (e.g. cooking classes, rural dinners, mountain adventures) in rural areas (ESRT and WWF-Vietnam, 2013). There is also a lack of attention paid to CBT at the national level:

> Although CBT is one of the four key pillars in Vietnam tourism development, CBT marketing activities still do not attract sufficient attention at the national level. The tourism sector and authorities still focus more on promoting general tourism and not CBT products and services.
>
> (Ngo, 2018, p. 61)

Capacity development

Capacity development is vital to ensuring that CBT products and services maintain consistent standards over time. However, CBT communities in Vietnam currently have many unmet needs in relation to comprehensive capacity development. Pham (2018) mentioned some barriers to capacity development for diverse CBT stakeholders (i.e. individual, community, local government and tourism authorities, and enterprises) (see Table 7.1). Common issues across stakeholders were lack of: (i) awareness of CBT's roles and impacts; (ii) sufficient CBT development plans; (iii) CBT or tourism expertise; and (iv) financial resources to invest or participate in CBT (Pham, 2018).

In terms of human resource skills, much of the current training for local communities to deliver tourism services still follows formal educational methods, with traditional classroom-style workshops in which theoretical information is provided. These methods are not sufficient for supporting the local community in building skills and knowledge for CBT operations, as mentioned in the proceedings:

Table 7.1. Barriers to capacity development for community-based tourism (CBT) stakeholders. Adapted from Pham, 2018.

Stakeholder	Barriers
Individual	Lack of time to participate
	Lack of skills to operate and manage a CBT project
	Foreign language barriers
	Lack of ability to adjust to the CBT environment
	Lack of interest in CBT or opposition to CBT development
Local community	Lack of CBT management skills
Local government and tourism authorities	Lack of capacity to connect with and support diverse stakeholders
Enterprises	Too much focus on product development and not enough attention to developing human resources for CBT

> The target audience for CBT training is people from rural communities. Thus, we need a vocational hands-on coaching approach instead of applying the same curriculum as in formal higher education.
>
> (Nguyen, 2018, p. 93)

An important part of coaching also involves increasing the awareness of locals regarding their assets. Often, when CBT is utilized as a part of poverty alleviation and development projects, there is a focus on what the community 'lacks' and existing resources are subsequently overlooked. As a visitor, the lead author found the local landscapes in many Vietnamese rural areas breathtaking and the local cultures unique. When local people were asked about tourism potential, most of them were not aware of what outsiders appreciate, simply because they grew up in the area and had become used to the existing landscapes and cultures.

The lack of opportunities to learn foreign languages, especially English, has created much frustration in CBT projects' daily operations and marketing as communities need to rely on external tour guides for translations. Language barriers also affect communities' ability to establish direct marketing relationships with tourists, external tour operators and tour companies. For instance, respondents from the Phú Minh CBT project in Hòa Bình province said:

> The minimum requirement is to be able to communicate in English. Second, we need to build good connections with tour companies and tour operators outside of the community. We also need to learn how to market our CBT to attract more tourists. For all of these, we are still lacking.
>
> (Giang T. Phi, 2014, unpublished data)

In a review of the existing CBT literature, Hamzah and Khalifah (2012) identified language, marketing skills, and connections with the private tourism sector as important elements for local communities to successfully operate CBT after the initial supporting organizations (e.g. NGOs) have withdrawn their involvement. Learning and capacity building, however, takes time. This is where most NGO-led CBT development often falls short, as support for local communities also disappears as the funding dries up for short-term projects.

Regulations

Although there exist policies that support CBT development, there is still a shortage of specific government regulations about CBT operations and development. This in turn creates negative impacts such as the commodification of culture, littering, deforestation and noise pollution in the locality. Furthermore, tourists' health and safety standards are also not established or enforced, leading to potential cases of food poisoning or injury. There are also differences between standards for domestic tourists versus international tourists. During a trip to Hòa Bình province, the lead author participated in rafting activities on a stream and noticed that domestic tourists tended to refuse to wear life jackets for their safety.

More importantly, due to a lack of CBT regulation frameworks and policies concerning local control and distribution of tourism profit, external investors are increasingly taking over local CBT resources and activities. These issues can be observed in the remote mountainous area of Pù Luông, Thanh Hóa province, where local land has been purchased en masse by investors from Hanoi and other parts of Vietnam to build commercial homestays. In Sapa, Lào Cai province, many external tour companies have also taken over existing hospitality and tourism enterprises, thus creating significant economic leakages in local CBT areas. This theme is reported in the 2018 conference on CBT:

> External tourism companies kept 80–90% profit from tourism has led to the destruction of many CBT models. Hence there is a heightened need for policies and regulations regarding profit distribution to CBT stakeholders.
>
> (Truong, 2018, p. 134)

Overall, the diverse challenges identified in this section (e.g. awareness, marketing and community capacity) tend to relate to the development of new CBT projects or the sustainability challenges of CBT activities self-initiated by local entrepreneurs. These challenges partly reflect the early stage of CBT growth and development in Vietnam and the need for specific CBT regulations and policies to direct this growth in a sustainable manner. The next section outlines and discusses key opportunities for CBT to address the above-mentioned challenges.

Opportunities for CBT

Three key factors were identified as having positive influences on the future sustainable development of Vietnam CBT. These are: (i) the increase in cross-sectoral collaboration; (ii) digitalization in tourism; and (iii) tourism policies (post)-COVID-19.

Cross-sectoral collaboration

Collaboration between the private, public, and not-for-profit sectors is essential to ensuring that tourism delivers benefits to local communities. For example, Mr Bình Dương, founder and director of CBT Travel, a consultancy and social enterprise, stated:

> The most important thing is to successfully persuade the residents, governments at various levels, NGOs, and private tourism companies to work together to develop CBT in Vietnam.
>
> (B. Dương, Hồ Chí Minh city, 2017, personal communication)

In recent years, traditional sectoral boundaries have been broken down, and new partnerships have been formed in Vietnam CBT thanks to the emergence of knowledge exchange initiatives by social enterprises and NGOs (Phi *et al.*, 2017). CBT Travel consultancy and social enterprise, for instance, has contributed to fostering diverse stakeholder collaboration through the organization of seminars, workshops and conferences to introduce CBT models and potential to national, province, commune and village levels of government, international NGOs operating in Vietnam, local or regional tour companies and local communities in rural and remote communities (Phi and Whitford, 2017). This social enterprise also encourages luxury hotels to directly provide pro bono training support in food and beverage services and donate old furniture to local CBT projects.

A number of nationwide CBT networks have also been established. For instance, the International Labour Organization has led the creation of a central support network for Vietnam CBT since 2016. Currently, in beta version (a piece of software) Vietnam Community Based Tourism Network is available at: https://vcbt-n.com (accessed 17 September 2021); the network's primary functions include capacity building, exchange of knowledge and resources, policy and advocacy, and communication and connection for CBT all over Vietnam. Another emerging CBT network is CTC – Community Tourism Club. Developed by the Vietnam Tourism Association, CTC contributes to promoting CBT products and destinations, as well as organizing familiarization trips and press trips for travel companies and journalists to experience local CBT projects directly (Vietnam Tourism Research Institute, 2018). These CBT social enterprises and networks can also collaborate with each community's local representatives and authorities to develop appropriate local rules and regulations for responsible CBT practices to minimize negative impacts on the community and the local environment. Furthermore, the cross-sectoral collaboration also opens up diverse financing opportunities for local communities. Beyond the standard self-funding option of local entrepreneurs or projects funded by development agencies and NGOs, there are now options for government-funded, low-interest partial loans from local government and development banks and partial in-kind support from local NGOs to support local CBT (Phi and Whitford, 2017).

The rise of digitalization in tourism

Vietnam has a relatively good communication infrastructure, and Internet service can now be easily assessable in remote and rural areas through mobile phones and 3G or 4G data services. Unsurprisingly, many tourism enterprises in CBT areas have used websites, social media and other online e-commerce and booking platforms to directly promote and sell their products and services. The *2018 Conference on Policies for CBT Development in Vietnam Proceedings*, for example, identified a number of new CBT websites such as thaihai.vn, kheroresort.com, yenducvillagetour.com and puluongretreat.com (Vietnam Tourism Research Institute, 2018). These websites, however, still require some initial funds and skills to create and maintain. Social media such as Facebook, Instagram, Flickr and Google Maps (including reviews) can be created for free and are more comfortable to set up and maintain. CBT communities in the northern mountainous areas – such as Mai Hịch CBT project or A Chu Homestay – are pioneers in effectively using social media while providing mentorship to new CBT projects and homestays that wish to use these digital tools to improve sales and marketing. Large online platforms such as Airbnb and booking.com have also started to dedicate staff to support locals in listing their small and medium-sized businesses, including homestays and tour providers in remote CBT areas. This is linked to these platforms' broader Corporate Social Responsibility initiatives, which aim to support local economic development and cultural preservation via tourism. For example, in 2017, Airbnb launched a €5 million global project, the Community Tourism Programme (Airbnb, 2017).

Also, modern communication technologies play an increasingly important role in CBT capacity building. Skype, phone calls, emails and Facebook messages have all been used as part of distance training and coaching. This is important as it significantly reduces the costs compared to having tourism experts physically visit the site, allowing organizations like the consultancy and social enterprise CBT Travel to provide pro bono training (B. Dương, Hồ Chí Minh city, 10 January 2020, personal communication). During the field trip to several local CBT projects in Mai Châu and Mộc Châu in January 2020, the authors also noticed that local people use Google Translate to bridge the language gap in real-time communication with foreigners, while learning introductory-level foreign languages through apps such as Duolingo.

Tourism policies in the (post)-COVID-19 period

With COVID-19 and the rise of domestic tourism, the potential for more CBT development targeting domestic tourists, especially urban residents, poses environmental challenges. In the (post)-COVID-19 period, CBT is a type of tourism that will help tourists to find the ideal place where they can 'return home'. For each person, the place of return is a place that creates a familiar feeling of home. Indeed, many South-east Asian areas are experiencing a rise in rural tourism and CBT as operators cater to demand socially distanced leisure and recovery from pandemic-related anxiety (Glusac, 2020; Phi, 2020b). A Vietnamese respondent noted that even Hợp Thành, a very remote commune in the northern mountains of Vietnam, still receives steady flows of tourists during the pandemic from the adjacent city of Lào Cai. He thus believed that 'there is unlimited potential to develop rural CBT targeting the urban citizens of Vietnam's 87 cities' (B. Dương, Hồ Chí Minh city, 2020, personal communication). At the same time, it is also noted that the domestic market is also very relevant in CBT and can be very important in locally originating – bottom-up – CBT ventures (see Zapata *et al.*, 2011).

Hence, despite encountering various challenges, the development of Vietnam CBT is also supported by many opportunities emerging in the broader CBT ecosystem, ranging from good infrastructure to support digitalization, to recent tourism policies. The next section will present two cases illustrating these challenges and opportunities.

CBT in Practice

Following theoretical analysis of challenges and opportunities to which CBT projects are facing, two case studies have been chosen. Each case sheds light on a different aspect of CBT in Vietnam. The Lác village CBT case enables readers to gain a deeper understanding of the challenges identified in the findings and the Sìn Suối Hồ CBT case discusses how the local community utilized opportunities to successfully overcome challenges in CBT development.

Lác village's CBT sustainability challenges

Hòa Bình province is located in the Northwest region of Vietnam, encompassing ten sub-districts and one provincial city. The majority of Hòa Bình's population consists of minority ethnic groups such as Mường, Thái, Tày, Dao and H'Mông. With its close proximity to the capital, Hanoi, beautiful mountainous landscapes and distinctive ethnic cultures, Hòa Bình became one of the first provinces to offer CBT in Vietnam. Towards the end of 1980s, CBT activities were only concentrated within Lác village in Mai Châu Commune, with the main visitors being Eastern European expatriates living in Hanoi who wished to relax and explore Thái culture over the weekend (Du lịch Hòa Bình, 2020). Over the years, Lác village has become a popular tourism destination for both domestic and foreign tourists. There are the following challenges:

- *Awareness challenges*: Despite its 'success' on the surface, many of the challenges for CBT development outlined in an earlier section are evident at Lác village. For instance, most of the tourism activities in the village are self-organized, individually initiated, and are designed without being mindful of community resources as a whole.
- *Marketing challenges*: In terms of marketing, the tourism products and services on offer at the village appear to be of lower quality and lacking in authenticity. Many authentic, traditional Thái wooden houses have been replaced by modern concrete houses. Hygiene and sanitation at the homestays, especially in toilets and bathrooms, are not guaranteed.
- *Lack of regulations*: Due to a lack of regulations, the village is dominated by countless souvenir shops selling inauthentic handicrafts that are mass produced and/or imported, while local traditional handicraft practices have almost disappeared (see Fig. 7.1). Non-traditional tourism products such as karaoke and electric cars also create noise and visual pollution and disturb the peacefulness of the whole village (Báo Hòa Bình, 2017). Concurrently, the average expenditure of tourists to Lác village remains relatively low, partly due to a lack of high-quality products and services but also because the main target segment of visitors is foreign backpackers on tight budgets.

Hence, without proper guidance from tourism experts to improve local capacity, as well as the absence of relevant regulations and management from the local authorities, self-organized CBT at Lác village has turned into typical mass tourism where the reduction in quality of life for local people starts to outweigh economic benefits brought about by tourism. As the number of visitors continues to increase in the future, Lác village will still face persistent challenges associated with sustainability, especially in environmental areas including the provision of clean water and waste treatment areas for both locals and tourists. To ensure sustainable CBT development, Lác village would also need to capitalize on opportunities discussed above, such as the integration of tourism with local agricultural value chains (e.g. via tourism experiences and the provision of local produce to tourists) and recovering traditional weaving and handicraft practices (which can be linked to existing local social enterprises such as Hoa Ban – a handicraft manufacturer that hires disabled and disadvantaged women in Lác village).

As of 2020, Hòa Bình has nine recognized CBT areas, with the involvement of 160 households. CBT development has been further

Fig. 7.1. A souvenir shop in Lác village. Photo courtesy of Giang T. Phi.

expanded to other communes outside Mai Châu, such as Đá Bạc, Cao Phong, Lạc Sơn, Tân Lạc and Lương Sơn. In March 2020, Hòa Bình's Provincial People's Committees issued Decree 446/QĐ-UBND that focuses on CBT development in conjunction with the New Rural Tourism policy, with the aim of developing 15 new CBT destinations in Hòa Bình. It remains to be seen whether other CBT development in the province can learn from the challenges of Lác village to ensure the sustainability of their future development.

Sìn Suối Hồ CBT: opportunities for awakening community spirit

Sìn Suối Hồ village is located around 30 km from the centre of Lai Châu, under Sơn Bạc Mây Mountain at an elevation of 1400 m above sea level, covering an area of 91.05 km², with a population of 2445. The village is populated by 127 households with 100% coming from the Flower H'Mông minority people. This is the first CBT officially recognized in 2015, created by H'Mông people based on the leadership of the village leader. As of 2014, the community successfully eliminated excessive drug and alcohol consumption rampant in the village for decades. Concurrently, Sìn Suối Hồ sought alternative livelihoods from farming (i.e. growing maize and rice) for their income. With a 'can do' attitude, Mr Chỉnh and Mr Xá initiated a tourism service by repairing and tidying up their houses for homestays. The villagers then followed their call to set up a tourism supply chain, avoiding competition by bringing villagers together playing different roles in the CBT project.

Sìn Suối Hồ CBT is systematically managed and operated via a CBT management board. Concurrently, Mr Chỉnh (the head of the village) has established a homestay team, and he is also the team's leader. The members of the homestay team actively invest resources in upgrading facilities and amenities for tourism, and ensure security and safety for tourists staying overnight. The village established a tour company to connect with tour operators in major cities. They also established a CBT fund, to which participants contribute 15% of their revenue for common facility development such as building trails, repairing roads, installing a signage system, public toilets, etc. Over time, Sìn Suối Hồ community has developed a wide range of authentic tourism products and services, including:

- *Homestays*: There are now 20 homestays in the village with special wooden welcome gates, listing the homestay owner's name, contact and their services for tourists, included Wi-Fi, food, hot water and a shower at a price of VND 80,000 per person/night (approximately US$4.00). All homestays are designed to be environmentally and naturally friendly styles with mountain-rock fences. The locals have decorated their houses and expanded their cymbidium orchid gardens to welcome visitors.
- *Tour guides and a traditional performance team*: There are ten local tour guides in the village, with two English-speaking guides, while the other eight are actively learning English. Tour guides are well trained in guiding skills by a small project, supported by the women's association. Local tour guides perform multiple tasks as they are also members of the traditional performance team of the village who perform at the Saturday market and other events.
- *Food and beverages*: Seasonal specialties made from local ingredients are prepared and monitored by hygienic standards to be served in homestays. There is one communal restaurant jointly owned by ten families in the village to serve large tour groups (see Fig. 7.2). These ten members receive 90% of the profits from the restaurant and 10% is contributed to the CBT fund.
- *Traditional handicrafts*: Traditional costumes are made from beeswax drawings on linen and are worn daily by villagers. Handicraft-making traditions such as embroidery, rice wine cooking, knitting, forging and orchid planting bring not only additional income for villagers but also showcase variations of CBT in the village. Tourists might experience making handicrafts while staying in the village.
- *Orchids*: Wild orchids were first used for house decoration, then tourists showed an interest in buying them. Gradually, the scale of the orchid farm business has grown from selling a little branch at the price of US$1.00–2.00 up to more than US$100 for a potted orchid. There are now 100 families cultivating orchids, producing over 20,000 orchids of different shapes and forms, leading to the village being named the 'Country of Orchids'.
- *Tourist attractions*: Traditional festivals include Gầu Tào – the biggest annual festival of the H'Mông people – in addition to the performance of folk games and celebration of the festival for the new rice season. There are also beautiful places for ecotourism such as waterfalls and flower valleys.
- *Saturday market*: This weekly market is held from 7 a.m. until 12 noon on Saturdays. Established in 2014, it is open to the village and the neighbouring villages for exchanging traditional products and necessities. A highlight in the market is that there is no wine or alcohol sold here because

Fig. 7.2. Local restaurant in Sìn Suối Hồ. Photo courtesy of Long H. Pham.

the community is well known for non-consumption of alcohol and tobacco as well as no theft or gambling.

Today, the village can accommodate 100 people/night at prices ranging from VND 70,000 to VND 100,000 per person/night (US$3.50–5.00). The village has received around 100,000 international and domestic arrivals. Average revenue for households generated from CBT and orchid planting ranges from VND 30 to VND 50 million/year (approximately US$1500–2500), with some exceptional households earning from VND 200 to 300 million/year (approximately US$10,000–15,000). The village's head earns about VND 250 million/year. Tourism contributes to poverty eradication by reducing the poverty rate from 60% to zero (Village head, Sìn Suối Hồ village, 2017, personal communication). The key success factors for this CBT model are discussed below.

- *Cross-sectoral collaboration*: The village's authorities always work with different residents to inform them of national and local policies on developing tourism, and work to enhance people's knowledge about protecting the environment at tourism spots. For instance, Chẻo Quẩy Hòa, Deputy Chairman of Sin Suối Hồ Village People's Committee said: 'Local authorities are teaching villagers to protect the forest, which provides a shady canopy for their lucrative cardamom and cymbidium orchids. Villagers now understand that protecting the forest will boost their income.' With the support of local authorities and private-sector tourism consultants, families repaired their own houses to meet homestay standards. In 2018, the villagers attended 4 days of CBT training on cooking and welcoming tourists in Bắc Hà, an initiative funded by the Centre for Social Initiatives Promotion (an NGO) and delivered by the social enterprise, CBT Travel.
- *Digitalization in tourism*: Concerning the role of digitalization in active learning and capacity development of the villagers, they all try their best to study English via YouTube, or visiting websites to learn how to work in CBT as a tour guide. The capacity for e-commerce, however, needs further training and development at Sìn Suối Hồ CBT to better market and connect the village with popular online booking platforms and both domestic and international companies.
- *Tourism policies*: Due to the success of Sìn Suối Hồ, the Lai Châu province has continued to expand the model to nine other CBT spots in other districts such as Lai Châu city, Phong Thổ district and Tam Đường district. Between 2017 and 2019, the total number of visitors to Lai Châu reached 350,000. Each foreign guest spent on average 1.6 days in the area, while domestic visitors stayed for 1.75 days. These policies continue to provide a favourable environment for CBT to flourish in both Sin Suối Hồ village and other parts of Lai Châu.

Throughout the two cases of CBT in practice, it is clear that the challenges and key success factors of CBT in Vietnam also largely depend on the CBT models that different communities adopt. For example, Sìn Suối Hồ CBT project represents a model of community-owned CBT. To awaken the community spirit and unify community efforts, the indigenous leadership has had a strong influence throughout the CBT development process. The village leader plays an important role in bringing local people together to develop and manage CBT in an environmentally, culturally and economically sustainable manner. Villagers are made aware that their H'Mông traditional identities are the key reason that draws tourists to the CBT. Hence each person actively contributes to preserving and sharing these cultural practices. Together, the villagers also develop various unique selling points (orchids, cleanliness, local food, authentic culture) of their village. Importantly, there is clear integration of existing livelihoods (e.g. horticulture and agriculture) of local people into the tourism value chain, in turn promoting more equality in the distribution of tourism profit. In contrast, Lác village's CBT development is characterized by model two, where CBT is individually developed and locally owned by the wealthier local entrepreneurs in the village. Without sufficient guidance from tourism experts and regulations, the adoption of this model not only leads to the social, cultural and environmental issues demonstrated above but also to a lack of equal distribution of economic benefits to the broader community. Without strengthening community spirit, it is unsurprising that the locals have gradually given up local assets and control of tourism to external investors. Many local homestays at Lác village are now rented out or sold entirely to investors from other provinces, mainly from Hanoi, indicating a gradual shift to model three – CBT externally owned or managed. Hence, interventions need to take place, supported by both the local community and the authorities, in order to ensure that Lác village can reduce the negative impacts associated with CBT development and expand positive impacts to local residents.

Conclusions and Recommendations

This chapter has contributed up-to-date information on the policies and practices of Vietnam CBT and identifies the key opportunities and challenges from its current development. Overall, Vietnam CBT has underdone significant expansion in the last few decades, and has great potential to contribute to the sustainable development of (ethnic) communities in the rural and remote areas of Vietnam. Vietnam CBT's fundamental challenges are identified in the areas of marketing, capacity development and regulation, while opportunities are found in the areas of cross-sectoral collaboration, digitalization in tourism, and tourism policies in the (post)-COVID-19 period. The cases of Lác village and Sìn Suối Hồ village have highlighted how these CBT challenges and opportunities take place in the context of the mountainous area of Vietnam. The cases also highlight the importance of leadership in uniting the community for tourism product and service development, along with the importance of capacity building for the villagers via leading by example, self-learning and, when possible, with guidance from tourism experts on site or via digital platforms. The establishment of management structures and proper regulation by the authorities is also key to ensuring the high standards of tourism products and services while ensuring that tourism brings more benefits than negative impacts to local people. Tourism policies in the (post)-COVID-19 period need to further support these local CBT projects to build human resources (especially in the areas of marketing and management) to better capture and serve high-value domestic tourists while still maintaining environmental sustainability and the authenticity and peacefulness of the villages.

Based on these findings, we propose the following practical recommendations for Vietnam CBT. First, there should be clear guidance for CBT product development (e.g. in the form of handbooks and expert training) that enables local communities to develop high-quality tourism products that can meet the needs of tourists. Government and private marketing agencies should also have specific strategies to promote the diverse range of tourism experiences offered by CBT. Second, in terms of capacity development, education and training programmes should be tailor made to suit the needs and strengths of local residents. Vocational visual training, hands-on coaching and on-the-job training, for instance, would better support local CBT than formal education programmes offered by higher education institutions. Funding for capacity development should also take into account long-term commitment and ongoing support for the local community. Finally, specific CBT policies and regulations should continue to be developed and adjusted to reflect the ongoing development of Vietnam CBT, with the goals of ensuring long-term social equality, as well as cultural and environmental protection. There should also be policies that further support cross-sectoral collaboration, CBT networks, and digitalization for CBT.

References

Airbnb (2017) Community Tourism Programme. Available at: https://www.airbnbcitizen.com/ctp/ (accessed 15 May 2021).

Báo Hòa Bình (2017) Giữ bản sắc du lịch cộng đồng [Conservation of community culture]. Available at: http://www.baohoabinh.com.vn/278/140577/Giu-ban-sac-du-lich-cong-dong.htm (accessed 20 February 2021).

Bauer, D.E. (2008) *Negotiating Development: Identity and Economic Practice in Coastal Ecuador*. ProQuest, Ann Arbor, Michigan.
Blackstock, K. (2005) A critical look at community-based tourism. *Community Development Journal* 40(1), 39–49. DOI: 10.1093/cdj/bsi005.
Braun, V. and Clarke, V. (2006) Using thematic analysis in psychology. *Qualitative Research in Psychology* 3(2), 77–101. DOI: 10.1191/1478088706qp063oa.
Britton, S.G. (1982) The political economy of tourism in the Third World. *Annals of Tourism Research* 9(3), 331–358. DOI: 10.1016/0160-7383(82)90018-4.
Chau, H.N.M. (2014) Stakeholder networks supplying rural tourism in Vietnam: with reference to a Japanese case. PhD dissertation, Okayama University, Okayama, Japan.
Dezsi, Ş. and Benedek, J. (2011) The role of rural tourism in the socio-economic diversification of rural space from Lăpuşului Land-Romania. *International Journal of Systems Applications, Engineering & Development* 2(5), 198–207.
Doan, M.C. (2018) Phát triển du lịch cộng đồng góp phần xóa đói giảm nghèo theo hướng bền vững [Community-based tourism for poverty eradication]. In *2018 Conference on Policies for CBT Development in Vietnam Proceedings*. Vietnam National Administration of Tourism, Hanoi, pp. 9–22.
Du lịch Hòa Bình (2020) B i học kinh nghiệm phát triển du lịch cộng đồng ở Hòa Bình [Lessons from community-based tourism development in Hòa Bình]. Available at: http://khudulichhohoabinh.vn/index.php/vi/tin-tuc/du-lich-hoa-binh/1989-ba-i-ha-c-kinh-nghia-m-pha-t-tria-n-du-la-ch-ca-ng-a-ng-a-ha-a-ba-nh (accessed 15 May 2021).
Ellis, F. (1999) *Rural Livelihood Diversity in Developing Countries: Evidence and Policy Implications*. Overseas Development Institute, London.
Ellis, S. (2011) Community-Based tourism in Cambodia: exploring the role of community for successful implementation in least developed countries. PhD thesis, Edith Cowan University, Perth, Western Australia, Australia.
ESRT (Environmentally and Socially Responsible Tourism) and WWF-Vietnam (2013) *Vietnam Community Based Tourism Handbook: a Market-based Approach*. ESRT Capacity Development Programme and WWF-Vietnam (World Wide Fund for Nature), Hanoi.
Flyvbjerg, B. (2006) Five misunderstandings about case-study research. *Qualitative Inquiry* 12(2), 219–245. DOI: 10.1177/1077800405284363.
Giampiccoli, A. and Mtapuri, O. (2017) Beyond community-based tourism. Towards a new tourism sector classification system. *Gazeta de antropologia* 33(1).
Glusac, E. (2020) The New Escapism: Isolationist Travel. *New York Times*, 24 June. Available at: https://www.nytimes.com/2020/06/24/travel/camping-outdoors-resorts-viru.html (accessed 20 December 2020).
Goodwin, H. and Santilli, R. (2009) Community-based tourism: a success. *ICRT* [International Centre for Responsible Tourism] *Occasional paper* 11(1), 37.
Haggblade, S., Hazell, P. and Reardon, T. (2010) The rural non-farm economy: prospects for growth and poverty reduction. *World Development* 38(10), 1429–1441. DOI: 10.1016/j.worlddev.2009.06.008.
Hamzah, A. and Khalifah, Z. (2012) Community capacity building for sustainable tourism development: experience from Miso Walai homestay. *Community Capacity Building* 2, 1–10.
Korten, D. (1980) Community organisation and rural development: a learning process approach. *Public Administration Review* 40(5), 480–511. DOI: 10.2307/3110204.
Kotler, P. (2012) *Kotler on Marketing*. Simon and Schuster, New York.
Lucchetti, V.G. and Font, X. (2013) Community based tourism: critical success factors. *ICRT* [International Centre for Responsible Tourism] *Occasional paper*, (OP27).
Mitchell, J. and Muckosy, P. (2008) *A Misguided Quest: Community-based Tourism in Latin America*. Overseas Development Institute, London.
Ngo, H.D. (2018) Công tác quảng bá, xúc tiến du lịch cộng đồng – thực trạng vàgiải pháp [Community-based tourism promotion: problems and solutions]. In *2018 Conference on Policies for CBT Development in Vietnam Proceedings*. Vietnam National Administration of Tourism, Hanoi, pp. 60–66.
Ngo, T. and Doan, T. (2014) Market access and the sustainability of community-based tourism enterprises: case study in Droong village, Dong Giang district, Quang Nam province, Vietnam. Paper presented at the International Conference on Sustainable Development and Ethnic Minority Poverty Reduction in Mountainous Regions of Vietnam, 11–13 June 2014, Thai Nguyen, Vietnam.
Ngo, T., Lohmann, G. and Hales, R. (2020) Integrating the third way and third space approaches in a post-colonial world: marketing strategies for the business sustainability of

community-based tourism enterprises in Vietnam. *Current Issues in Tourism* 23(15), 1914–1932. DOI: 10.1080/13683500.2019.1694494.

Nguyen, V.L. (2018) Chú trọng đào tạo phát triển nguồn nhân lực để du lịch cộng đồng phát triển bền vững, có trách nhiệm. In *Proceedings of Conference on Policies for CBT Development in Vietnam, 2018*. Vietnam National Administration of Tourism, Hanoi, pp. 88–95.

Pelto, P.J. (2017) *Mixed Methods in Ethnographic Research: Historical Perspectives*. Taylor & Francis, New York.

Pham, H.L. (2018) Những khó khăn, rào cản đối với việc phát triển nhân lực phục vụ du lịch cộng đồng hiện nay [Barriers to the development of human resources for community-based tourism]. In *2018 Conference on Policies for CBT Development in Vietnam Proceedings*. Vietnam National Administration of Tourism, Hanoi, pp. 96–102.

Phi, G.T. (2017) A critical analysis of microfinance tourism and poverty alleviation: characteristics, opportunities and constraints.PhD thesis, Griffith University, Brisbane, Queensland, Australia.

Phi, G.T. (2020a) Development tourism and poverty alleviation: towards an integrative framework. In: Brandão, F., Breda, Z., Costa, R. and Costa, C. (eds) *Handbook of Research on Role of Tourism in Achieving the Sustainable Delopment Goals*. IGI Global, Hershey, Pennsylvania, pp. 20–41.

Phi, G.T. (2020b) Staycation & Backyard Tourism: Silver Linings for Tourism Recovery in the COVID-19 Pandemic? Available at: https://destination-review.com/en/staycation-backyard-tourism-silver-linings-for-tourism-recovery-in-the-COVID-19-pandemic/ (accessed 20 September 2020).

Phi, G.T. and Whitford, M. (2017) Social entrepreneurship in Viet Nam. In: *Managing Growth and Sustainable Tourism Governance in Asia and the Pacific*. UNWTO (World Tourism Organization), Madrid, pp. 132–139.

Phi, G.T., Whitford, M. and Dredge, D. (2017) Knowledge dynamics in the tourism–social entrepreneurship nexus. In: Sheldon, P. and Roberto, D. (eds) *Social Entrepreneurship and Tourism*. Springer, Cham, Switzerland, pp. 155–172.

Pro-poor Tourism Partnership (2005) *Pro-Poor Tourism: Annual Register 2005*. Pro-Poor Tourism Partnership, London.

Saayman, M. and Giampiccoli, A. (2016) Community-based and pro-poor tourism: initial assessment of their relation to community development. *European Journal of Tourism Research* 12, 145–190.

Scheyvens, R. (2012) *Tourism and Poverty*. Routledge, New York.

Sharpley, R. and Telfer, D. (eds) (2008) *Tourism and Development: Concepts and Issues*. Channel View Publications, Clevedon, UK.

Truong, S. V. (2018) Đề xuất mô hình du lịch cộng đồng tại làng nghề [A proposal for establishment of community-based tourism in craft villeges]. In *2018 Conference on Policies for CBT Development in Vietnam Proceedings*, pp. 130–135.

United Nations (2016) Transforming our world: the 2030 Agenda for Sustainable Development. Available at: https://sustainabledevelopment.un.org/post2015/transformingourworld (accessed 20 September 2020).

UNWTO (World Tourism Organization) (2002) *Tourism and Poverty Alleviation*. UNWTO, Madrid.

van Beers, T. (2011) Discourse in tourism development: a narrative of inclusive business. Master's thesis. Wageningen University, Wageningen, the Netherlands.

Vietnam National Administration of Tourism (2020) Tourism Statistics. Available at: https://vietnamtourism.gov.vn/english/ (accessed 21 May 2021).

Vietnamtourism (2017) 2017 Vietnam Tourism Law. Available at: https://vietnamtourism.gov.vn/index.php/docs/853 (accessed 17 September 2021).

Vietnam Tourism Research Institute (2018) *2018 Conference on Policies for CBT Development in Vietnam Proceedings*. Vietnam National Administration of Tourism, Hanoi.

Yin, R. (2009) *Case Study Research: Design and Methods*, 4th edn. Sage, Thousand Oaks, California.

Zapata, M.J., Hall, M., Lindoc, P. and Vanderschaeghe, M. (2011) Can community-based tourism contribute to development and poverty alleviation? Lessons from Nicaragua. *Current Issues in Tourism* 14(8), 725–749. DOI: 10.1080/13683500.2011.559200.

8 Heritage Conservation and Tourism: the Present and Future of the Past

Huong T. Bui[1], Andrew Le[2]* and Thomas E. Jones[1]
[1]Ritsumeikan Asia Pacific University (APU), Japan; [2]International College of Management Sydney (ICMS), Sydney, Australia

Introduction

Tourism is perceived as an important opportunity for sustainable heritage conservation. Heritage tourism can transform an historic city into a touristic-historic city (Ashworth and Tunbridge, 2000), when the dual roles of heritage conservation and tourism attraction might assist the city to accomplish multiple goals, including education, recreation, financial and community engagement (Timothy and Boyd, 2006). However, 'modern mass tourism blights the aura of all notable relicts' (Lowenthal, 2015, p. 425), and thereby 'may dilute both [the] speciality and specificity' (Ashworth and Tunbridge, 2000, pp. 55–56) of a historic site owing to the double forces of commodification and globalization driven by both tourism and conservation.

Concern for urban heritage conservation has proliferated across Asia in recent years as rapid economic growth has caused large-scale transformation in physical landscapes and has disconnected people from their past and their sense of place (Winter and Daly, 2012). Within the theoretical debate on an Asian approach to heritage conservation, there are critiques on the limitations of a Eurocentric approach to heritage conservation and heritage tourism in Asian cities (Logan, 2002; Winter, 2014). Within this debate, Vietnam sets an interesting example of its Western European colonial legacy and contemporary Eastern European influence combined with the country's long tradition as a South-east Asian society. Having been colonized by the French for nearly a century, many concepts and practices for heritage conservation in the country were inherited from the colonial past and have carried on to the current time. The process of UNESCO World Heritage designation guided by the dominant systemic universal values further supports the Eurocentric nature of the approach to conservation in the country.

Having analysed Hội An Ancient Town, a living urban heritage site designated by UNESCO in 1999, and one of the most popular touristic destinations in Vietnam, the authors present findings derived from a longitudinal study over 5 years, before and during the COVID-19 pandemic in 2020–2021. The authors first review existing research on heritage conservation and tourism in both colonial and contemporary Vietnam. Then we centre our analysis on the case of Hội An Ancient Town by looking at the 'frozen in time and space' conservation practices, under pressure from the exponential growth of mass tourism in the past three

*Corresponding author: ale@icms.edu.au

(eds Huong T. Bui *et al.*)
DOI: 10.1079/9781789242782.0008

decades and the recent suspension of tourism since early 2020. Although the suspension of mass tourism negatively impacts the economic wealth of local residents and businesses, it offers an opportunity for administrators to rethink and revise the existing tourism development policy.

Heritage Conservation and Tourism

Philosophical grounds

Asian philosophies of heritage reorient the focus from a primarily physical historical connection with the past to an ongoing cultural and metaphysical connection that is reflected in three arguments. First, Asian heritage is valued for its spiritual significance rather than its material authenticity. Heritage is constantly maintained and even facelifted by people without concern for the original material. An old structure may receive significant improvements and additions to other facilities, but such practice may be defined as 'vandalism' by international agencies (Wijesuriya, 2001). UNESCO also emphasizes the importance of continuity in the original fabric of physical structures in defining the authenticity of built heritage. In particular, the Nara Document on Authenticity, which originated in Japan, stresses the continuity of use in heritage buildings rather than structure and materials, because their fabric is usually periodically renewed or refurbished (Long and Sweet, 2006, p. 447). That leads to the second point, where Asian heritage needs to be understood in its totality and continuity rather than as being representative of a particular objective style and 'frozen' in a particular time (Chapagain, 2013). Third, unique local practices exist in many traditional communities. However, these practices often conflict with the internationally proliferated standards. In other words, heritage in an Asian context often differs from the Eurocentric norm that prioritizes historical monuments and 'high culture', instead embracing the meaning of sites in ordinary, everyday landscapes (Taylor, 2012).

The paradigm shift in heritage conservation is marked by introducing a 'living heritage approach' (Poulios, 2014), valuing spiritual and intangible beliefs even at the expense of the holistic conservation of tangible or material heritage. For instance, Japan has long had a robust national system for protecting tangible and intangible heritage with an emphasis on 'living culture'. In other words, the ways buildings are put to use is more important than their fabric (Aygen and Logan, 2015). Similarly, China's recently developed heritage management system has been strongly influenced by Western thinking and practices, but clearly indicates that China's heritage serves the national interest (du Cros and Lee, 2013). Other Asian countries, such as Indonesia and India have also developed national charters to guide professional practice. The Vietnamese are determined to maintain control and selectively use Western principles with respect to heritage conservation, reflecting the notion of syncretism in which successive incoming new elements add to, rather than replace, what exists (Aygen and Logan, 2015). The holistic notion of the 'living heritage approach' offers valuable lenses to examine tourism phenomena. Accordingly, the authors adopted the definition of heritage as that which 'embodies competing notions of the unchanging and authentic past and the consciously constructed and transformed present and future; and it is bound up with issues of local, ethnic, provincial, national and global identities' (Hitchcock *et al.*, 2010, p. 11).

Historical grounds

Colonial and de-colonial approaches to heritage conservation coexist in contemporary Vietnam. Dating back to the colonial period, spanning the mid-19th to mid-20th centuries, heritage conservation and tourism were intertwined with French colonial policy in Indochina (the present-day Vietnam, Laos and Cambodia). The preservation of monuments served dual purposes: (i) helping to legitimize French colonialization by providing a scientific sanction; and (ii) ensuring the suitability of tourism operations. First, the preservation of historical monuments was guided by colonial scientists in the belief that local populations were unable to conserve their own environment and heritage (Demay, 2014). From the beginning of colonization in the late 19th century, heritage preservation was gradually initiated and it was legalized in the late 1920s and early 1930s. Scientific research undertaken to classify and preserve sites allowed France to establish its prestige vis-à-vis other colonial empires. The French scientific approach heavily concentrated archaeological efforts on the restoration of physical monuments, which was supported by non-scientific reasons including the substantial funds invested in scientific excavation, examination and preservation (Demay, 2014). Second, the overlap between tourism and scientific preservation allowed tourism to be justified as a reason for the protection of heritage (Demay, 2014). Natural and historical riches were an integral part of tourist sites, being a tool for communication and a way to present a playful and exotic Indochina through its monuments, landscape and animals, which, as a consequence, enhanced France's effort to preserve them. The old sacred sites were to be incorporated into the map of the colony, including their ancient prestige (Anderson, 1991).

After Vietnam gained its independence from the French in 1954, heritage became a highly politicized project concerned with the construction of national identity (King and Parnwell, 2011). This mission was for the newly independent state to select and deploy archaeological discoveries and heritage sites to present national resilience, unity and innovation, often in the context of an 'imagined' glorious age which was subsequently eclipsed by colonialism (Glover, 2003, p. 17). In this respect, post-independent governments have often had to reposition their archaeological sites to express indigenous achievement and demonstrate the legitimacy conveyed by ancient genealogy, in contrast to Western interpretations of the sites (Hitchcock *et al.*, 2010). This process of identity construction involved the study and preservation of local heritage (Long and Sweet, 2006, p. 463). The excavation, conservation and designation of the Central Sector of Thăng Long Citadel as a UNESCO Cultural Heritage site on the occasion of the 1000th anniversary of Thăng Long – Hà Nội in 2010 served the reconstruction of the national identity (Bui and Lee, 2015). This practice echoes Ian Glover's argument as it represents an 'attempt to create discourses with the past in order to legitimize and strengthen the position of the state and its dominant political communities' (Glover, 2003, pp. 16–17).

Following decades of military conflict, international sanctions and struggles for post-war economic and socio-cultural recovery, the country has reformed through the *Đổi Mới* economic policy since 1986. Cultural heritage plays an especially important role in this transitional period, serving as a cultural 'anchor' for contemporary Vietnam (Suntikul *et al.*, 2010). Following international standards, the Vietnamese classification of heritage divides into tangible and concrete elements (buildings, monuments, artefacts, sites and constructed landscapes), and 'intangible cultural heritage' (behaviour, action, performance, etc.). Heritages are governed by three levels in accordance with their significance: (i) prefectural level: heritage with local value; (ii) national level: heritage with national value; and (iii) outstanding national level: heritage with outstanding national value. As of 2020, eight sites of cultural and natural heritage are listed as UNESCO World Heritage Sites (WHS) (see Table 8.1.).

In addition to the UNESCO heritage sites, Vietnam possesses a unique mixture of different types of 'heritage' sites, appealing to different groups of visitors. These include sites representing traditional Sino-Vietnamese culture and the romantic relics of French colonialism as well as less conventional types of attractions that capitalize on the controversial trend of 'war tourism' (Suntikul *et al.*, 2010).

Tourism grounds

Heritage has become a political tool in negotiations over identity, but it is also part of an 'industry'. The interpretation of cultural heritage is contested among different groups of visitors, site managers, and levels of governance as found from previous writing on heritage tourism in Vietnam. First, Henderson (2000) notes that '[h]eritage attractions in Vietnam and more widely remain a highly political issue and the influence of government policy cannot be ignored' (p. 276). For instance, the Central Sector of the Imperial Citadel of Thăng Long, a UNESCO WHS located in the capital city of Vietnam has been politicized and commodified for tourism

Table 8.1. The eight UNESCO World Heritage Sites (WHS) in Vietnam. Compiled by the authors.

Name	Province	National classification	UNESCO designation	Year of designation
Complex of Huế Monuments	Thừa Thiên – Huế	Historical cultural, art and architecture	Cultural	1993
Hôi An Ancient Town	Quảng Nam	Art and architecture	Cultural	1999
Mỹ Sơn Sanctuary	Quảng Nam	Art and architecture	Cultural	1999
Hạ Long Bay	Quảng Ninh	Natural landscape	Natural	1994/2000
Phong Nha – Kẻ Bàng National Park	Quảng Bình	Historical cultural and natural landscape	Natural	2003/2015
Central Sector of The Imperial Citadel of Thăng Long – Hà Nội	Hà Nội	Historical and archaeological	Cultural	2010
Citadel of the Hồ Dynasty	Thanh Hóa	Historical, art, architecture and archaeological	Cultural	2011
Tràng An Landscape Complex	Ninh Bình	Natural landscape and historical cultural, art and architecture	Mixed	2014

with that purpose (Bui and Lee, 2015). Second, heritage tourism develops under the jurisdiction of conservation, but also stays at the forefront of commodification. The negotiation between conservation and commodification is observable in the two contradicting phenomena of gentrification and museumification (Jones *et al.*, 2020) that affect the demography of the local residents inhabiting heritage sites. Commodification drives the development of sectors supporting tourism, such as souvenir vendors in heritage cities (Bui and Jolliffe, 2013; Trinh *et al.*, 2014). There are also warnings that heritage sites such as Hội An may be moving off the track of sustainable development (Avieli, 2015), encountering various socio-economic and environmental issues resulting from tourism development (Bui *et al.*, 2018). Third, heritage tourism is an object of the 'tourist gaze', in which the disparity between international and domestic visitors' imagination and perception exists. For instance, Bui and Le (2016) examine Hạ Long Bay and discover that international visitors have higher standards and are more critical than domestic visitors when judging the destination and its associated services. Gillen (2010), by examining the operations of travel agencies, found that domestic tourists are marginalized. Such businesses segment their markets, not according to the familiar language of cultural, eco- or special interest tourists, but via a reading of their clients' everyday spending and behavioural habits. Similarly, in the Huế Complex of Royal Monuments, Johnson (2010) finds an increasing presence of frivolous Vietnamese visitors – the nouveaux riche – who are diluting the meaning and authenticity of the site.

In order to deepen our understanding of the trilogy of philosophy, history and tourism grounds shaping heritage tourism in contemporary Vietnam, it is necessary to explore a heritage city with significant tourism activities that involve different sectors and to follow its historical development. The case of Hội An Ancient Town is ideally justified for these purposes.

The case of Hội An Ancient Town

Hội An Ancient Town is located in the central coastal region of Quảng Nam province, approximately 30 km from Đà Nẵng city and 812 km south of Hà Nội and 845 km north of Hồ Chí Minh city. From the 2nd to 15th centuries CE, native settlements in the area (of the Cham ethnic group) earned their living through fishing and trade. During the 9th and 10th centuries Hội An became an important commercial port and in the 16th and 17th centuries Vietnamese settlers engaged in farming, fishing and handicraft production there. Hội An's decline as a trading port began in the early 18th century when its rulers implemented a closed-door policy that restricted marine trade. In addition, the siltation of the Thu Bồn River forced the subsequent French colonial administrators to designate neighbouring Đà Nẵng city as the main regional port in the early 20th century, ending the golden age of Hội An as an international trading port. Throughout the first French era (1858–1945), the French return to Indochina (1946–1954), and the partition into North and South Vietnam (1959–1975), Hội An remained a quiet, isolated rural town focused on small-scale fishery and handicraft industries, which continued until the wake of the *Đổi Mới* economic reforms initiated by the central government in 1986.

The conservation of Hội An Ancient Town can be divided into four periods (Ando *et al.*, 2015). The 'preparation period' (1976–1986) taking place in the first 10 years after the end of the Vietnam War, was initiated with the management of abandoned houses and saw the start of the designation and protection of the national cultural heritage. Conservation of the historic core caught the attention of Polish architect Kzimierz Kwiakowski (Kazik) in 1983, raising international awareness of the traditional wooden houses and monuments of the area (Di Giovine, 2009, p. 263). The 'formulation period' between 1987 and 1996 established a system and a framework to protect national cultural heritage with international support. Due to fiscal and technical issues, the restoration work languished until the involvement of overseas development agencies and universities, beginning in 1992. The 'implementation period' was from 1997 to 2005, with building restoration work funded by the national government and revenue from tourism. Since then, revenue from entrance fees has assisted local residents in restoring their homes by supporting 80–100% of the total cost of restoration (Di Giovine, 2009). In 1997, the second conservation plan formalized core and buffer zones, with the former remaining virtually unchanged to this day. We are currently in 'the management period', which started in 2006, and marks the progress of restoration work and infrastructure development in line with tourism growth (see Fig. 8.1).

Tourism has brought dramatic change to Hội An's economy and population since the early 1990s. The first coverage of Hội An in the 1991 budget guidebook *Lonely Planet* attracted backpackers in such numbers that a tourism programme was established immediately (Di

Fig. 8.1. Hội An Ancient Town. Photo courtesy of Long H. Pham.

Giovine, 2009). In 1999, Hội An Ancient Town was designated as a UNESCO World Cultural Heritage site under criterion (*ii*), the fusion of culture over time, and criterion (*v*) the well-preserved culture of an Asian trading port. A mere 4000 visitors came in 1991, but by 2017 the city had more than 1 million visitors overnight in the ancient city.

Due to a strictly regulated heritage law, new lodging facilities cannot be established in the core zone. Hotels and resorts therefore cluster in the buffer zones, surrounding the Minh An Ward, on the islet of Hội An River (Cẩm Nam Ward), along the road connecting the city and Cửa Đại beach along the Đế Võng River (Cẩm Châu Ward) and in the areas of Tân An, Cẩm Phô and Sơn Phong that ring the core. The core zone is also surrounded by a ring of five revitalized craft villages for cultural tourism (Trà Quế – horticulture; Cẩm Nam – fisheries; Thanh Hà – pottery; Kim Bồng – carpentry and woodwork; and Hội An silk village), and a beach tourism area along the seafront (Hội An People's Committee, 2017). The five areas of tourism development in Hội An are: (i) heritage tourism in the core restricted area; (ii) beach tourism along the coast expanding from Cửa Đại to An Bàng; (iii) the ecotourism zone on Chàm Island; (iv) ecotourism mixing with cultural tourism in the Cẩm Thanh area; and (v) cultural tourism in the suburban area, where there are clusters of handicraft villages. The distribution of hotels is concentrated in the buffer zone surrounding the restricted zone of the ancient streets and along the coast (see Fig. 8.2).

After three decades of rapid tourism development, from the beginning of 2020, international tourism has been suspended due to travel restrictions issued as countermeasure to prevent the spread of coronavirus. The strict lockdown policies implemented in Đà Nẵng – Hội An in the summer of 2020 further cancelled out the domestic tourism. Having depended heavily on international tourism, Hội An has undergone a critical period that demands a rethinking of a development model highly contingent on external forces. Our analysis, driven from a longitudinal, multi-source and methodological study, enriches the discussion on the symbiosis of heritage conservation and tourism in the

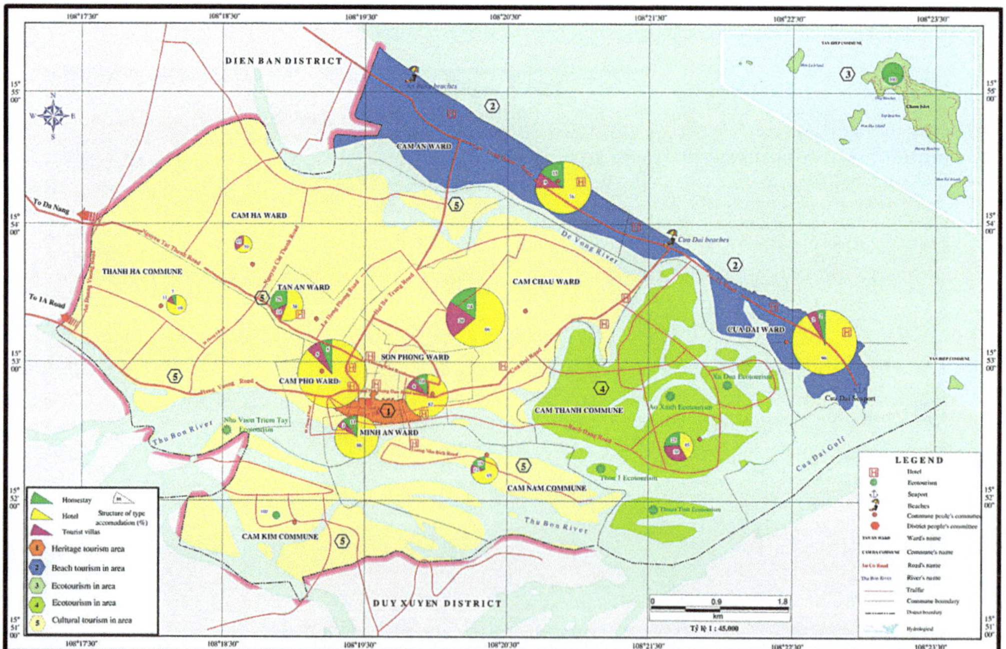

Source: Author's compilation based on a tourist map provided by Hoi An Department of Trade and Tourism

Fig. 8.2. Tourism resources and hotel distribution of Hội An. Map courtesy of Thuy T. Duong.

context of contemporary Vietnam, especially under the period of the global health crisis in 2020–2021.

Study Methods

The case of Hội An has been explored in full by combining different methods and information sources, enabling an examination of competing forms of evidence to elucidate more convincing findings (Yin, 2009). The first stage of the fieldwork was conducted from 2015 to 2018 using both primary and secondary sources of information. Primary data came from two dozen interviews with provincial and municipal government officers, international consultants, non-governmental organization (NGO) staff, business owners (hotels and shops), local residents and researchers. Another source of primary data is the opinions of domestic and international tourists visiting the site surveyed by the consultant team of a European-funded programme for Environmentally and Socially Responsible Tourism (ESRT). Answers to a self-administered questionnaire were collected from international visitors who understand English, as well as domestic visitors. The questionnaire had four sections collecting information on visitors': (i) travel arrangements and spending; (ii) evaluation of destination attributes; (iii) perception of destination images; and (iv) demographic information. The questionnaire also included three open-ended questions on positive aspects, negative aspects as well as recommendations, all of which provided data for qualitative analysis in this study. After two phases of data collection, 662 completed questionnaires were collected from 301 domestic and 361 international visitors. Descriptive statistics were used to generate profiles of the visitors, their travel arrangements and visitation to attractions. To compare means of related dimensions between domestic and international groups, one-way analysis of variance (ANOVA) was employed. To assist an in-depth analysis of visitor experiences, the authors used content analysis to extract major themes from visitors' comments (text data).

The second stage of data collection was carried over 2 years, between 2019 and 2020. Data sources consisted of statistics collected from governmental departments and policy papers concerning the governance and administration of economic and social activities in addition to numerous reports and documents published by international NGOs, and central and local governments. Media sources and some online interviews on the recent COVID-19 pandemic were also collected and analysed using critical discourse methods to explore the dialectic link between texts and practice (Fairclough, 2003). By employing multi-source data collection and multi-method data analysis at various data collection points, the authors could triangulate findings across sources and time, and contrast between groups of domestic and international tourists, to draw a holistic picture of heritage tourism in Hội An in particular, and Vietnam in general.

Findings

Findings emerged from a quantitative analysis of visitor surveys, compared to the themes extracted from interviews and visitors' opinions. Contrasting views of domestic and international tourists are also presented. The temporal dimension of before and during the COVID-19 pandemic is also considered for data comparison.

Visitors' survey

The visitor profile shows that domestic tourists are from younger age groups as those under 44 years old account for 90% of the respondents, who are largely civil servants (29%) and students (25%), mainly from Hà Nội (24%) and Hồ Chí Minh city (28%). Contrastingly, 31% of international visitors are in the senior age group (44–64 years old) and 51% are of the middle age group (25–44 years old), primarily from Australia (22%), the UK (20%), other European countries (27%), North America (8%) and other countries in Asia (24%). International visitors are employees (41%), self-employed (14%), students (13%) and retirees (11%).

With reference to the context of their trips, the majority of international visitors visited Hội An for the first time (84.8%), while only 34.9% of domestic travellers belong to this category. International visitors stay in upper-class hotels (resorts and luxury hotels above three-star) while domestic travellers prefer a cheaper range of hotels (one- or two-star) or hostels. The most striking difference between international and domestic visitors is their spending and length of stay at the destination. While more than 90% of domestic visitors are either day trippers or stay 1–2 nights, international tourists stay for 3–4 nights (40%), for approximately a week (17.8%) or even for more than a week (16%). Longer stays result in much higher spending among international tourists at around VND 12 million (US$650), approximately 20-fold the domestic market average of VND 747,500 (US$32) (see Table 8.2).

Visitation to cultural heritage (73.1%) and tailor-made clothing shops (57.2%) are among the top services utilized by international visitors. Top services used by domestic visitors are local specialties (75.1%), visitation to cultural heritage (55.4%) and scuba diving (41.8%) (see Table 8.3). Handicraft villages in the suburban areas are essential to retaining traditional craftsmanship and are an integral part of cultural heritage tourism, however, they contribute marginally to the product portfolio of the site. Ten years since the establishment of the handicraft villages for tourism in 2005, less than 5% of international tourists have visited these villages. Among four handicraft village sites, the non-authentic silk village, located on the main road connecting Đà Nẵng and Hội An, attracts higher frequencies of visitation. Another three villages (Thanh Hà, Trà Quế and Kim Bồng) located within 5 km of the restricted zone of the historic town, are accessible both by boat and road, yet they are not appealing to visitors (Table 8.3).

The ancient town atmosphere has been perceived as artificial with a mean score of 4.69 for domestic and 4.84 for international tourists on the scale from 1 to 6. International tourists are also sceptical in their evaluation of the destination being environmental friendly and respecting local culture, rating these features as 4.53 and 4.43, respectively, which were lower than the domestic group's ratings of 4.95 and 5.15, respectively (Table 8.4).

Visitors' opinions

Thematic coding from comments on open-ended questions aid in furthering our understanding of visitors' perceptions of the destination. Major themes from the comments to the question '*What did you like the most in your visit to Hội An?*' left by 75 international visitors and 60 domestic tourists were coded to generate text clouds based on the frequency of keywords. Domestic and international visitor's comments are relatively similar. Both groups highlight the valued atmosphere, the ancient town and friendly people; these themes show the highest frequencies in the text clouds. In other words, both tangible elements (the ancient town) and intangible features (atmosphere, people) of Hội An impressed tourists the most. International visitors rated 'food' highly positively, followed by 'beach' and 'shopping' among tourism products offered at the destination. For example, international visitors wrote:

> 'Relaxed atmosphere, friendly people especially at the resort. Our tour guide, Tommy, was also excellent.'
>
> 'Everything! The ancient town is stunning, and the beach is beautiful: accommodation, service in the hotel and all restaurants, we went to was outstanding.'

Positive comments from the survey respondents include key words such as 'old town', 'atmosphere', 'ancient' and 'landscape' appearing most frequently (see Fig. 8.3A and B). This implies that efforts at making conservation are recognized by tourists.

For the questions '*What did you like the least?*' and '*What could be improved?*' there were 75 comments from domestic visitors and 150 from international visitors. Both domestic and international visitors are concerned with issues of transportation; no public buses, unclear instruction of road signage, and restricting motorbikes were among the top comments of domestic visitors. Other comments included:

> 'Hard to walk on footpaths, too many bikes parked. [It's] hard to cross the road; very dangerous.'
>
> 'Needs better traffic control – motorbike, cyclists and cyclos should be restricted in areas and cannot hear as they come up behind you.'

Table 8.2. Characteristics of visitors' trips to Hội An. From authors' data analysis.

	Domestic visitors (N = 301)		International visitors (N = 361)	
Travel services	Frequency	Percentage	Frequency	Percentage
Type of travel				
Package tour	77	25.6%	72	19.9%
Self-organized	221	73.4%	275	76.2%
Other	2	0.7%	14	3.9%
Length of stay in Hội An				
Day trip	101	34.4%	8	2.2%
1–2 nights	161	54.8%	86	23.9%
3–4 nights	23	7.8%	144	40.0%
5–6 nights	5	1.7%	64	17.8%
> 7 nights	4	1.4%	58	16.1%
Frequency of visit to Hội An				
First	105	34.9%	306	84.8%
Second	77	25.6%	27	7.5%
Third	37	12.3%	17	4.7%
> Three times	81	26.9%	11	3.0%
Main travel purpose				
Leisure/vacation	241	80.1%	346	95.8%
Business	12	4.0%	1	0.3%
Conference, workshop	9	3.0%	1	0.3%
Visiting family, friends	17	5.6%	8	2.2%
Other	21	7.0%	5	1.4%
Accommodation				
Resort	24	8.0%	63	17.8%
Five-star hotel	8	2.7%	25	7.1%
Four-star hotel	12	4.0%	67	19.0%
Three-star hotel	71	23.6%	81	22.9%
One- or two-star hotel	36	12.0%	35	9.9%
Guest house	16	5.3%	16	4.5%
Lodge/hostel	51	16.9%	9	2.5%
Homestay	21	7.0%	36	10.2%
Other	37	12.3%	21	5.9%
Total expenditure in Hội An	VND 747,500		VND 12,595,100	

Table 8.3. Awareness and consumption of tourism products by domestic and international visitors. From authors' data analysis.

	Domestic (N = 301)		International (N = 361)	
Travel services	Frequency	Percentage	Frequency	Percentage
Consumption				
Visit world cultural heritage sites	118	55.4%	207	73.1%
Making traditional lanterns	51	23.9%	29	10.2%
'One day as a farmer' Programme	11	5.2%	16	5.7%
Scuba diving/snorkelling at Cham Island	89	41.8%	48	17.0%
Traditional folk music and dance	32	15.0%	31	11.0%
Tailor-made clothing	33	15.5%	162	57.2%
Local specialties	160	75.1%	133	47.0%
Other	3	1.4%	19	6.7%
Visits to attractions				
Hội An Ancient Town	289	96.7%	322	95.8%
Mỹ Sơn holy land	85	28.4%	79	23.5%
Cửa Đại beach	166	55.5%	148	44.0%
Thanh Hà ceramic village	36	12.0%	17	5.1%
Hội An silk village	67	22.4%	45	13.4%
Cù Lao Chàm village	155	51.8%	49	14.6%
Núi Thành mountain	34	11.4%	35	10.4%
Trà Quế village	25	8.4%	16	4.8%
Kim Bồng carpentry village	15	5.0%	14	4.2%
Other	113	37.8%	24	7.1%

Table 8.4. Destination image perception and evaluation by domestic and international visitors. From authors' data analysis.

	Means			
	Domestic (N = 301)	International (N = 361)	Student's *t*-test value	Significance
Image				
Authentic/artificial	4.69	4.84	–1.48	0.14
Modern/traditional	3.32	3.01	2.57	0.01
Diversified/simple	2.39	2.69	–2.75	0.01
Evaluation				
Environmentally friendly destination	4.95	4.53	4.16	0.00
Respect the environment and local culture	5.15	4.43	8.11	0.00
Overall satisfaction	4.97	5.34	–5.37	0.00

Fig. 8.3. Positive comments by (A) domestic visitors; (B) international visitors. From authors' data analysis.

'Taxi drivers need to be more considerate on everyone when driving on the road. Too fast, no need to rush!'

Among 75 domestic visitors, there were 35 comments about the environment of Hội An. They are concerned about the hygiene due to rubbish on the street, in the river and in the ocean. As noticed among visitors 'litter and rubbish is getting worse'. The outskirt beach area is notified with dirty water together with pollution, for instance, 'dirty water at Cửa Đại beach, more variety in shop[s] they all sell the same; pushy sellers'. High prices for food, souvenirs and entrance tickets are the most frequent comments from domestic tourists. Some noticed that the price for services has increased dramatically while service quality has declined: 'food is getting more expensive while quality is getting poor' or 'unfriendly services in many souvenir shops selling at high price'. Similarly, sellers and unfriendly selling behaviour of shopkeepers are top concerns of international tourists. A wide variety of words were used to describe this unfriendly selling behaviour of shopkeepers, including 'aggressive', 'annoying', 'forced', 'badgered', 'pressured' and 'hassled'. Comments included:

'We felt a little pressed for the sellers, some were a little bit angry if we didn't buy anything.'

'feeling of being ripped off, impossible to just look around, without being pressured to buy something.'

The most frequent issues raised by international tourists focused on the behaviour of sellers (see Fig. 8.4A and B), although some acknowledged increasing commercialization and erosion of

Fig. 8.4. Negative comments by (A) domestic visitors; (B) international visitors. From authors' data analysis.

cultural identity: 'It's beginning to look like Disneyland, in a bad way, too many shops, they all sell the same things, it doesn't feel like a Vietnamese town anymore.'

Administrators' viewpoints

Visitor viewpoints on overcrowding and commercialization of the town are somewhat similar to administrators' views. The majority of touristic activities and facilities are concentrated in the core zone of the town, which is too small to allow a spatial distinction between the conservation area and touristic facilities. As a result of an adaptive re-use policy of conservation, commercial and domestic vernacular heritage have been turned into tourist facilities, such as shops, coffee houses or galleries, attracting more tourists to the core zone. Hội An is becoming a victim of its own success as an administrator reveals:

> In the evening, the central streets [in Minh An Ward] are very crowded ... Hội An means 'Meeting place of Peacefulness'. Hội An now is no longer peaceful and quiet. Hội An is beautiful only after 9 or 10 p.m. when streets become quiet.
>
> (Officer of Hội An Center for Conservation of Monuments, March 2018)

Analysis of the economic structure of Hội An reveals some insights on developmental imbalances. First, hotels represent the fastest growing economic sector. There have been turning points in the regulation of touristic accommodation, leading to different growth paths for the tourism industry. The policy was to convert Hội An from a backpacking budget destination to a 'high-quality' resort city defined by an average spending target of US$200/day initiated in the year 2000. The requirement for minimum investment into hotels, to be US$1 million meant that only strong external financial investors would be able to develop resorts and hotels, as very few people in Hội An had the required capital (Avieli, 2015). Along with this policy, luxury resorts, hotels and holiday homes flourished along the 7 km road connecting Hội An, Cửa Đai beach and nearby An Bàng beach, which has transformed the ancient town from an historic cultural destination into a resort destination (Parnwell, 2016). While the 'quality' tourism started showing limitations because it was exclusive to external investors and international tourists, opportunities for local homestays accelerated from 2009 to 2010, following a popular cooking series; this has been institutionalized since 2012. The local government approved the homestay business and officially allowed households to lease out unused rooms to tourists. From 2012 to 2017, homestays proliferated unevenly across the city, accounting for the largest share of accommodation in the ecotourism and cultural tourism zone outside the heritage centre (Bui *et al.*, 2020). On the one hand, it is a sign of increasing community resilience and participation in tourism-led growth. On the other hand, it signals the shift towards an inclusive notion of tourism growth.

Second, the overall growth of the touristic hotel sector should be examined in relation to other economic sectors. Tourism development has sparked the prosperity of other tourism-related services, creating a whole economic system centring around tourism, while handicrafts and agriculture have been relatively neglected. Fisheries and seafood processing primarily serve local restaurants and hotels. The tourist market mainly drives demands for handicrafts and souvenirs. The traditional craft villages surrounding Hội An, such as Kim Bồng (carpentry and woodwork), Thanh Hà (pottery) and Trà Quế (horticulture), as well as traditional fishing villages (Cẩm Nam), have been on tourist maps since 2005 to showcase traditional skills and revitalize the intangible cultural heritage in these villages (Hội An Center for Monuments Management and Conservation, 2008), though they no longer retain their original functions for production (Bui and Jolliffe, 2013). In other words, agriculture and fisheries have been changed into performative staging activities for tourist consumption.

> Agriculture and handicraft in Hội An are small in scale, and outputs are mainly made for tourism, therefore, these industries cannot develop independently. The agricultural activities in Hội An are no longer agriculture for production. It is arranged and performed as an artistic type of agriculture. In the long

> term, agriculture and handicraft cannot be independent from tourism.
>
> (Officer of Hội An People's Committee – interview posted on Hội An's World Heritage Portal, 2020)

Tourism has been a driving force for Hội An's economy, contributing to more than 70% of the city's GDP. Since March 2020, the COVID-19 pandemic has severely impacted all aspects of the tourism business in Hội An, resulting in the temporary closure of 90% of touristic shops. This poses a question about the opportunity cost of focusing on tourism development at the expense of other traditional productive sectors. The suspension of international and domestic tourism pushes tourism-related households into a critical condition. In the meantime, other sectors where locals retain their traditional production, as with agriculture and fisheries, have remained relatively stable.

> Many locals live and work in agriculture in the suburban areas of Cẩm Kim, Cẩm Hà, Cẩm Thanh and Cẩm Nam and do not entirely depend on tourism, they retain the tradition of agriculture. For many years, administrators and other sectors have undervalued their contribution. When tourism is suspended, it is the time to revive and renovate these traditional production systems for local consumption. Agriculture might not yield a very high income like tourism, but it has proven to be a sustainable livelihood for locals whereas the pandemic might have prolonged effects.
>
> (Former secretary of the local cabinet, interview posted on Hội An's World Heritage Portal, 2020)

Tourism development has transformed the landscape of the coastal city where many coastal areas for fisheries have been replaced by hotels, resorts and supporting services, while fishermen have gradually integrated into the tourism economy. Prior to the pandemic, a fisherman shared: 'It was a happy time being home with family. My income from my home service seafood restaurant is threefold the income derived from fishery'. Having gone through two strict lockdown periods in March and July 2020, his family restaurant had no customers. When their savings were almost gone, the fishermen had to return to the sea to fish (interview posted by The New York Times, 2021 (republished on *VnExpress*)).

Third, the emergence of the domestic market has shifted the orientation for tourism services and products. The number of visitors to Hội An in 2020 hit a record low of less than 1 million, 80% fewer than in 2019 (Hội An's World Heritage Portal, 2021). Switching from the international market to the domestic market became a national strategy initiated by the Ministry of Culture, Sport and Tourism as part of a campaign called *Vietnamese travel in Vietnam* from 1 June to 31 December 2020. The campaign covered promotional packages with high quality and reasonable prices showing strong commitment from accommodation providers, entertainment facilities and airlines to provide holiday packages that were compliant with safety regulations and social distancing.

To attract Vietnamese visitors, a series of cultural and recreational events were promoted to enhance the travel experience and generate revenue to complement the loss of income from admission fees, food, drinks and accommodation. Hội An put on a trial a show named *Waking Up Hội An* for Christmas 2020 and New Year 2021. The initiative marked intensive participation of the local communities and local administrators in partnership with private business to renew the atmosphere and scenery of the ancient town. The key attraction was *Light Up Hội An* held every full moon together with the regular *Lantern Night*. The world-renowned show *The Memoir of Hội An* has been the largest real-world show globally since March 2018, and celebrated its 3 years of inauguration on 18 March 2020. It has attracted around 1 million visitors. In addition, there has been *Hội An Show* with a storytelling format for reinventing Hội An as an international trading port of the 16th to 19th centuries (Hội An's World Heritage Portal, 2021).

Discussion

Hội An tourism at its peak of development has shown excessive commercialization and has become an 'artificial' destination. The commodification of heritage leads to the creation of landscapes that are described as 'elitist' and removed from the everyday experiences of the local residents, causing the displacement of low-income groups by an incoming 'gentry'

(Davidson and Lees, 2010). It should be noted that gentrification has prevailed at the site following policies to depopulate the core zone and adaptively reuse heritage sites for commercial purposes (Jones *et al.*, 2020). These, in fact, are consequences of a major shift in city planning, focusing on upgrading Hội An from a small-scale historic town to an expanded resort city since 2000. Along with the gentrification process is the arrival of outsiders, such as traders, new owners of properties and investors, representing an external driver for tourism growth. Tourism has developed at the expense of traditional agriculture, fisheries and handicrafts. The concentration of investment from outsiders in the fast-growing accommodation and service sectors to cater to the needs of the international tourist markets, has been in question. The suspension of tourism demand for more than a year since early 2020 has virtually sent the town back to a time before tourism development in the early 1990s. In this situation, the role of agriculture, fisheries and handicrafts have been included in discussions about balancing the economic structure of the town.

Another point that needs rethinking on the external-led growth of Hội An tourism is the market aspect. The growth policy in favour of the international market has resulted in negligence towards domestic and local needs. As shown from the survey data, Vietnamese visitors tend to have short stays and do not use accommodation and touristic services as much as international tourists. Their spending, therefore, is much less. In order to attract domestic tourists from nearby regions, Hội An has moved towards an experience-based economy by revitalizing cultural traditions through various forms of performance arts. The revitalization of tradition triggers the participation of local communities, leading the destination towards an endogenous growth model with the mobilization of internal sources.

The boom and bust of heritage tourism in Hội An analysed above presents two layers of development curves. Initially, small-scale tourism in the early days utilized local resources and investors serving both domestic and international markets, representing endogenous growth from within. Tourism in Hội An has long been led by international tourists, although the early days of investment mainly came from domestic investors. The international tourist market has been a growth engine for both endogenous and exogenous growth curves. From a small-scale independent destination to a large-scale resort destination, international market-led growth has resulted in a radical shift from production to consumption; from production-oriented activities to a tourism-service-based economy. This external driver of tourism growth represents the exogenous growth curve, which is larger in size and scope. However, the externalization of the resources and markets has proved to be too unstable to withstand sudden changes in the socio-economic environment such as the COVID-19 pandemic starting in the year 2020. Showing its pitfalls, the exogenous growth model demands a shift in development towards an inclusive growth model, that integrates domestic and local resources for sustainable future growth. Over 30 years of tourism development, the seed of growth has been planted in the community. Recently, the evidence of social memory and community-led growth and traditions as foundations and engines for sustainability and resilience of the town have been discussed (Bui *et al.*, 2020). That might suggest the emergence of an inclusive growth model incorporating the active role of locals. Development towards an inclusive growth model is in favour of domestic markets, encouraging multiple investors to include local initiatives, and shifting from a tourism-based to an experience-based economy. Three growing curves of endogenous, exogenous and inclusive growth models are presented in Fig. 8.5.

Overall, our research furthers the understanding of the risks and consequences of an international tourism-led growth model. This study exemplifies the notion that when a tourism-led model transforms places to meet international tourists' tastes and preferences, local inhabitants and domestic tourists are alienated. Within a broader context of heritage conservation, the findings corroborate Chapagain's (2013) contention that the beauty of Asian heritage lies in its cultural pluralism and continuity, with people, nature and culture as integral parts of a holistic concept. The 'continuity' and 'pluralism' argument cannot be overlooked as Hội An reclaims vestiges of its cosmopolitan history as an international hub of trade and commerce. Also, questions

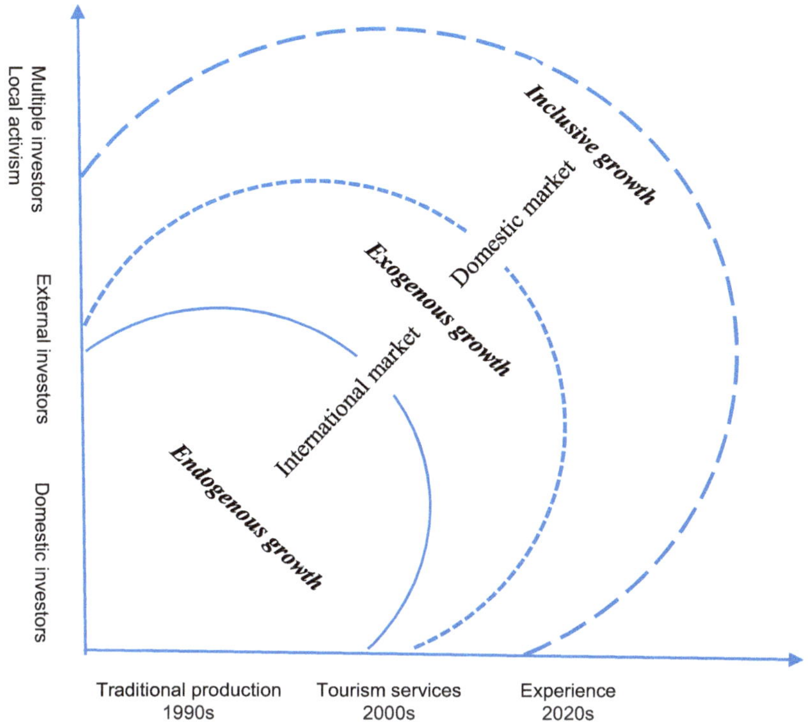

Fig. 8.5. An integrated growth model of Hội An. Compiled by the authors.

remain about revitalizing cultural traditions and history through various shows and performances. On the one hand, it shows diversification and shifting towards experience-based tourism, while on the other hand, it might just decorate the town with trinkets, potentially making it appear even more artificial.

Conclusions

This in-depth case study of Hội An examines the relationship of heritage conservation and tourism development from physical, spatial and social perspectives. The analysis contributes to the current literature by highlighting distinctive Vietnamese practices of heritage conservation, a syncretic approach to reconcile the complexity of the tourism–conservation relationship. This syncretism advocates both uniqueness and universality of the WHS and applies to both tourism and conservation

Practically, the research highlights important issues in managing conservation and tourism at a living WHS. The living heritage approach to site management would encourage the role of the local community and better integrate them into overall economic and social change at the heritage site. Also, researchers have raised critical questions over the perennial planning dilemma of core and buffer zones for tourism development (Jones *et al.*, 2020). Excessive commercialization of the core zone and unregulated development of the buffer zone might offer a warning for the Vietnamese tourism authorities to slow down development. In the meantime, during the COVID-19 pandemic the paucity of tourism offers an ideal thinking time for the city and its administrators to find better ways to sustain the growth of tourism and benefits for locals that have not been proven in the endogenous growth model.

A number of limitations, however, are inherent in this chapter. First, the research employs a single case study, raising questions of representativeness (Yin, 2009). In order to have a broader view of the topic, a series of case studies or cross-cultural comparison to other cases would yield better transferability. Second, a full-scale quantitative survey of local residents and tourists to compare and contrast their perception of the current practice of conservation and tourism development is another option for future research. The limitations, however, provide opportunities to investigate unique concepts of Asian heritage management and the symbiosis of tourism and conservation particular to the value system of the society.

References

Anderson, B. (1991) *Imagined Communities: Reflections on the Origin and Spread of Nationalism*. Verso, London.

Ando, K., Fukukawa, Y. and Tomoda, H. (2015) The formulation process and characteristics of historic conservation system in Hoi An, Vietnam. *Journal of Architecture Planning* 80(708), 379–389. DOI: 10.3130/aija.80.379.

Ashworth, G.J. and Tunbridge, J.E. (2000) *The Tourist-Historic City*. Arnold, London, UK.

Avieli, N. (2015) The rise and fall (?) of Hoi an, a UNESCO world heritage site in Vietnam. *SOJOURN: Journal of Social Issues in Southeast Asia* 30(1), 35–71. DOI: 10.1355/sj30-1b.

Aygen, Z. and Logan, W. (2015) Heritage in the 'Asian century': responding to geopolitical change. In: Logan, W., Carith, M. and Kockel, U. (eds) *A Companion to Heritage Studies*. Wiley Blackwell, Chichester, UK, pp. 410–425.

Bui, H.T. and Jolliffe, L. (2013) Souvenir production and attraction: Vietnam traditional craft village. In: Cave, J., Jolliffe, L. and Baum, T. (eds) *Tourism and Souvenirs: Global Perspectives from the Margins*. Channel View Publications, Bristol, UK, pp. 161–175.

Bui, H.T. and Le, T.-A. (2016) Tourist satisfaction and destination image of Vietnam's Ha Long Bay. *Asia Pacific Journal of Tourism Research* 21(7), 795–810. DOI: 10.1080/10941665.2015.1075564.

Bui, H.T. and Lee, T.J. (2015) Commodification and politicization of heritage: implications for heritage tourism at the Imperial Citadel of Thang Long, Hanoi (Vietnam). *Austrian Journal of South-East Asian Studies* 8(2), 187–202.

Bui, H.T., Le, T.A. and Ngo, P.D. (2018) Managing UNESCO World Heritage in Vietnam: visitor evaluation of heritage mission and management of Trang An Landscape Complex. In: *Managing Asian Destinations*. Springer, Singapore, pp. 89–105.

Bui, H.T., Jones, T., Weaver, D. and Le, A. (2020) The adaptive resilience of living cultural heritage in a tourism destination. *Journal of Sustainable Tourism* 28(7), 1022–1040. DOI: 10.1080/09669582.2020.1717503.

Chapagain, N. (2013) Introduction: contexts and concerns in Asian heritage management. In: Silva, K.D. and Chapagain, N.-K. (eds) *Asian Heritage Management: Context, Concerns, and Prospects*. Routledge, London, pp. 1–30.

Davidson, M. and Lees, L. (2010) New-build gentrification: its histories, trajectories, and critical geographies. *Population, Space and Place* 16(5), 395–411. DOI: 10.1002/psp.584.

Demay, A. (2014) *Tourism and Colonialization in Indochina (1898–1939)*. Newcastle, Cambridge Scholars Publishing.

Di Giovine, M.A. (2009) *The Heritage-scape: UNESCO, World Heritage and Tourism*. Lexington Books, Plymouth, Massachusetts.

Du Cros, H. and Lee, Y.-S. (eds) (2013) *Cultural Heritage Management in China: Preserving the Cities of the Pearl River Delta*. Routledge, London.

Fairclough, N. (2003) *Analysing Discourse: Textual Analysis for Social Research*. Routledge, London.

Gillen, J. (2010) Tourism and entrepreneurialism in Southeast Asian cities. *Geography Compass* 4(4), 370–382. DOI: 10.1111/j.1749-8198.2010.00322.x.

Glover, I. (2003) National and political uses of archaeology in South-East Asia. *Indonesia and the Malay World* 31(89), 16–30. DOI: 10.1080/13639810304439.

Henderson, J.C. (2000) War as a tourist attraction: the case of Vietnam. *International Journal of Tourism Research* 2(4), 269–280. DOI: 10.1002/1522-1970(200007/08)2:4<269::AID-JTR219>3.0.CO;2-A.

Hitchcock, M., King, V.T. and Parnwell, M. (2010) *Heritage Tourism in Southeast Asia*. NIAS Press, Copenhagen.
Hội An Center for Monuments Management and Conservation (2008) *Cultural Tourism and Heritage Management in the World Heritage Site of Ancient Town of Hoi An, Viet Nam*. United Nations Educational, Scientific and Cultural Organization, Bangkok.
Hội An People's Committee (2017) *Plan for Development of Tourist Accommodation Network of Hoi An City period 2017–2020 (Kế hoạch phát triển mạng lưới lưu trú du lịch Thành Phố Hội An giai đoạn 2017–2020)*. Hội An People's Committee, Hội An, Vietnam.
Hội An's World Heritage Portal (2020) Available at: http://www.hoianworldheritage.org.vn/vi/news/Dat-va-nguoi-Hoi-An/bao-ton-va-phat-huy-du-lich-di-san-trong-boi-canh-dich-covid-19-thach-thuc-tao-ra-co-hoi-2032.hwh (accessed 20 August 2020).
Hội An's World Heritage Portal (2021) Available at: http://www.hoianworldheritage.org.vn/vi/news/Du-lich-Hoi-An/hoi-an-show-la-noi-de-du-khach-dien-vien-ke-ve-thuong-cang-xua-2050.hwh (accessed 20 January 2021).
Johnson, M. (2010) Aspiring to the 'tourist gaze': selling the past, longing for the future at the World Heritage Site of Hue, Vietnam. In: Hitchcock, M., King, V. and Parnwell, M. (eds) *Heritage Tourism in Southeast Asia*. NIAS Press, Copenhagen, pp. 173–201.
Jones, T., Bui, H. and Ando, K. (2020) Zoning for World Heritage Sites: dual dilemmas in development and demographics. *Tourism Geographies* (Advanced online). Available at: https://doi.org/10.1080/14616688.2020.1780631
King, V.T. and Parnwell, M.J. (2011) World Heritage Sites and domestic tourism in Thailand: social change and management implications. *South East Asia Research* 19(3), 381–420.
Logan, W. (2002) *The Disappearing 'Asian' City : Protecting Asia's Urban Heritage in a Globalising World*. Oxford University Press, Oxford, UK.
Long, C. and Sweet, J. (2006) Globalization, nationalism and world heritage: interpreting Luang Prabang. *South East Asia Research* 14(3), 445–469.
Lowenthal, D. (2015) *The Past is a Foreign Country – Revisited*. Cambridge University Press, New York.
Parnwell, M. (2016) Heritage management and tourism development in Hoi An, Vietnam. In: King, V. (ed.) *UNESCO in Southeast Asia: World Heritage Sites in Comparative Perspective*. NIAS Press, Copenhagen.
Poulios, I. (2014) Discussing strategy in heritage conservation: living heritage approach as an example of strategic innovation. *Journal of Cultural Heritage Management and Sustainable Development* 4(1), 16–34.
Suntikul, W., Butler, R. and Airey, D. (2010) Vietnam's heritage attractions in transition. In: Hitchcock, M., King, V. and Parnwell, M. (eds) *Heritage Tourism in Southeast Asia*. NIAS Press, Copenhagen, pp. 202–220.
Taylor, K. (2012) Heritage challenges in Asian urban cultural landscape settings. In: Daly, P. and Winter, T. (eds) *Routledge Handbook of Heritage in Asia*. Routledge, New York, pp. 266–280.
The New York Times (2021) Interview republished on *VnExpress*, March. Available at: https://vnexpress.net/mot-nam-the-gioi-khong-co-du-lich-4247489.html (accessed 20 March 2021).
Timothy, D.J. and Boyd, S.W. (2006) Heritage tourism in the 21st century: valued traditions and new perspectives. *Journal of Heritage Tourism* 1(1), 1–16. DOI: 10.1080/17438730608668462.
Trinh, T.T., Ryan, C. and Cave, J. (2014) Souvenir sellers and perceptions of authenticity – the retailers of Hội An, Vietnam. *Tourism Management* 45(2), 275–283. DOI: 10.1016/j.tourman.2014.05.005.
Wijesuriya, G. (2001) Protection of sacred mountain – towards a new paradigm in conservation. In: *Report of the Thematic Expert Meeting on Asia-Pacific Sacred Mountains, Wakayama, Japan, 5–10 September 2001*. UNESCO, World Heritage Committee, Tokyo, pp. 47–62.
Winter, T. (2014) Beyond Eurocentrism? Heritage conservation and the politics of difference. *International Journal of Heritage Studies* 20(2), 123–137. DOI: 10.1080/13527258.2012.736403.
Winter, T. and Daly, P. (2012) Heritage in Asia: converging forces, conflicting values. In: Daly, P. and Winter, T. (eds) *Routledge Handbook of Heritage in Asia*. Routledge, New York, pp. 1–35.
Yin, R.K. (2009) *Case Study Research: Design and Method*, 5th edn. Sage Publications, London.

9 Dark Tourism in the Former Demilitarized Zone (DMZ)

Phuong M. Ngo[1], Huong T. Bui[2]* and Alexandru Dimache[2]
[1]*Investment, Trade and Tourism Promotion Agency, Quảng Trị, Vietnam;*
[2]*Ritsumeikan Asia Pacific University (APU), Japan*

Introduction

Research on dark tourism has proliferated in the past two decades. The production of dark tourism is not only driven by the supply side in the broader political and cultural climate (Seaton, 1999; Stone, 2013) but also by the demand side through consumer preferences (Stone and Sharpley, 2008). The majority of the studies of dark tourism at war sites, however, look at places in Europe, Australia, the Pacific and the USA (Light, 2017). Such sites are associated with World War I (WWI), World War II (WWII) or the American Civil War. This approach, however, has its limitation as those who directly experienced these wars are now mainly lost due to the passage of time. In contrast, the war in Vietnam took place in Asia in the latter half of the 20th century, so the memories of that war still live among the soldiers and citizens. The Vietnam War is one of the most controversial wars of the 20th century, thus its sites provide an opportunity for studying the battlefield tourism. Research on the war in Vietnam within a tourism context has been of increasing interest (Henderson, 2000; 2007; Schwenkel, 2006; Lema and Agrusa, 2013; Gillen, 2014; 2018; Upton *et al.*, 2018). However, the viewpoint of the war from the Vietnamese perspective has been rather limited in its availability in English literature (Le, 2014; Ngo and Bui, 2019).

The current study was conducted in the former Demilitarized Zone (DMZ). The DMZ area in Quảng Trị has long been considered a sacred site for domestic visitors. Examination of the structure of dark tourism industry, issues of governance, promotion and production of dark tourism as well as the involvement of tourists and tour guides in historic sites of Quảng Trị further our understanding on the form of dark tourism in post-war Vietnam.

Literature Review

Conceptualization of dark tourism

Dark tourism is defined as 'travel to a location wholly, or partially, motivated by the desire for actual or symbolic encounters with death' (Seaton, 1996, p. 240). Dark tourism has been used interchangeably with war/battlefield tourism and thanatourism. Originating from the Greek word *Thanatos*, thanatourism conveys the meaning of the personalization of death, as derived from thanatopsis – death-related tourist activity (Seaton, 1996). In common, thanatourism and dark tourism 'encounter with

*Corresponding author: huongbui@apu.ac.jp

 Vietnam Tourism: Policies and Practices (eds Huong T. Bui *et al.*)
DOI: 10.1079/9781789242782.0009

remembrance with death, rather than death itself' (Seaton, 2018, p. 14). Alternatively, war/battlefield tourism (Smith, 1998) takes place at the sites where conflicts occurred but also war-related places such as cemeteries, monuments, museums and other infrastructure. Travel to the battlefield is 'collectively embraced by the distinction between battlefield tourism and pilgrimage' (Baldwin and Sharpley, 2009, p. 189). Battlefields have been used as venues to convey the official narrative of national history through a cultural process embedded in 'site sacralisation' (Miles, 2012, p. 8). Owing to the official historical narrative at battlefield sites, interpretations presented at the sites do not only deal with psychological reactions, but also carry 'multiple truths' shaped by personal stories and social ideologies for interpretation of the past, present and future (Ryan, 2007, p. 4). Encountering differences in regional, social, cultural and political background among tourists, the 'art of interpretation' (Miles, 2012, p. 1) is manifested in many war-related sites, and has been polarized into 'hot and cold', 'poly-vocal and uni-vocal' and 'complex and simple', among others (Ryan, 2007, p. 5). Accordingly, conflict sites are open for commemoration, entertainment, education and pilgrimage (Dunkley *et al.*, 2007).

Theorizing the relationship between demand and supply in dark tourism, Sharpley (2009b) proposes that visitor engagement at the dark site ranges from the *palest* (i.e. consumption) to the *darkest* that 'seek to integrate themselves with death, either through witnessing violent or untimely deaths or, in the extreme perhaps, travelling in the knowledge or expectation of death' (Sharpley, 2009b, p. 18). On the supply side, there are two different forms of consumption: *accidental* supply and *purposeful* supply, which link the site to the visitors' fascination with death. Another profound theory on the shade of darkness is the dark tourism spectrum (Stone, 2006) that classifies visitation to sites of death and suffering with six levels of darkness and 'seven dark suppliers' based on this darkness scale: (i) dark-fun factories; (ii) dark exhibitions; (iii) dark dungeons; (iv) dark resting places; (v) dark shrines; (vi) dark conflict sites; and (vii) dark camps of genocide (p. 152).

In terms of politics, Light (2017) contends that dark tourism studies offer a different significance limited to 'places of death and suffering by different "users" and has highlighted issues of tension or dissonance among those users' (p. 284). Nationalism is another argument that emerges from visitation to the sites of death and suffering. Travelling to dark sites generates a chance to negotiate senses of nationhood (Seaton, 1999; Light, 2017). Moreover, most of the war heritage sites in Western countries are related to the legacy left by WWI and WWII, which are distant from the present. Consequently, significant historical events and places have significant political roles as places of national memory (Winter, 2009; Dunkley *et al.*, 2011; Stone, 2012). Besides, the experience created by dark tourism products seems to be less about 'living memories' as the generations who used to be involved in those events have mostly passed away.

Major criticism of the dark tourism conceptualization centres on the limitations of shades of darkness (Sharpley, 2009b), based on the level of visitors' interest in the dark sites as the basis for classification. For instance, visitors who travel to a particular dark site repeatedly due to different values and attachments; their trips are not determined by physical artefacts at the destination, but by their beliefs and emotions. It is worthwhile to point out that Stone's scale (2006) only captures a dualistic conceptualization – a dark site either has an educational or an entertainment orientation. The problem is that when the function of the dark site could be located in between these two poles, such as places for remembrance and mourning, the dualism of the education-entertainment poles fails to accommodate the complex features and multidimensional conceptions of the war-related sites.

Dark tourism: supply and demand

Dark tourism is driven by specific supply and demand (Farmaki, 2013). While the study of demand for dark tourism relates to tourists' motivation to visit the dark sites (Dunkley *et al.*, 2011; Podoshen, 2013), the supply side is defined as 'commodities' (Smith, 1998, p. 183) manifested by a wide range of political factors governing attractions. The notion of demand for dark tourism is multifaceted. Tourists might seek *spiritual motivation* (Ashworth and Hartmann, 2005; Cohen, 2011), *educational motivation* (Cohen, 2008, pp. 8–9), or *pleasure motivation* (Chen and Tsai, 2019).

The supply side in dark tourism involves various stakeholders such as governments, tourists, host communities, tourism businesses and other sectors. The human factor is a major element in supplying the dark tourism experience, as 'the key to success in managing dark tourism attractions is to engage the staff in storytelling' (Wiltshier, 2016, p. 44); thus, interactions between visitors and staff at the site 'determines relevance and eventually satisfaction to all engaged in the supply chain' (p. 53). Therefore, the supply-side managerial unit or tour operators should 'encourage an understanding of the moral and ethical values attached to a place in order to allow it to shape the minds of the tourist rather than the tourist shaping the identity of a place' (Johnston *et al.*, 2016, p. 165). The supply-side actors play an important role in the delivery of dark experiences and in connection from customer-demand to attraction-supply.

An Asian context of dark tourism

Dark tourism has been 'primarily a Western phenomenon' (Stone, 2006, p. 149). In the Western-centric viewpoint, the tourist is motivated by 'the contemplation of their own mortality' (Cohen, 2018, p. 169). The concept of dark tourism and thanatourism have sometimes been used uncritically in non-Western contexts (Light, 2017). The Western-centric perspective of dark tourism has been challenged as the number of studies on dark tourism/thanatourism in non-Western regions has increased significantly. It is important to recognize that the relationships between societies and their dead take very different forms in Asian society. This means that the use of Western frameworks for understanding the tourism–death relationship in other parts of the world may not be applicable (Lee *et al.*, 2012; Yoshida *et al.*, 2016). In Asian countries, the existence of dark tourism/thanatourism seems to manifest in various ways. There is a need for alternative ways of theorizing tourism–death relationships by scholars working in non-Western contexts. A number of researchers in Asia have begun to address this challenge (Zhang *et al.*, 2016; Zheng *et al.*, 2017; Weaver *et al.*, 2018).

Dark tourism in Vietnam

There are several alternative concepts that have been associated with dark tourism in Vietnamese, such as '*Du lịch hoài niệm*' – nostalgia tourism, '*Du lịch chiến trường xưa*' – battlefield tourism, and '*Du lịch tâm linh*' – spiritual tourism. Meanwhile, dark tourism in Vietnam is examined by international researchers through the different lenses of: (i) ideology and politics (Alneng, 2002); (ii) multifaceted practices of memory and knowledge production (Schwenkel, 2006); (iii) the visitor experience at battlefields (Upton *et al.*, 2018); (iv) migrant experiences of war sites (Le, 2014); (v) landscapes of war (Lema and Agrusa, 2013); and (vi) wartime attractions (Henderson, 2000; 2007; Gillen, 2018). In particular, the 'spiritual process' in the Vietnamese tradition 'broke with state ideologies of collective moral heroism to also recognize individual loss and grief, as well as the important role that cultural beliefs and values play in commemorating and communing [with] the war dead' (Schwenkel, 2013, p. 110). In the interpretation of war sites in Vietnam, tour guides play a significant role in conveying meaningful messages to tourists by enhancing their cognitive and affective access to historical sites (Ngo and Bui, 2019). War heritage or war-related sites widely remain a highly political issue (Henderson, 2000). With many battlefield sites left over from various wars in the 20th century, Vietnam provides an ideal setting for the study of dark tourism. However, research on war-related sites and dark tourism in Vietnam is very limited.

Study Site and Methods

The site

Located in Central Vietnam, Quảng Trị is well known for its DMZ tour. The Bến Hải River became the temporary border dividing the country following the Geneva Accord in 1954; thereafter Quảng Trị functioned as a DMZ positioned on the frontlines of North and South Vietnam (see Fig. 9.1). The DMZ was at the frontline of two distinct ideologies; the battlefields of armies in the past are places of contested memories, and eventually become places of dissonant touristic interpretations in the present. The landscape, artefacts and remnants that remain at the sites evoke memories, imbue the historical narratives and transform into touristic narratives in the present. 'The past is as explosive as any of the remnants buried in this land' (Nguyen, 2016, p. 46). Close to 50 years after the ending of the Vietnam War, those first generations who directly participated in this struggle and their

Fig. 9.1. The location of Quảng Trị in Vietnam. Map courtesy of Thuy T. Duong.

authentic experiences still exist; they can still tell their stories, in other words, the 'living experience' of the 'first-hand generation', which contributed invaluable sources of data that make the study of battlefield tourism in the DMZ both distinctive and unique.

The methods

First, the researchers conducted various trips to historic sites for observation in 2019 by participating in activities, rituals and interactions with locals and tourists. Participant observation is a strategy where the researcher directly involves themselves in a given social group for a prolonged period, creating a personal relationship with its members, describing their actions and motivations through a process of identification (Corbetta, 2003). The participant observer serves two purposes: (i) to engage in activities appropriate to the situation; and (ii) to observe the activities, people and physical aspects of the situation.

Second, semi-structured interviews were employed during the research phase because this type of method permitted the collection of a set of data on research-related topics, while also allowing a certain degree of flexibility for understanding participant behaviour. The researchers employed a purposeful sampling technique to capture a range of specific populations. The selection of samples for this study basically reflects the relationship among stakeholders who are involved in tourism products and tourism activities at the research sites. The questions for the interview were designed to capture several objectives of both the public and private sectors. The interviewees were visitors, tour operators, tour guides and government officers. Interview questions covered information about the site, tour routes, tour guides, narratives and interpretation at the site. For tour guides and tour operators, the researchers inquired about product packaging. The researchers also aimed attention at attraction-supply or government officers who provided information regarding tourism policy and strategy. Tour guides were asked to provide information with regards to the kind of tourists they usually meet, why tourists visit the site, and what type of questions or comments tourists often ask or say when visiting the site. The researchers conducted 17 interviews lasting about 15–30 minutes in early April 2019. Interview recordings were transcribed and analysed along with observations and field notes. In addition to informal interviews, the researchers also conducted several informal conversations to collect a variety of information resources necessary for the research. Profiles of the interviewees are shown in Table 9.1, including six visitors, five on-site guides, three tour guides from tourism agencies, two tour operators and one government officer.

Based on transcription and translation, the data was reduced to descriptive themes. The purpose is to highlight the most relevant and meaningful phrases or passages of text. These coding phrases were collected according to themes of what the informants mentioned regarding the dark sites. The researchers analysed discourses of groups of tour guides, tourists, and tour operators separately. Then, the themes that emerged from various groups were collected, linked and formulated into dominant themes that emerged from several sides, which centred on the customer-demand and attraction-supply of dark tourism.

Findings

Tourism policies

The Department of Culture, Sports and Tourism of Quảng Trị is a government agency under the provincial People's Committee, and is in charge of governing and managing tourism in the province. The department is under the jurisdiction of the Ministry of Culture, Sports and Tourism and the Vietnam National Administration of Tourism. From the administrative viewpoint and policies, war-related heritage sites are the key elements in the overall strategic tourism development plan.

> Towards 2025, tourism will become a key economic sector, accounting for over 10% of GDP. Quảng Trị tourism is branded with four major themes 'War memories – The aspiration for peace', 'The gateway to the East-West Economic Corridor', 'The Road of Heritage Sites in Central Vietnam', and 'The Legendary Hồ Chí Minh Trail'. With its special geographical position and unique tourism resources, it is an

Table 9.1. Profiles of the interviewees. Compiled by the authors.

Respondents	Gender	Occupation	Residence
Tourist 1	Female	Student	Phú Yên
Tourist 2	Male	Veteran	Hà Nội
Tourist 3	Female	Unknown	Quảng Bình
Tourist 4	Male	Officer	Thanh Hóa
Tourist 5	Male	Veteran	Hà Nam
Tourist 6	Female	Unknown	Quảng Bình
On-site guide 1	Female	Tour guide at Vĩnh Mốc Historical Relic	Quảng Trị
On-site guide 2	Male	Tour guide at Vĩnh Mốc Historical Relic	Quảng Trị
On-site guide 3	Female	Tour guide at the Quảng Trị Ancient Citadel	Quảng Trị
On-site guide 4	Male	Tour guide at the Quảng Trị Ancient Citadel	Quảng Trị
On-site guide 5	Male	Tour guide at Hiền Lương Historical Relic	Quảng Trị
Tour guide 1	Male	Independent tour guide	Huế
Tour guide 2	Male	Tour guide	Hà Nội
Tour guide 3	Female	Independent tour guide	Quảng Bình
Tour operator 1	Male	Director	Quảng Trị
Tour operator 2	Male	Director	Quảng Bình
Government officer 1	Female	Tourism Division of Department of Culture, Sports and Tourism of Quảng Tri	Quảng Trị

> important advantage for Quảng Trị's tourism economic development.
>
> (Government officer 1)

Within this strategy, priority for investment in tourism development has been attributed to the DMZ:

> The DMZ tour is a top priority and leaves a strong impression on tourists, especially veterans who used to fight in Vietnam. Revenue from tourism is mainly due to this source of visitors.
>
> (Government officer 1)

However, the policy has proved inadequate in utilizing and promoting the available advantages of dark sites. The gap between policy proposed by governmental planning and expectations from private tourism enterprises has yet to be closed by effective implementation.

> Governmental support for businesses is basically little to none. The latter independently operate and market themselves ... I have directly sent my inquiry about entrance tickets to historical sites to the Executive Management of Natural-Historical Landscape Conservation Center but the only response has been: 'To be reconsidered later' ... Governmental authorities have been fulfilling their responsibilities. However, there has not been a more proactive environment, more progress, or any project to encourage firms. The promotion of the tourism industry is still a paper-based plan.
>
> (Tour operator 1)

War heritage sites are providing great opportunities to attract tourists thanks to their cultural and historical values. However, in recent years, visits by veterans to the sites have declined because they are getting old. Another key reason is that investment into the dark heritage sites has been stopped, and there is a lack of specific design or clear concepts for the DMZ tours.

Current state of the dark sites

Quảng Trị has an important economic and geopolitical location on the national transport routes including the highway, the north–south railway and the seaway to the east. Major war heritage sites are located close to the main road.

For example, the Hiền Lương Bridge on National Highway 1A connecting the north and the south, is easily accessible to visitors. The dark sites in Quảng Trị are located throughout the province with a high density so that it is convenient to move from place to place:

> This site is quite close to Quảng Bình. My guests are also quite old, and want to learn about history and culture rather than entertainment activities.
>
> (Tourist 6)

Informal talks with tour guides revealed the fact that most accommodation in the province is small scale (with fewer than 50 rooms) and belong to private small- and medium-sized enterprises. The accommodation quality is generally poor and only meets the minimum requirements of tourists.

> The touring area is not too large; there was no problem with food and accommodation. Hotels have recently been renovated; there are even three- and four-star hotels now. The food meets the tourists' appetites. Although the services are not as luxurious and high quality as those of other places, they meet our customers' basic demands.
>
> (Tour operator 1)

Facilities in some old hotels are out of date and need upgrading. In addition to accommodation, the researchers recorded several informants' comments about other services and facilities, for instance, in Vĩnh Mốc Tunnels:

> I hope this site will attract more tourists and the facilities and services for food, beverages and accommodation will be upgraded. After visiting Vĩnh Mốc Tunnels, there should be more interesting sites to attract the tourists to stay in Quảng Trị. Currently, the quality of service is too poor. If it is improved, it can keep guests staying for more days.
>
> (On-site tour guide 1)

Indeed, the limitations of the services and facilities, on the one hand, limited visitors' expenditures, and on the other hand, limited the duration of their stays at the war heritage sites. This is a reason why visitors only spend 1 day for DMZ tours. Some visitors consider Quảng Trị as a 'transit' spot rather than a 'destination' as showed in tour guides' comments:

> [Visitors] still stay in Quảng Trị but little to none.
>
> (On-site tour guide 3)
>
> Rarely do [visitors] remain in Quảng Trị afterwards.
>
> (On-site tour guide 4)

Tour operators have also noticed many shortcomings during their operations:

> A standardized tour includes the following services: meals, accommodation, and destination, transportation, which are still limited. Currently the Quảng Trị Department of Industry and Trade is promoting several agricultural goods, which shows certain efforts, though not much.
>
> (Tour operator 1)

Although most war-related sites are managed by the local government itself, the renovation and renewal of a part or the whole of a monument requires serious scientific and historical research and cannot be carried out by individuals or organizations. This process requires the coordination of several units, across both private and public sectors. Therefore, the development of tourism, especially for cultural and historical sites, needs to be carefully researched to meet the changing needs of tourists while still retaining the authentic value of the sites.

Dark tours' itineraries

The itinerary for visitation of war-related sites in Quảng Trị is strikingly different for international and domestic groups. The international and domestic visitors choose different ways to book tours based on the level of tour information and convenience:

> International tourists often book through a tour company at their location. Then, these companies forward the booking to ground-handling agents for the necessary preparation. Domestic tourists either choose a tour provided by travel companies or they travel on their own and then buy a tour at a travel agent in Quảng Bình.
>
> (Tour guide 3)

For the daily tours, the international visitors often depart from a major hub of international tourism in the centre of Vietnam, Huế city. A day tour visiting the DMZ includes sites such as Khe Sanh Base, the Rockpile, the McNamara Line, Cồn Tiên Base and Dốc Miếu Base. Eventually, it has become a standardized tour offered by many travel agencies. A 'DMZ tour' designed for the international tourist is described as follows:

> Along Highway 9, the bus traveled to Camp Carroll ... and then to the 'Rockpile', a 230-meter-high limestone formation that served as a firebase and observation point ... The next destination was an impoverished 'Montagnard' community of the Bru-Vân Kiều people ... The last stop on Highway 9, close to the international border with Laos, was the Khe Sanh Combat Base, site of the infamous seventy-seven-day siege by a PAVN force in 1968. After lunch ... tourists made brief stops at the 'skeleton cathedral' (the ruins of a bombed Catholic church) and the Hiền Lương Bridge, which spans the Bến Hải River and marked the 'former border' at the Seventeenth Parallel. Tourists then briefly traveled to 'the other side', to visit the Vịnh Mốc tunnels, where they edged their way slowly through the narrow underground pathways built by villagers to escape US bombing.
>
> (Schwenkel, 2013, pp. 119–120)

The DMZ tour has expanded to include new historical sites outside the normal tour routes including many sites of memory tailored to domestic tourist demands. For veterans or elders, the DMZ tour is usually packaged to include a number of historical destinations. Veterans return to Quảng Trị to recall old memories and to carry out rituals for the fallen soldiers. Their trips are very special, including visits to former battlefields, cemeteries and other sites of personal memories.

> My group consists of 40 people; we are Trường Sơn veterans visiting Quảng Trị. Our tour includes Đồng Lộc Junction, Yến Island – the Grave of General Võ Nguyên Giáp, Highway 9 Martyrs' Cemetery, Trường Sơn Martyrs' Cemetery, [and] Quảng Trị Ancient Citadel. Tomorrow, we will visit Hiền Lương Bridge.
>
> (Tourist 2)

Visitors' experiences at dark sites

From 2005 to 2017, the total number of visitors to Quảng Trị war sites reached more than 9,000,000 arrivals. Travel agencies in Quảng Trị organized for 4.5 million visitors with social revenue of VND 2140 billion; The Centre for Natural-Historical Landscape Conservation welcomed and served 1.8 million visitors, and the revenue from ticket sales and donation sources reached nearly VND 35 billion (Quảng Trị Department of Culture, Sports and Tourism, 2018).

Vietnamese veterans visit former battlefields through regular programmes delivered by the Veterans Association, such as the 'Memories of the Old Battlefields and Comrades' (*Hoài niệm về chiến trường xưa và đồng đội*), and the 'Legendary Road' (*Con đường huyền thoại*) on the occasions of national commemorative events. Visitors to Quảng Trị war heritage sites often perform a kind of ritual practice and pray for those who sacrificed their lives. On-site tour guides at the Quảng Trị Ancient Citadel noted various emotional moments when veterans and their relatives visited the dark site:

> They come here to visit their old battlefields. Someone who used to fight here, someone who used to live in this tunnel. Many people come here to find their memories and tell their stories of the past.
>
> (On-site guide 1)

Several visitors expressed their emotions when coming to this war heritage site. It is worth mentioning that they do not only come once or twice, but repeat that trip to the same location many times:

> We occasionally organize [a trip] back to the old battlefield tour if guests request it. The veterans visit these sites every year. They mainly come here to honour and perform memorial rituals to their comrades.
>
> (Tour guide 2)

When being asked why the Vĩnh Mốc Tunnels, and Hiền Lương Bridge are indispensable destinations in the tour itinerary, a tour guide from Huế said, 'these are typical and significant places in the Vietnam War' (Tour guide 1). Indeed, the historical sites in Quảng Trị are iconic destinations for many visitors. It is as if coming to Quảng Trị without visiting such a war site, would be an omission and source of regret. Visitors to Quảng Trị have a variety of motivations, which will be clarified in the next section of this study. However, there is a characteristic that numerous tourists mention: the 'sacredness' of the war heritage:

Fig. 9.2. Tà Cơn airfield. Photo courtesy of IPA Quảng Trị.

> The tourists are highly impressed by the Citadel for its 'sacredness' and the historical meaning associated with the 81-day battle.
> (On-site tour guide 4)

On-site tour guides at war heritage sites emphasize that international tourists often watch documentary films, see the artefact and narrative in the museum, visit the remnants of the war (i.e. aircraft corpses at Tà Cơn (see Fig. 9.2) and underground tunnels in Vĩnh Mốc), and listen to tour guides' interpretation. For domestic visitors, the time spent on experiencing the relics is often shortened or skipped, as most will practise a worshipping ritual for the deceased. This marks a distinctive feature compared with the usual forms of tourism.

> Veteran groups come back to visit the old battlefield every year to honour their comrades with their deep gratitude.
> (On-site guide 2)

Domestic and international visitors prefer to visit different sites as their interest in the dark sites is completely different. Even visiting the same place, they had different activities and behaviours:

> International tourists don't perform any ritual at the memorial sites but come to the museum for the tour guide's interpretation of Quảng Trị Citadel's history and the 81-day long battle.
> (On-site guide 4)

> [Domestic tourists] consistently ask about the history of the 81-day battle. The tourists pay attention to the meaning behind the common grave/Memorial; the location of the 81-day battle and the value of Thạch Hãn River.
> (On-site guide 4)

> [For American tourists] [t]he most frequent questions being asked is why American forces wanted to destroy Vĩnh Linh? How was life in the tunnel? Where did they find food? How did they light up the tunnel? What are Vietnamese attitudes towards Americans? Do Vietnamese hate Americans?
> (On-site guide 2)

In between demand and supply sides: tour guides' roles

Domestic and international tourists also hold very different viewpoints of what happened at the dark sites, given their background. While tour guides are seldom involved in political debate regarding historical events, on-site guides often follow the official narrative.

> The DMZ tour is usually chosen for learning about history ... Tourists have often learned about these sites before arriving. Additional information was collected through the [on-site] guide's interpretation at the site. In my personal experience, I seldom express my political viewpoints as tourists might have a very different view on the related events I mention. I let the on-site guide explain the national narrative of the events as they are authorized to do so.
>
> (Tour guide 1)

Interpretation of historical events presents one of the greatest challenges for tour guides who are without local and knowledgeable skills. The symbolic meaning can only be conveyed by a skilful guide. It is easy to be omitted from the tour itinerary if a tour guide is short on site information and personal experience. For example, being one of the most important sites for pilgrimage, Quảng Trị Ancient Citadel (see Fig. 9.3) is sometimes skipped because its symbolic meaning and narrative are not easy to present, especially when the tour guide is not experienced enough.

> Quảng Trị Ancient Citadel is officially included in the package tour, but is often omitted from the itinerary for international travellers. Firstly, it is not easy to interpret as it requires deep and specialized historical knowledge. Secondly, the Citadel has a high symbolic meaning with very few tangible exhibits to show visitors. Apart from the common grave and museum, it is difficult to explain historical events that occurred [there] in the past.
>
> (Tour guide 2)

The political system cannot be separated from education in Vietnam. Political education is embedded in the official narrative at historic sites, and the on-site tour guides become teachers. Interaction with history at physical settings (the site) and symbolic interpretation (by the guides) engages students and makes them more excited in historical lessons. Historical sites and events are live learning experiences and materials. The young students have visited Quảng Trị to understand the wartime events that happened in the past.

Fig. 9.3. Quảng Trị Citadel. Photo courtesy of IPA Quảng Trị.

> Students are from universities nationwide, mainly from Hà Nội and Hồ Chí Minh city, from the departments of history and tourism. There are also students from the Đà Nẵng University of Science and Technology. Local school and college students visit the site for the educational program on the occasions of Youth Day (March 26th) and National Reunification Day (April 30th).
>
> (On-site guide 1)

The educational programme targets young visitors, to nurture a sense of pride and appreciation among the young generations, from the preschool and kindergarten level. Owing to the fact that the entire province is covered with many historical sites, the tradition of paying tribute to national martyrs in war-related sites has mainly been carried out frequently by school children from primary to high school. Children from local schools also participate in various national commemorative events. These activities raise the sense of national pride and responsibility for future generations. Even tourists can recognize the educational functions of the sites as revealed in the interviews:

> Coming here, you will feel both the national pain and pride very clearly that become the best place for young generation education.
>
> (Tourist 6)

Several veterans come to Quảng Trị to visit the former battlefield and their comrades with different moods. Some of them stand sadly on this heroic land to remember a past time of loss and sacrifice. Other veterans want to talk about their feelings, the meaning of life and their aspirations. International tourists are among those who visit old battlefields, carrying with them a different understanding of history, and consequently different motivations and feelings. Therefore, the on-site tour guide plays a peacemaker role, acting as a mediator to connect the past to the present and relieving the physical and mental wounds caused by the war.

Various shortcomings related to human resources at dark sites in Quảng Trị were identified from interviews with tour guides. For instance, on-site guides have good professional knowledge but lack some professional skills to serve tourists.

> I found the guide's language and knowledge is fine, yet their style is not as professional as in Quảng Bình. For example, the day I went to Hiền Lương, no guide appeared there so that I had to find them. The off-peak period might cause this problem.
>
> (Tour guide 3)

In particular, the number of English-speaking tour guides is limited. This restriction in supply which has not been able to meet the demand of international visitors as of yet, affects both English and non-English speaking tourists.

> In fact, our agency has some training courses for the employees. However, in addition to specialized knowledge, foreign language is a major obstacle. Most employees do not study foreign languages. Although some foreign language classes are held, only [a] low level of English is offered. Some tourists require French [or] Italian speaking guides, which is impossible to be met.
>
> (On-site guide 1)

In summary, the analysis provides an overview of the supply-side factors in dark tourism. The DMZ tour in Quảng Trị is associated with the Vietnam War and the reunification of the nation. Tourism products are still poor; the infrastructure is underdeveloped; the quality of tourism services has not been focused on improvement; and policies to support tourism development are not very effective.

Discussion

War heritage is 'where history and heritage are contested between different groups of stakeholders' (Timothy and Boyd, 2006, p. 3), and it is where there is 'a political construction of what is remembered' (Richter, 1989, p. 109). In Vietnam, the war heritage sites 'remain a highly political issue with the influence of government policy' (Henderson, 2000, p. 276). Findings from the current study on political embeddedness in the war heritage sites challenge the notion of dark tourism typologies, proposed by Sharpley (2005), particularly the notion of 'fascination with death' on the demand side and supply in correlation to two poles, 'accidently' and 'purposefully' (p. 225) created attractions. War sites in Vietnam are packaged for political purposes, therefore, the argument that 'battlefields that have, frequently, become tourist attractions by accident' (pp. 224–225) does not hold true in this case. Also, conventional motivations such as 'entertainment' (Sharpley, 2009a, p. 14) are largely missing from the empirical evidence shown in this study. Owing to the fact that the Vietnam War is still quite a recent major historical event of the 20th century, many who were directly involved in the war are still alive and continue to tell stories of the war to their next generations. Therefore, both primary and secondary memories of the war come into existence. This special feature enables the visitor experience to be fueled by their living memories (both primary and secondary), which has an influence on the way visitors perceive and connect to 'death' meaningfully, not simply as 'death of others' with little connection. Death in the case of the Vietnam War is the death of friends and relatives, of comrades who fought together. Thus remembrance and spiritual ceremonies dedicated to the dead have much deeper meaning, and connect generations, where remembrance of the past is to pray for the present and the future.

In the context of dark tourism, on-site tour guides create a very valuable performance, not simply through storytelling and meaning-making, but by creating 'emotions' that 'colour our memories and experience' (Uzzell and Ballantyne, 1998, p. 152). As a peace ambassador at the dark sites, on-site tour guides are important links to connect the present and the past, between visitors and the site, helping those who seem to be enemies in the past to reconcile in the present. The on-site tour guide is an important link in conveying the sense of nationalism and desire for happiness and peace at the war heritage sites. Although interpretation of battlefield sites has proved to be important (Miles, 2012; Hooper and Lennon, 2017), very little attention is paid to tour guide contribution in mediated interpretation.

Variation of dark tourism is contextualized to a particular location which presents its ethics, politics, religion, society, culture, history, geography, demography and psychology. In her writing about living memory of the war in the book *The Country of Memory: Remaking the Past in Late Socialist Vietnam*, the historian Hồ Tài, H.-T (2001) mentioned various ways by which the living remember the dead on national, community and familial scales. The dead join the altar with family ancestors. Vietnamese believe in two worlds of a person: (i) the living moment with the body as a tangible world; and (ii) after-death life with the soul as an intangible world. Taking care of this type of spiritual life is related to spiritual and secular practice rather than religion. In other words, there is an interreligious encounter among gods, heroes and ancestors in Vietnam (Tran, 2018). The tradition shows the love for the ancestors and the country and the younger generation are educated about such Vietnamese customs through years of: '*Uống nước nhớ nguồn*' (drinking water, thinking of its source).

The findings of the study draw important implications for tourism product development and managerial practices in dark tourism in the former DMZ in Quảng Trị. First, dark tourism in Quảng Trị needs a system of sufficient infrastructure and facilities to attract visitors, thereby extending the time that visitors stay. With currently limited capital investment, the province can choose a few 'focal points' for investment and renovation. Second, improving service quality and developing associated products requires cooperation among stakeholders: investors, suppliers, tour operators, tour agencies and government. Third, the human factor is the most important factor contributing to the quality of tourism products. It is necessary to improve the qualifications of tour guides and staff at historical relics. Tour guides and staff should be trained frequently in both professional skills and language skills. Training can be conducted in place or linked to major training centres throughout the country. In addition, it is necessary to strengthen the capacity of management staff by attracting highly qualified human resources and by transferring technological policies.

Conclusions

The legacy of the Vietnam War means the central region of Vietnam is known for dark tourism. Among former battlefields of the Vietnam War, Quảng Trị province and the former DMZ are well known for fierce battles. The current study emphasizes a unique perspective, namely Vietnamese interpretation of the heritage of the Vietnam War, where on-site guides perform multiple political and educational functions as well as mediating roles between the past and the present while also being ambassadors for peace. It is argued that the motivation, perception and experience of visitation to the dark sites is more complicated than the conventional dualistic dark–light spectrum. The findings of the study provide insights into understanding the structure and characteristics of dark tourism in the Asian context. The study highlights the gap between demand and supply and implications for dark-tourism product development in Quảng Trị. The research also advances our understanding of the dark tourism experience in general and in the context of Vietnam, where understanding the war from a Vietnamese perspective has been neglected in the tourism literature.

The outcomes of the study have a number of limitations that serve as a starting point for future research. First, the number of informants in the interview phase is limited and only captured domestic visitors. Expanding the interviews to international visitors and other institutions under the various viewpoints (i.e. international tour operators, international travel agencies, tourism managers, government officials) will contribute to enhancing the understanding of dark sites. In addition, the study analyses the data from a Vietnamese perspective and quite deeply focuses on domestic tourists. A comparative approach of the balance between domestic and international visitors is recommended. Second, analysis from qualitative data collection cannot forecast for the future, yet predicting the tourism market is necessary for practical applications related to tourism product development. Market consumer research based on a quantitative approach is suggested for better measurement of tourist demand in both domestic and international segments.

Acknowledgements

The work presented in this chapter was supported by Japan Society for the Promotion of Sciences (JSPS) research grant number 18H03462.

References

Alneng, V. (2002) 'What the fuck is a Vietnam?': touristic phantasms and the pop colonization of (the) Vietnam (War). *Critique of Anthropology* 22(4), 461–489.

Ashworth, G. and Hartmann, R. (2005) *Horror and Human Tragedy Revisited: the Management of Sites of Atrocities for Tourism*. Cognizant Communication Corporation, Elmsford, New York.

Baldwin, F. and Sharpley, R. (2009) Battlefield tourism: bringing organised violence back to life. In: Sharpley, R. and Stone, P. (eds) *The Darker Side of Travel: the Theory and Practice of Dark Tourism*. Channel View Publications, Bristol, UK, pp. 186–206.

Chen, C.-M. and Tsai, T.-H. (2019) Tourist motivations in relation to a battlefield: a case study of Kinmen. *Tourism Geographies* 21(1), 78–101. DOI: 10.1080/14616688.2017.1385094.

Cohen, E. (2008) *Youth Tourism to Israel: Educational Experiences of the Diaspora*. Vol. 15. Channel View Publications, Bristol, UK.

Cohen, E. (2018) Thanatourism: a comparative approach. In: Stone, P.R., Hartmann, R., Seaton, T., Sharpley, R. and White, L. (eds) *The Palgrave Handbook of Dark Tourism Studies*. Palgrave Macmillian, London, pp. 157–172.

Cohen, E.H. (2011) Educational dark tourism at an in populo site: the Holocaust Museum in Jerusalem. *Annals of Tourism Research* 38(1), 193–209.

Corbetta, P. (2003) *Social Research: Theory, Methods and Techniques*. SAGE Publications, Oxford, UK.

Dunkley, R.A., Morgan, N. and Westwood, S. (2007) A shot in the dark? Developing a new conceptual framework for thanatourism. *Asian Journal of Tourism and Hospitality* 1(1), 54–63.

Dunkley, R., Morgan, N. and Westwood, S. (2011) Visiting the trenches: exploring meanings and motivations in battlefield tourism. *Tourism Management* 32(4), 860–868. DOI: 10.1016/j.tourman.2010.07.011.

Farmaki, A. (2013) Dark tourism revisited: a supply/demand conceptualisation. *International Journal of Culture, Tourism and Hospitality Research* 7(3), 281–292. DOI: 10.1108/IJCTHR-05-2012-0030.

Gillen, J. (2014) Tourism and nation building at the war remnants museum in Ho Chi Minh City, Vietnam. *Annals of the Association of American Geographers* 104(6), 1307–1321. DOI: 10.1080/00045608.2014.944459.

Gillen, J. (2018) It begins with the bombs: operationalizing violence at a Vietnamese dark tourism site. *Cultural Geographies* 25(4), 519–536. DOI: 10.1177/1474474018762810.

Henderson, J.C. (2000) War as a tourist attraction: the case of Vietnam. *International Journal of Tourism Research* 2(4), 269–280. DOI: 10.1002/1522-1970(200007/08)2:4<269::AID-JTR219>3.0.CO;2-A.

Henderson, J.C. (2007) Communism, heritage and tourism in East Asia. *International Journal of Heritage Studies* 13(3), 240–254. DOI: 10.1080/13527250701228197.

Hooper, G. and Lennon, J.J. (eds) (2017) *Dark Tourism: Practice and Interpretation*. Routledge, London.

Hồ Tài, H.-T (2001) *The Country of Memory: Remaking the Past in Late Socialist Vietnam*. University of California Press, Berkeley, California.

Johnston, T., Tigre-Moura, F. and Mandelartz, P. (2016) Welcome to the Home of Auschwitz Tours: the online marketing of genocide tourism. In: Mandelartz, P. and Johnston, T. (eds) *Thanatourism: Case Studies in Travel to the Dark Side*. Goodfellow, Oxford, UK, pp. 155–170.
Le, K.L. (2014) Cu Chi tunnels: Vietnamese transmigrant's perspective. *Annals of Tourism Research* 46(4), 75–88. DOI: 10.1016/j.annals.2014.02.007.
Lee, C.-K., Bendle, L.J., Yoon, Y.-S. and Kim, M.-J. (2012) Thanatourism or peace tourism: perceived value at a North Korean resort from an indigenous perspective. *International Journal of Tourism Research* 14(1), 71–90. DOI: 10.1002/jtr.836.
Lema, J. and Agrusa, J. (2013) Revisiting the war landscape of Vietnam and tourism. In: Butler, R. and Suntikul, W. (eds) *Tourism and War*. Routledge, London, pp. 245–253.
Light, D. (2017) Progress in dark tourism and thanatourism research: an uneasy relationship with heritage tourism. *Tourism Management* 61(2), 275–301. DOI: 10.1016/j.tourman.2017.01.011.
Miles, S. (2012) Battlefield tourism: meanings and interpretations. PhD thesis, University of Glasgow, Glasgow, UK.
Ngo, P.M. and Bui, H.T. (2019) Contested interpretation of Vietnam war heritage: tour guides' mediating roles. *Journal of Tourism & Adventure* 2(1), 61–84. DOI: 10.3126/jota.v2i1.25933.
Nguyen, V.-T. (2016) *Nothing Ever Dies: Vietnam and the Memory of War*. Harvard University Press, Cambridge, Massachusetts.
Podoshen, J.S. (2013) Dark tourism motivations: simulation, emotional contagion and topographic comparison. *Tourism Management* 35(4), 263–271. DOI: 10.1016/j.tourman.2012.08.002.
Quảng Trị Department of Culture, Sports and Tourism (2018) Quang Tri – tourism potentials. Available at: https://dulich.quangtri.gov.vn/en/news/Tourism/quang-tri-tourism-potentials-36.html (accessed 25 May 2018).
Richter, L.K. (1989) *The Politics of Tourism in Asia*. Univerisity of Hawaii Press, Honolulu.
Ryan, C. (ed.) (2007) *Battlefield Tourism: History, Place and Interpretation*. Elsevier, Amsterdam.
Schwenkel, C. (2006) Recombinant history: transnational practices of memory and knowledge production in contemprary Vietnam. *Cultural Anthropology* 21(1), 3–30. DOI: 10.1525/can.2006.21.1.3.
Schwenkel, C. (2013) Post/socialist affect: ruination and reconstruction of the nation in urban Vietnam. *Cultural Anthropology* 28(2), 252–277. DOI: 10.1111/cuan.12003.
Seaton, A.V. (1996) Guided by the dark: from thanatopsis to thanatourism. *International Journal of Heritage Studies* 2(4), 234–244. DOI: 10.1080/13527259608722178.
Seaton, A.V. (1999) War and thanatourism: Waterloo 1815–1914. *Annals of Tourism Research* 26(1), 130–158. DOI: 10.1016/S0160-7383(98)00057-7.
Seaton, T. (2018) Encountering engineered and orchestrated remembrance: a situational model of dark tourism and its history. In: Stone, P.R., Hartmann, R., Seaton, T., Sharpley, R. and White, L. (eds) *The Palgrave Handbook of Dark Tourism Studies*. Palgrave Macmillian, London, pp. 9–31.
Sharpley, R. (2005) Travels to the edge of darkness: towards a typology of dark tourism. In: Ryan, C., Page, S. and Aicken, M. (eds) *Taking Tourism to the Limits: Issues, Concepts and Managerial Perspectives*. Elsevier, Oxford, UK, pp. 217–228.
Sharpley, R. (2009a) Dark tourism and politial ideology: towards a governance model. In: Sharpley, R. and Stone, P. (eds) *The Darker Side of Travel: the Theory and Practice of Dark Tourism*. Channel View Publications, Bristol, UK, pp. 145–163.
Sharpley, R. (2009b) Shedding light on dark tourism: an introduction. In: Sharpley, R. and Stone, P. (eds) *The Darker Side of Travel: the Theory and Practice of Dark Tourism*. Channel View Publications, Bristol, UK, pp. 1–22.
Smith, V.L. (1998) War and tourism: an American ethnography. *Annals of Tourism Research* 25(1), 202–227.
Stone, P. (2006) A dark tourism spectrum: towards a typology of death and macabre related tourist sites, attractions and exhibitions. *Tourism: an International Interdisciplinary Journal* 54(2), 145–160.
Stone, P. (2012) Dark tourism and significant other death: towards a model of mortality mediation. *Annals of Tourism Research* 39(3), 1565–1587.
Stone, P. (2013) Dark tourism scholarship: a critical review. *International Journal of Culture, Tourism and Hospitality Research* 7(3), 307–318. DOI: 10.1108/IJCTHR-06-2013-0039.
Stone, P. and Sharpley, R. (2008) Consuming dark tourism: a thanatological perspective. *Annals of Tourism Research* 35(2), 574–595. DOI: 10.1016/j.annals.2008.02.003.
Timothy, D.J. and Boyd, S.W. (2006) Heritage tourism in the 21st century: valued traditions and new perspectives. *Journal of Heritage Tourism* 1(1), 1–16. DOI: 10.1080/17438730608668462.

Tran, A.Q. (2018) *Gods Heroes, and Ancestors: an Interreligious Encounter in Eighteenth Century Vietnam*. Oxford University Press, New York.

Upton, A., Schänzel, H. and Lück, M. (2018) Reflections of battlefield tourist experiences associated with Vietnam War sites: an analysis of travel blogs. *Journal of Heritage Tourism* 13(3), 197–210. DOI: 10.1080/1743873X.2017.1282491.

Uzzell, D. and Ballantyne, R. (1998) Heritage that hurts: interpretation in a post-modern world. In: Uzzell, D. and Ballantyne, R. (eds) *Contemporary Issues in Heritage and Environmental Interpretation: Problems and Prospects*. The Stationery Office, London, pp. 152–171.

Weaver, D., Tang, C., Shi, F., Huang, M.-F., Burns, K. *et al.* (2018) Dark tourism, emotions, and postexperience visitor effects in a sensitive geopolitical context: a Chinese case study. *Journal of Travel Research* 57(6), 824–838. DOI: 10.1177/0047287517720119.

Wiltshier, P. (2016) Site management and consuming death: the attraction of death, disaster and the macabre. In: Mandelartz, P. and Johnston, T. (eds) *Thanatourism: Case Studies in Travel to the Dark Side*. Goodfellow, Oxford, UK, pp. 43–59.

Winter, C. (2009) Tourism, social memory and the Great War. *Annals of Tourism Research* 36(4), 607–626. DOI: 10.1016/j.annals.2009.05.002.

Yoshida, K., Bui, H.T. and Lee, T.J. (2016) Does tourism illuminate the darkness of Hiroshima and Nagasaki? *Journal of Destination Marketing & Management* 5(4), 333–340. DOI: 10.1016/j.jdmm.2016.06.003.

Zhang, H., Yang, Y., Zheng, C. and Zhang, J. (2016) Too dark to revisit? The role of past experiences and intrapersonal constraints. *Tourism Management* 54(2), 452–464. DOI: 10.1016/j.tourman.2016.01.002.

Zheng, C., Zhang, J., Zhang, H. and Qian, L. (2017) Exploring sub-dimensions of intrapersonal constraints to visiting 'dark tourism' sites: a comparison of participants and non-participants. *Asia Pacific Journal of Tourism Research* 22(1), 21–33. DOI: 10.1080/10941665.2016.1175489.

10 Vietnam Tour Operators and Responsible Tourism: a Behavioural Approach

Quynh N. Bui[1]*, Giang T. Phi[2] and Andrew Le[3]
[1]*Vietnam National University, Hanoi, Vietnam;* [2]*VinUniversity, Hanoi, Vietnam;*
[3]*International College of Management Sydney (ICMS), Sydney, Australia*

Introduction

'Responsible tourism' (hereinafter referred to as RT) is considered highly applicable (Neto, 2003; Goodwin, 2011) among terminologies in relation to sustainable tourism. Although some RT practices have been implemented by the not-for-profit sector in Vietnam, there needs to be strong involvement from the private tourism sector, especially from tour operators. The central role of tour operators in RT implementation has been mentioned in many statements, including the 2002 Cape Town Declaration, the 2008 Kerala Declaration and the 2011 Alberta Declaration (Chettiparamb and Kokkranikal, 2012).

The concept of RT originated as early as the 1980s but has only gained global traction in the last two decades. The first workshop on RT was held in Cape Town in 2002 alongside the World Summit on Sustainable Development in Johannesburg (Goodwin, 2014). This workshop led to the creation of the Cape Town Declaration, in which the definition of RT has three dimensions: (i) RT development helps improve the quality of life of the community in the tourism destination; (ii) RT helps create better business opportunities; and (iii) RT brings quality experiences for visitors. Echoing Goodwin's (2014) opinion, Spenceley *et al.* (2002) defined RT based on its positive impact on bringing better travel experiences to tourists, better business opportunities for tour operators while also ensuring a better life for local communities. Indeed, the World Tourism Organization (UNWTO) also emphasizes how the tourism industry works and operates in a responsible way towards sustainable development, in which tourism 'provides good working conditions for all, promotes local culture and products, protects natural resources, and thus preserves the unique attraction of each tourism destination' (UNWTO, 2016, p. 5).

The essence of RT is that it is not merely a solution to sustainable development either by the types of products or by the scale of operation, but a fundamental change in a set of values and actions that results in an improvement of tourism activities involving all stakeholders including tourists, tourism service suppliers, the local community and tourism managers. It is the awareness and responsible behaviour of stakeholders in tourism activities, with the greatest responsibility on tour operators and tourists, to ensure 'nice places for people to live

*Corresponding author: buinhatquynh1512@gmail.com

DOI: 10.1079/9781789242782.0010

and visit' (Goodwin, 2011, p. 102). Similarly, Leslie (2012) believes that RT should be interpreted as a behavioural characteristic, taking responsibility for the consequences of tourism production and consumption. Thus, RT is not a specific type of tourism, nor a specific tourism product/brand, but rather represents tourism development plans, policies and strategies that ensure that benefits are distributed optimally among stakeholders such as local communities, authorities, tourists and tour operators (Husbands and Harrison, 1996, cited in Leslie, 2012, p. 20). Furthermore, RT is not about how to do small-scale or alternative tourism; instead, the emphasis is on individual behaviour, more specifically being aware of the importance of having a sense of responsibility in tourism development with other stakeholders (Gilbert *et al.*, 1994).

With their unique position in the tourism system, tour operators serve as the hub for connecting diverse stakeholders in tourism, ranging from tourists to the local community to local destination authorities. A tour operator can offer RT products and services to motivate and educate tourists in consuming tourism in responsible ways and can concurrently improve awareness in local communities as well as contributing benefits to the tourism destination across the three aspects of the economy, society and the environment (Vu, 2015; Camilleri, 2015). Furthermore, these tourism enterprises know how to reach visitors through publications and the media, thereby connecting sustainable tourism destinations with customers. Despite tour operators' essential role, there exists a gap in the literature regarding their RT behaviours and implementation in the context of Vietnam. This chapter thus seeks to bridge this gap through a qualitative research approach.

Theoretical Approaches to RT

A number of different approaches are pertinent in the research on RT. The first focuses on the consumption and production of tourism products and services, or in other words, associating RT with the behaviour of purchasing and consuming tourism products by tourists (Budeanu, 2005; 2007; Kasim, 2007; Caruana *et al.*, 2014; Stanford, 2014). The second approach centres on the stakeholders, namely enterprises, employees in enterprises, tourists, local communities and tourism authorities. The interrelationships among stakeholders are among the top concerns of sustainable tourism development, as they mediate between interests, needs, supply, consumption and production (Driha, 2018). Camilleri (2015) suggested that the common interests for both enterprises and stakeholders could be created by tourism and responsibility initiatives. Along with that, RT also helps cement the relationship between enterprises and the social community and tourism authorities, sharing benefits in human resource management, market development, cost savings and business operation. Meanwhile, Byrd (2007) emphasized the role of tourists and host communities in RT development at a destination, particularly in RT product development. Godfrey (1998) found out the attitudes of tourism managers towards RT implementation and concluded that the attitude of managers would affect their position on implementation and the success of RT implementation. The third approach stems from the choices in the analysis of the responsible and ethical behaviours of the individual or the responsible and ethical behaviours of the organization. Individual ethical values and social standards are mutually interrelated. According to Fennell and Malloy (2007), ethical standards are the basis for the intentions of RT. Researchers also focused on analysing attitudes, behaviours and decision making of individual tourists, also known as 'responsible tourists' (Goodwin and Francis, 2003; Caruana *et al.*, 2014; Stanford, 2014; Lee *et al.*, 2017).

In recent years, researchers have also approached RT regarding behaviours on tourism activities, from awareness, attitude to action. This research approach is partly influenced by the definitions of RT from Goodwin (2011) and Leslie (2012). Advancing the theoretical aspect of the behavoural approach to RT, Mihalic (2016) proposed an RT model consisting of three factors – awareness, agenda and action – explaining the process of evolution from awareness to behaviour in RT.

Linked to tour operation, RT is viewed as an essential process of behaviour, from awareness to intention and decision making. Frey and

George (2010) found that even though tourism enterprises have a positive awareness and attitude towards RT, implementation of the principle of RT is facing a lack of resources, high costs of operation, lack of support from the authorities, and a fiercely competitive environment. In a similar vein, Sheldon and Park (2012) pointed out that although there are implementation guidelines, tour operators still face obstacles in their practices. The successful implementation rate is very low, accounting for only 30% of the total number of enterprises that support RT behaviour. Another barrier to the implementation of RT is the behaviour of tourists. Tourists' actions are not always consistent with their awareness of RT (Goodwin and Francis, 2003). Although tourists always expect tour operators to implement RT, they are unwilling to pay the price for RT products. For this reason, enterprises seem not to be wedded to RT implementation. To support the implementation of RT by tour operators, it is crucial to find out what motivates tourism enterprises in Vietnam, to identify their awareness, attitudes and behaviours, and find out the influencing factors for these enterprises to commit and act in respect to RT.

The Vietnamese Context

Concerns for RT have emerged in Vietnam since the early 2000s. To set up the framework for RT implementation, organizations from the not-for-profit sector (e.g. the Netherlands Development Organization (SNV), the World Wide Fund for Nature (WWF) and the European Union-funded Environmentally and Socially Responsible Tourism (ESRT) Capacity Development Programme) have initiated and invested in many practical projects throughout the country. As a result, RT guidelines are primarily based on foreign RT practices, which challenge the Vietnamese private sector to adopt owing to the country's different political and economic system. The implementation of RT in the private tourism sector is led by the Responsible Travel Club of Vietnam (RTC Vietnam), an association established in 2009 that unites tourism organizations, individuals and enterprises to promote RT implementation.

There are five major forms of RT practices in Vietnam, which are quite similar to the five area actions of operator business activities defined by the United Nations Environment Programme (UNEP, 2005). These forms are:

- RT product development and management (such as prioritizing local suppliers, altering irresponsible elements on tour);
- responsible supply chain management (namely cooperating with responsible suppliers, training suppliers about the way to conduct RT);
- responsible cooperation with the local community at the tourism destination (such as purchasing local products, respecting local cultural norms);
- performing corporate social responsibility as a way to provide information about appropriate behaviour (donation and charity); and
- responsible internal management (such as achieving RT criteria, delivering green office principles, offering good employment and working conditions).

Vietnamese researchers advocate that RT is an orientation that stakeholders in tourism need to be aware of, which will help them behave adequately for the long-term purpose of sustainable development (Nguyen, 2012; Duong, 2013; Trang, 2013; Vu, 2015). Further elaborating RT in the context of Vietnam, Pham contended:

> Responsible tourism comes from the awareness to behaviour and responsibility of each individual, enterprise, organisation towards the natural and social environment, responsible tourism gives specific requirements for the stakeholders involved in tourism activities. The responsible tourism implementation process is not only temporary actions but also a process of changing from awareness to behaviour in tourism activities of all stakeholders.
>
> (Pham, 2015, p. 42)

To further our understanding of RT in the context of Vietnam, researchers discuss the current state of and barriers to implementation of RT among tour operators. To fulfil the aim of this research, we seek to address the following research questions: (i) How do tour operators perceive and implement RT?; and (ii) What factors affect the RT behaviour of tour operators in Vietnam?

Research Method

This research employed a qualitative approach that allows discovering the nature of different group behaviours by involving an interpretive and naturalistic approach (Denzin and Lincoln, 2008). Within this research approach, semi-structured in-depth interviews guide the conversation with a pre-prepared set of questions while allowing certain flexibility during the interview situation (Lindlof and Taylor, 2010). The research question is also framed within a constructivist paradigm, which views 'the truth' and 'reality' as being socially constructed in the minds of individuals, and is also frequently negotiated and communicated between them (Denzin and Lincoln, 2011). A qualitative research approach utilizing semi-structured interviews is thus suitable to address the above-mentioned research questions as there is not much existing literature regarding tour operator's awareness, attitudes and behaviours in relation to RT.

Within a wide array of qualitative research methods, in-depth interviews were adopted. In the interview phase, the authors used a list of exploratory questions to ask respondents (see Table 10.1). The interviews were conducted with ten international travel businesses and two tourism experts. Among the operators, six are current members of the RTC Vietnam, and four are non-RTC Vietnam members. The reason why the interviewed participants were dominated by the members of RTC was that this NGO is acting with the aim to promote the sustainable growth of tourism in Vietnam. RTC Vietnam consists of inbound companies who are concerned the most about RT and attempt to act for sustainable tourism. The majority of tour operators affiliated with the RTC are small and medium-sized, and their business partners are mainly from developed countries. In addition, to ensure the representativeness and generalization and avoid bias, the researchers selected another four international tour operators. These non-RTC tour operators may reveal different perceptions and implementation towards RT compared with those of RTC Vietnam and thus could act as a benchmark/comparison check. A series of 30–90 minute in-depth interviews were conducted in-house with the tour operators, directors, vice-directors, or heads of management (for participants' profiles see Table 10.2). The researchers also have more than 10 years' experience working and conducting research in the tourism field.

The interview process was flexible. The interviewer could change the order of the interview questions to suit the context and answers of the interviewee so that the conversation could take place naturally to collect as much information as possible. The researchers employed several interviewing skills suggested by Kvale (1996), including memoing, note taking, and special attention to not imposing subjective inferences and interpreting the interviewee's ideas. Oral data from interviews were recorded and transcribed to text for analysis.

The method of thematic coding was utilized for data analysis. This method of coding is defined as 'a set of techniques for the systematic analysis of texts of many kinds addressing not only manifest content but also the themes and core ideas found in texts as primary content' (Drisko and Maschi, 2015, p. 85). The researchers selected themes that most frequently appeared in the text. However, some unique themes were presented. Themes were then arranged into categories of relevant issues. The findings from interviews are presented in the following section.

Findings

Three categories emerged from thematic coding by aligning themes by similarities and connections. The first category reveals tour operators' awareness of RT as an industry practice. The second cluster of themes is a collection of current RT practice implemented by tour operators. The final one addressed major barriers hindering the effectiveness of RT implementation in tourism businesses.

Awareness

Overall, awareness of the nature of RT remains weak, inconsistent among managers and employees of tourism businesses, as stated by a tour operator's representative:

> [...] Awareness of responsible tourism is not consistent; managers and employees should have been aware of (responsible) products, customers, management, but not yet.
>
> (Interviewee 1)

Table 10.1. Interview questions. Compiled by the authors.

No.	Main questions	Suggestions
Assessment of the perception and implementation of responsible tourism (RT)		
1	Could you please describe the responsible tourism activities that your business has initiated or implemented?	• Reason for doing? • How much time? • Activities performed?
2	How do you rate the current state of responsible tourism implementation of tour operators in Vietnam?	• Result? The remaining problem? • What are the advantages and disadvantages of implementation? • Mechanisms and policies related to responsible tourism? • Benefits obtained?
3	In your opinion, what activities does the responsible tourism behaviour of tour operators include?	• What is responsible tourism behaviour? • What are the activities associated with responsible tourism development?
Assessment of the impact of organizational factors on RT behaviour		
4	In your opinion, do organizational factors affect the responsible tourism behaviour of tour operators? How is the impact?	• Which factors affect the responsible tourism behaviour of tour operators? • When you make decisions on responsible tourism, do all of your employees support it? • When you make decisions on responsible tourism, do all of your employees agree to implement it? • Is teamwork effective in your tour operator business? Do groups have an impact on your decision to implement tourism? How is the impact? • Does your tour operator have a reward and sanction policy for the implementation of responsible tourism? • Is the management policy for the implementation of responsible tourism strictly enforced in your enterprise?
5	Please indicate the reason why your business does not provide responsible tourism products.	• Give reasons. Explain • What are the difficulties encountered by the tour operator itself, by partners, and by tourists?
6	In your opinion, apart from organizational factors, are there any external factors that affect the responsible tourism behaviour of tour operators?	• Does the management policy have an impact? • Does the tourists' needs/travelling trends have an impact? • Do the partners or service providers of your business have an impact? • Does the local community at the tourist destination have an impact? • Does the community, in general, have an impact? • Do your competitors have an impact? • Do the media agencies, the daily press which report on the implementation of responsible tourism have any impact? • Do other international tour operators who are implementing responsible tourism have an impact?

Table 10.2. Participants' profiles. From fieldwork data.

Interviewee	Position	Organization
Interviewee 1	Director	Tour operator
Interviewee 2	Vice-director	Tour operator
Interviewee 3	Director	Tour operator
Interviewee 4	Director	Tour operator
Interviewee 5	Director	Tour operator
Interviewee 6	Director	Tour operator
Interviewee 7	Director	Tour operator
Interviewee 8	Director	Tour operator
Interviewee 9	Head of management office	Tour operator
Interviewee 10	Director	Tour operator
Interviewee 11	Lecturer/researcher	University
Interviewee 12	Manager	Vietnam National Administration of Tourism

This weak awareness is partly attributed to the emphasis on mass tourism in the Asian tourist market. In the meantime, RT is a relatively new concept that has not been widely adopted by a large number of tour operators and other service providers. The concept and implementation of RT newly emerge among small groups of service providers but perhaps needs more time to become a norm for the industry. For example, interviewees 9 and 10 said:

> [...] Many managers are promoted from tour guides. [They] do not learn the management structure, so they only sell what is short term. [They] do not understand the internal resources to make a business model or satisfy the customer needs, which is the current situation of Vietnamese tourism businesses.
>
> (Interviewee 10)

> [...] Regarding responsible tourism products, tour operators have not done much as they are in the middle, linking existing products and clients but not making products. To change, [we] must start from tourism service providers such as hotels. However, mainly five-star hotels would provide responsible tourism products, while most three-star hotels have not yet implemented responsible tourism. Tour operators have to identify [proper providers] but [there are] not enough resources. For instance, if the Asian tourist market focuses on mass tourism, the tour operator must focus its resources on that market.
>
> (Interviewee 9)

Furthermore, interviewees often mentioned the lack of information and communication on RT. Many enterprises have not been informed about the definition as well as the significant advantages of RT development, causing them not to properly understand RT and not to take actions leading to RT. Even when several enterprises gradually become aware of RT practices, they still lack the specific tools, knowledge and methods to develop tourism in a responsible way:

> [...] The barrier is that we have just started implementing responsible tourism. We have to widely disseminate the concept and raise awareness of the community because it directly affects the interests of all relevant stakeholders ... We do not have effective media channels so that all stakeholders in the tourism business, from the government to enterprises, can be fully aware of this issue and concept. They may appear somewhere, several actions that people or enterprises already did for themselves, arising but not in a systematic manner.
>
> (Interviewee 12)

Current practices

The lack of awareness has led to many problems in RT implementation among tour operators. Researchers have identified five main forms of RT which encourage tourists, suppliers and workers to comply with RT principles among RTC members.

(1) Introducing RT practices in operation

Tour operators mentioned some practices introduced to tourists in order to promote environmental and social sustainability. These practices aim to create some changes in tourist behaviours. For example, the practice called 'Refill My Bottle', which encouraged tourists to refill water at water stations instead of using plastic bottles, is applied widely among businesses involved in this study. Additionally, several tour operators set up a 5-gallon water bottle on travelling buses so that their tourists are able to refill their bottles while on the move. Some tour operators attempt to guide their tourists to reduce using plastic and wet tissues on their tours:

> [...] We want to replace plastic bottles with inox bottles on tours and give them to our customers as a gift.
>
> (Interviewee 2)

(2) Encouraging tourists to make a donation or volunteer

Another form of RT is to provide an act of donation. Donation could be included on a package tour or given separately as a way to enhance social responsibility.

> [...] The company has set up a charity fund called ... Foundation, bought books for the library, supported orphans in 64 provinces in Vietnam, supported [by] the Vietnam National Administration of Tourism (VNAT), Institute of Tourism Development Research (ITDR).
>
> (Interviewee 5)

A few tour operators design tour programmes to include socially responsible actions. They contact some non-profit organizations which aim to help disadvantaged groups in the society and provide tourists with an opportunity to be involved in some volunteer activities.

> [...] Our customers could visit the orphanage in a pagoda and donate money. We may assist our tourists to buy clothes, foods and other school things in order to donate to those orphans ... [Tourists] can also opt to travel to several destinations and take a great deal of effort to support the local community in building/restoring elementary schools. They also organize skill training courses for children ...
>
> (Interviewee 6)

(3) Integrating suppliers' compliance to RT

The integration of suppliers who commit to RT into a tourism supply chain creates an entirely responsible packaged tour. Therefore, tour operators committed to RT select to work with suppliers, such as those supplying accommodation, restaurants, catering and transport agencies, who share the same vision.

> [...] We explain to our suppliers about responsible tourism, clarify the importance/long-term advantages of implementing responsible tourism, and encourage them to care more about the environment and business ... We always cooperate with the suppliers which share the same vision in implementing responsible tourism.
>
> (Interviewee 4)

(4) Supporting the local community

Utilizing local suppliers helps the tour providers to reduce their carbon footprint and bring a unique experience and indigenous culture to tourists. Tour operators have engaged with local value chains to purchase and promote local products, namely handicraft products or local specialties:

> [...] We purchase seafood from fishermen in Nam Dinh, Thai Binh and stock to restaurants located in Hanoi.
>
> (Interviewee 5)

In addition, responsible tour operators follow the 'fair trade' principle when cooperating with the local community, offering them price protection even in the off-peak season:

> [...] We will not use 'price squeezing' when doing business with the local community during their low season ...
>
> (Interviewee 3)

(5) Implementing responsible internal management

An additional set of practices which are chosen by tour operators is implementing responsible management within their enterprises. This may include employee compensation and incentive policies (such as the organization of yoga classes for employees after their working hours, training programmes and creating green offices, namely the planting of lots of trees, saving electricity and other forms of energy saving):

> [...] As a director of an enterprise, I have to take responsibility with all of my employees for how the business generates benefits, how to control the consumption level properly. In this case, there is a need to save electricity and water and then the provision of a compensation policy, which are internal issues.
>
> (Interviewee 2)

While the above list of RT implementation is extensive, these actions are only taken by a small numbers of tour operators in Vietnam, as stated by Interviewee 12:

> [...] However, there are not many tour operators understanding and involving responsible tourism activities into their every tour today. It is possible that only the RTC could have included them in their tours.
>
> (Interviewee 12)

Barriers to implementation

Barriers to RT implementation are identified including internal factors such as the administrative and managerial structure of the companies, and external factors attributed to norms, social pressure and culture.

(1) Administrative structure

Tour operators perceived the implementation of RT should have been legitimized. The higher level of state and provincial administrators should be aware of RT and support responsible practices. The awareness of other stakeholders such as regulators, partners, local communities and service providers should also be considered because of their impact on the behaviour of the travel business, as well as their willingness to implement RT in operations:

> [...] In the case of Ben Tre, for example, the secretary of the provincial Party Committee attaches great importance to these responsible tourism programmes. Following the issue of their order, the implementation of the [responsible] activities will be very convenient. Perceptions of partners are also important, especially of accommodation establishments.
>
> (Interviewee 4)

> [...] If the highest level of authorities is aware of RT, they will always remind enterprises and local people about the implementation of RT, even this conception must be imbrued with members of the State Committee on tourism. If the highest level can do it, it will affect the remaining level, the central level will spread to all levels.
>
> (Interviewee 10)

(2) Management policy

From the tour operators' point of view, sustainable tourism must be embedded in business management and execution. The management orientation aims to ensure that employees are fully aware of and firmly grasp sustainable development criteria to take action and operate the business according to set goals.

> [...] Sustainable tourism involves management, legislation, and action. Being a member of RTC, tour operators must commit to ensuring sustainability in the environment, society, and internally, in terms of management, awareness, and ensuring everyone in the business must firmly grasp the criteria. Tour operators also focus on business performance, so they focus more internally.
>
> (Interviewee 5)

In particular, mechanisms for rewards and sanctions have influenctial effects. Interviewee 8 said:

> [...] The reward and punishment policy is not related to responsible tourism behaviour, and related to revenue. This behaviour has more to do with individual perception.
>
> (Interviewee 8)

(3) Group dynamics

Group dynamics are associated with the organization and are considered to be an important factor affecting the RT behaviour of businesses along with leadership and management policies:

> [...] Teamwork differs from organization behaviour and from personal behaviour. Teamwork is considered as one of the

> characteristics of organizational behaviour, so team and even leadership factors are the most impactful factor on the enterprise operation.
>
> (Interviewee 5)

In a group, the socialization process will help the individual acquire the knowledge, behaviour and standards necessary to fully become part of the business.

> [...] In my tour operator, I see that my colleagues, and the company's business strategy have never been related to responsible tourism. In addition, when my tour operator alliances with other businesses, I also participate in seminars and meetings with them, and then I realize that they often focus on how to attract more and more tourists and earn money, without caring about responsibility. My readiness or willingness to promote responsible tourism is affected to some extent.
>
> (Interviewee 7)

(4) Social pressure

Pressure from the broader society is also considered as both an enabler and a barrier to corporate behaviour. Businesses tend to follow what is considered the social norm, based on the perception of society to perform or not perform a particular behaviour. Therefore, some businesses perform responsible behaviour to meet societal pressures:

> [...] The most crucial goal of the business is to create profits, maximize profits. This profit is not only money but also the image and brand of the business. ... Today, businesses want to build images, use images to attract visitors. Therefore, businesses practise social responsibility and follow what the law says to build a good image in society.
>
> [...] Visitor pressure creates general change in the tourism system. Reflected by customers, many boats in Ha Long do not allow discharge ...
>
> (Interviewee 3 and Interviewee 4)

Local communities also have power in pressuring tourism enterprises and tour operators to adopt RT practices. Interviewee 10's quote perhaps best sums up the local community's power:

> [...] A criterion of responsible tourism is that local people have power. They can offer suggestions, participate in defending their opinions and keep developing ... So how to let local people participate? Having a sense of ownership from tour operators and local people will make them feel the benefits brought to them. When the stakeholders have deeply embraced the responsible tourism mindset, responsible tourism will go deeper. This is different to implementation based on the definition, which responsible tourism will only be on paper, not going into details ...
>
> (Interviewee 10)

(5) Cultural philosophy

Besides social pressure, cultural philosophy is another external factor which plays an important role in the process of fostering RT awareness and ethics of responsibility issues, thereby forming the responsible behaviour of individuals and organizations. Cultural philosophy is the cultural ideology that affects the thinking process and perceptions of business managers and each employee of the enterprise, thereby affecting the business management, operations and behaviour in the external environment.

> [...] Culture has a direct impact on that behaviour. Culture is the root ... Enterprise envisions as a person. People are influenced by natural culture, social culture and family culture. Culture forms the core of the nation.
>
> (Interviewee 11)

In Vietnam, enterprises always seek to build their own identity and brand in relation to their competitors, but there is still mutual coordination among enterprises in doing business:

> The management, the self-governance, the boundary line, the thing that makes the difference, the own business ethics in that company, all the peculiarities in that company, create a competition between one enterprise and another, that is autonomy. But they are together, they still have the link of community. That enterprise cannot do it alone. 'Birds of a feather, flock together' is a very famous thing in Vietnam. In other countries, there is no such thing as one street trades the same product, but Vietnam is willing to do it in one street.
>
> (Interviewee 12)

However, this mutual coordination sometimes leads to the adoption of bad practices among businesses:

> [...] the Westerners are still very surprised about the fact that in Vietnam if one enterprise makes fake goods, they will influence the whole group to do the same. The reason is that if two enterprises A and B both trade in the same product, enterprise A intends to produce fake goods but is afraid that if business B is uncompromising, enterprise B will denounce enterprise A. Therefore, enterprise A will invite enterprise B to counterfeit together, to increase their income, and if they are exposed, they will all go to court together instead of standing trial alone. There is still sense of community in court.
> (Interviewee 11)

In summary, this chapter presents the findings related to awareness and current forms of RT practices, and barriers to RT implementation among tour operators in Vietnam. RT practices are attributed to tourists, suppliers and the internal policies of tour operators. Barriers to RT have their roots in internal factors relating to the organization such as policies and structure, as well as the external environment from the society and culture within which the tour operators function.

Discussion

Emerging from our data analysis is the fact that responsible tourism (RT) is in its infancy in Vietnam. There appears to be a lack of awareness of RT and a complete absence of sharing knowledge between managers and employees within the enterprise. This leads to the fact that the involvement of tour operators in Vietnam has not been fully embedded and practised as RT. Notably, there are businesses that only use the image of RT to gain benefits rather than actually applying responsible actions in their tourism activities. This is reflected in our review of current RT practices among tour operators, which remains discrete, small scale and driven by individual businesses' motivations.

The major contribution of the current study to the understanding of RT in Vietnam is its identification of barriers hindering the effectiveness of RT practices. The research results show that the group dynamic may infleunce the tendency to implement RT. This is in line with the findings of previous research by O'Leary and Pangemanan (2007) and Steiner (1972). These authors argue that groups are created to make more collective decisions, to achieve greater consensus, possibly due to the pressure to get along with each other and respect each other within the same group. The effectiveness of group decision making over individual decision making is explained through factors such as the group members' interactive relationships, hence it is a more objective and perceptive approach to the issue; tightening quality and error control in operation, prompting new initiatives (Steiner, 1972). The feeling of fear of being out of sync with other team members, prevents individuals from expressing their individual opinions differently from decisions agreed to by most of the group. This affirms that the group creates a strong incentive to change individual thoughts and actions, as demonstrated in the research of McGrath (1984).

Findings from the current research demonstrate two contrary views on the impact of the management system on the intention to implement RT. On the one hand, businesses emphasize that the management policies of their enterprises, particularly the reward policy and sanctions, shape behaviour in fulfilling corporate, environmental and social responsibility. This is because management decisions are often well thought out and deliberated about before they are made, and they comply with the laws, and respect rights in order to avoid sanctions (Premeaux, 2004). Businesses are more likely to practise RT if the incentives and penalties are made public and transparent. On the other hand, there is a view that management policy has no impact on the RT practices of the business, because in the process of employee selection, businesses have given criteria for attitudes and behaviours and only select employees who match these evaluation criteria. Therefore, the admitted employees will not violate the principles set by the business, so the reward and punishment policy will not affect the behaviour much.

In addition, the research results also indicate that social pressure from stakeholders has a strong influence on corporate decision making. Businesses in the market must satisfy the desired needs of tourists, and at the same time, to get favourable conditions for business development, businesses need to follow legal regulations and support the community and society. This implies that before making a decision, it is necessary to anticipate the impact of that decision on the environment, the economy and society.

Furthermore, the expert interviews also pointed out the importance of cultural factors influencing RT implementation, particularly collectivism and community spirits. Collectivism promotes cohesiveness among individuals and the prioritization of the group over the individual, manifesting in loyalty and long-term commitment to the group (Schwartz, 1990). Vietnam has long been known as a collectivistic society (Tran, 1996). As demonstrated in the findings above, collectivism might have both positive and negative influences on the responsible behaviour of Vietnamese businesses.

Conclusions

This chapter has clarified the concept of RT behaviour of tour operators. The individual's RT behaviour has been studied and verified in many situations, but there are very few studies on the RT behaviour of tour operators. Scholars and researchers are still trying to explain RT as behaviour and how tourism activities are carried out, so more research assessments and more general empirical studies (Leslie, 2012; Burrai *et al.*, 2019) are required. The researchers have explained to some extent the RT behaviour of businesses in Vietnam, and show that it is a decision-making process that leads to participation in RT action. This process is mainly influenced by personal factors (awareness), internal factors (management policy, group dynamics) and external factors (social and cultural pressure).

For tour operators in Vietnam, the research results have contributed to pointing out the role and the importance of social pressure coming from stakeholders related to the business's implementation of RT. From this position, it proposes policies that contribute to building and promoting relationships with the stakeholders ranging from employees in the business to suppliers, partners, to the state authorities in tourism, to communities in the tourist destination, and to press/media agencies, while at the same time, enlisting the stakeholders' support and help in the operation and growth of the tourism business in the face of unpredictable changes in markets and society. In addition, the management policies of enterprises also contribute to promoting the implementation of RT behaviour of individual employees. Hence businesses need to pay attention to build a clear and transparent management policy system to promote the implementation of RT, while limiting any irresponsible behaviour by employees.

For tourism authorities, the results obtained from this study will help managers better understand the level of awareness, actual implementation, and determinants of the tourism behaviour of tourism companies in Vietnam; thereby making necessary adjustments in planning/development strategies, creating opportunities for small and medium-sized tour operators to participate in building mechanisms and policies to promote the implementation of RT, and contributing to the sustainable development goals of the tourism industry.

Limitations

Although the research was carefully implemented based on a detailed research process, there are still limitations. First, this study is exploratory based on the perceptions of just 12 interviewees. Future research can verify the factors identified in this study on a larger scale, with the possible use of a quantitative approach. Second, this research's sample involves mainly small and medium-sized businesses. There is potential to further explore the viewpoints of larger tour enterprises and broaden the geographical research area to increase reliability. In addition, future research can focus on assessing the factors affecting the employee's responsible behaviour in the enterprise. Along with that, other businesses in the tourism sector, including hotels, restaurants, other types of accommodation, and transportation service providers also need to be analysed to find out the main factors affecting their performance in tourism activities.

References

Budeanu, A. (2005) Impacts and responsibilities for sustainable tourism: a tour operator's perspective. *Journal of Cleaner Production* 13(2), 89–97. DOI: 10.1016/j.jclepro.2003.12.024.

Budeanu, A. (2007) Sustainable tourist behaviour – a discussion of opportunities for change. *International Journal of Consumer Studies* 31(5), 499–508. DOI: 10.1111/j.1470-6431.2007.00606.x.

Burrai, E., Buda, D.M. and Stanford, D. (2019) Rethinking the ideology of responsible tourism. *Journal of Sustainable Tourism* 27(3), 1–16. DOI: 10.1080/09669582.2019.1578365.

Byrd, E.T. (2007) Stakeholders in sustainable tourism development and their roles: applying stakeholder theory to sustainable tourism development. *Tourism Review* 2(62), 6–13. DOI: 10.1108/16605370780000309.

Camilleri, M.A. (2015) Responsible tourism creates shared value among stakeholders. *Tourism Planning and Development* 13(2), 219–235. DOI: 10.1080/21568316.2015.1074100.

Caruana, R., Glozer, S., Crane, A. and McCabe, S. (2014) Tourists' accounts of responsible tourism. *Annals of Tourism Research* 46(5), 115–129. DOI: 10.1016/j.annals.2014.03.006.

Chettiparamb, A. and Kokkranikal, J. (2012) Responsible tourism and sustainability: the case of Kumarakom in Kerala, India. *Journal of Policy Research in Tourism, Leisure and Events* 4(3), 302–326. DOI: 10.1080/19407963.2012.711088.

Denzin, N. and Lincoln, Y. (2008) *Collecting and Interpreting Qualitative Materials*. Sage Publications, Thousand Oaks, California.

Denzin, N.K. and Lincoln, Y.S. (eds) (2011) *The Sage Handbook of Qualitative Research*. Sage Publications, Thousand Oaks, California.

Driha, O. (2018) Stakeholders and indicators of sustainable tourism: an introduction to sustainable tourism. Presentation at workshop '1st TOURIST Training at Hue University, Vietnam', 2–5 October 2018.

Drisko, J.W. and Maschi, T. (2015) *Content Analysis*. Oxford University Press, Oxford, UK.

Duong, Q.N. (2013) Phát triển du lịch có trách nhiệm. *Tạp chí Du lịch Việt Nam* 11, 22–23.

Fennell, D. and Malloy, D. (2007) *Codes of Ethics in Tourism. Practice, Theory, Synthesis*. Channel View Publications, Clevedon, UK.

Frey, N. and George, R. (2010) Responsible tourism management: the missing link between business owners' attitudes and behaviour in the Cape Town tourism industry. *Tourism Management* 31(5), 621–628. DOI: 10.1016/j.tourman.2009.06.017.

Gilbert, D.C., Penda, J. and Friel, M. (1994) Issues in sustainability and the national parks of Kenya and Cameroon. In: Cooper, C.P. and Lockwood, A. (eds) *Progress in Tourism, Recreation, and Hospitality Management*. Wiley, Chichester, UK, pp. 30–45.

Godfrey, K.B. (1998) Attitudes towards 'sustainable tourism' in the UK: a view from local government. *Tourism Management* 19(3), 213–224. DOI: 10.1016/S0261-5177(98)00020-X.

Goodwin, H. (2011) *Taking Responsibility for Tourism*. Goodfellow Publishers, Oxford, UK.

Goodwin, H. (2014) What is responsible tourism? Responsible Tourism Partnership, 24 September. Available at: https://responsibletourismpartnership.org/what-is-responsible-tourism/ (accessed 12 June 2020).

Goodwin, H. and Francis, J. (2003) Ethical and responsible tourism: consumer trends in the UK. *Journal of Vacation Marketing* 9(3), 271–284. DOI: 10.1177/135676670300900306.

Kasim, A. (2007) Corporate environmentalism in the hotel sector: evidence of drivers and barriers in Penang, Malaysia. *Journal of Sustainable Tourism* 15(6), 680–699. DOI: 10.2167/jost575.0.

Kvale, S. (1996) *Interviews: an Introduction to Qualitative Research Interviewing*. Thousand Oaks, California, Sage Publications.

Lee, H.Y., Bonn, M.A., Reid, E.L. and Kim, W.G. (2017) Differences in tourist ethical judgment and responsible tourism intention: an ethical scenario approach. *Tourism Management* 60(4), 298–307. DOI: 10.1016/j.tourman.2016.12.003.

Leslie, D. (2012) The responsible tourism debate. In: Leslie, D. (ed.) *Responsible Tourism: Concepts, Theories and Practice*. CAB International, Wallingford, UK, pp. 17–42.

Lindlof, T.R. and Taylor, B.C. (2010) *Qualitative Communication Research Methods*, 2nd edn. Sage Publications, Thousand Oaks, California.

McGrath, J.E. (1984) *Groups: Interaction and Performance*. Prentice Hall, Englewood Cliffs, New Jersey.

Mihalic, T. (2016) Sustainable-responsible tourism discourse – towards 'responsustable' tourism. *Journal of Cleaner Production* 111(Part B), 461–470. DOI: 10.1016/j.jclepro.2014.12.062.

Neto, F. (2003) *A New Approach to Sustainable Tourism Development: Moving Beyond Environmental Protection*. United Nations, New York.
Nguyen, V.L. (2012) Developing responsible tourism in community-based tourism. *Vietnam Tourism Review* 11, 20–21. [in Vietnamese].
O'Leary, C. and Pangemanan, G. (2007) The effect of group work on ethical decision-making of accountancy students. *Journal of Business Ethics* 75(3), 215–228. DOI: 10.1007/s10551-006-9248-5.
Pham, T.H. (2015) Responsible tourism: from perception to behaviour. *Vietnam Tourism Review* 1, 42–43, 52. [in Vietnamese].
Premeaux, S.F. (2004) The current link between management behaviour and ethical philosophy. *Journal of Business Ethics* 51(3), 269–278. DOI: 10.1023/B:BUSI.0000032495.53772.42.
Schwartz, S.H. (1990) Individualism-colletivism: critique and proposed refinements. *Journal of Cross-Cultural Psyhology* 21(2), 139–157.
Sheldon, P.J. and Park, S.-Y. (2012) An exploratory study of corporate social responsibility in the US travel industry. *Journal of Travel Research* 50(4), 392–407. DOI: 10.1177/0047287510371230.
Spenceley, A., Relly, P., Keyser, H., Warmeant, P. and McKenzie, M. (2002) *Responsible Tourism Manual for South Africa*. Department of Environmental Affairs and Tourism, Pretoria.
Stanford, D. (2014) Responsible tourist behaviour. *Tourism Management* 43, 4–5. DOI: 10.1016/j.tourman.2014.01.006.
Steiner, G.A. (1972) Social policies for business. *California Management Review* 15(2), 17–24. DOI: 10.2307/41164414.
Tran, N.T. (1996) *Find Out About Vietnam Cultural Identity*. Ho Chi Minh City, Vietnam: Ho Chi Minh City Publisher. [in Vietnamese].
Trang, L. (2013) Actions for responsible tourism. *Vietnam Tourism Review* 4(47), 16–17. [in Vietnamese].
UNEP (United Nations Environment Programme) (2005) *Integrating Sustainability into Business: a Management Guide for Responsible Tour Operators*. UNEP, Paris.
UNWTO (World Tourism Organization) (2016) *The Tourism Sector and the Sustainable Tourism Goals – Responsible Tourism, a Global Commitment*. United Nations Global Compact Network, Madrid.
Vu, Q.T. (2015) Formulating a legal framework for resonsibe tourism. *Vietnam Tourism Review* 2, 24–25. [in Vietnamese].

11 Public Opinion on Quarantine Hotels: a Question of Corporate Social Responsibility

Binh Nghiem-Phu[1]* and Hang T.B. Tran[2]
[1]University of Hyōgo, Kobe, Japan; [2]Thuong Mai University, Hanoi, Vietnam

Introduction

The accommodation sector, as part of the hospitality and tourism industry, is an important economic sector in Vietnam. In 2019, there were 5880 hotels of various classes (one to five stars) providing 262,300 rooms (Vietnam National Administration of Tourism, 2020). The average occupancy rate of the four- to five-star hotels in particular in Vietnam was approximately 60% with an average room rate of US$60 per room/night (CBRE, 2020). However, the global COVID-19 pandemic has interrupted the growth of the hotel industry. The negative impacts of the pandemic have been prolonged for the year 2021. Several relief measures have been introduced, such as financial assistance and incentives given to both hotel employees (US$43–77/month) and businesses (e.g. 0% interest, collateral-free loans and suspension of contributions to their retirement funds). In addition, credit support, deferred tax and land rental payment, and electricity bill reduction have also been granted to the hotels (CBRE, 2020).

To prevent the transmission of COVID-19, the Vietnamese government has been imposing a strict border control policy which orders overseas returnees and visitors to quarantine in designated facilities from 14 to 21 days upon their arrival. Resulting from this policy, certain hotels have transformed their lodgings into quarantine facilities for those entering Vietnam on permitted flights (Hanoi Department of Health, 2021; Ho Chi Minh City Department of Health, 2021). Although this repurposing of hotel facilities for health and safety reasons can be justified as an action towards corporate social responsibility (CSR) (Vietnam Investment Review, 2020; Teng *et al.*, 2021), a risk is that brand delusion (deteriorated perceptions of hotel brand or hotel image) may arise. None the less, from a business perspective, utilizing the lodging facilities for quarantine generates additional revenues for hotels in compensation for the loss of regular revenue from international-visitor and domestic markets.

Under these circumstances, the current study was carried out to answer two research questions: (i) What are the public opinions (i.e. images or impressions) about hotels offering a quarantine service at their facilities?; and (ii) To what extent will the image of such hotels be affected by these practices? This chapter employed an analysis of online data about public opinions including Internet-based press content and user-generated content. Following the analysis of these data we discuss their implications

*Corresponding author: binhnghiem@gmail.com

DOI: 10.1079/9781789242782.0011

for hotel businesses in managing their CSR and business activities during and after the COVID-19 pandemic.

Literature Review

Business and CSR activities in the hotel industry

The main function of hotels is to provide guests with a place to stay and rest (Alexander *et al.*, 2010; Ling *et al.*, 2012; Cvelbar *et al.*, 2021). Hotels also offer additional services to their customers, including those of restaurants, bars, sports and entertainment facilities, and meditation and cultural experiences (Stringam, 2008; Lin, 2009; Wu and Liang, 2009; Hung *et al.*, 2015; Lu, 2020; Sun *et al.*, 2021). These offerings are in the hotels' product profiles for guests to evaluate when reserving accommodation (Sohrabi *et al.*, 2013; Wang *et al.*, 2020), determine their satisfaction levels (Zhou *et al.*, 2014; Liu *et al.*, 2017), and influence their intention to recommend friends to stay at, and themselves to return to, a particular hotel (Tanford *et al.*, 2012; Yang and Lau, 2015).

In the era of increasing awareness and implementation of social and environmental sustainability, many hotel managers are concerned about their responsibilities to the society that are defined by the CSR approach (Holcomb *et al.*, 2007; Calveras, 2015). The specific CSR activities may be divided into five domains, covering environmental, ethical, legal, social or philanthropic, and financial issues (Wong and Kim, 2020). Perceptions of CSR practices can affect hotel employees' perceived quality of life, job satisfaction and work engagement (Gürlek and Tuna, 2019; Kim *et al.*, 2020). Furthermore, they can also influence hotel guests' perceived hotel image, satisfaction and loyalty (Palacios-Florencio *et al.*, 2018; Latif *et al.*, 2020), and determine local residents' attitudes towards the hotel industry (Boğan *et al.*, 2020). Therefore, CSR has become a significant strategy to improve the overall performance of hotel businesses (Benavides-Velasco *et al.*, 2014; Ghaderi *et al.*, 2019).

Recent context of hotel business and CSR

First reported in late 2019, the global health crisis brought about by the COVID-19 virus has tremendously affected the socio-economic activities of many countries (Aristovnik *et al.*, 2020; McKibbin and Fernando, 2021). The hotel industry is one of the most severely impacted business sectors (Anguera-Torrell *et al.*, 2021). Thus, many alterations to the normal business activities have been implemented for the industry to survive and thrive. For example, hotels in Macao focused on improving pricing and maintenance practices in the initial stage of the pandemic (Lai and Wong, 2020) to maximize yield from diminishing sales. In the USA, many hotels shut down their peripheral or supporting products and services, such as swimming pools, business centres and fitness centres, during the peak period of the pandemic. These businesses emphasized measures to guarantee social distancing and good sanitation conditions of their core product, the room service (Smart *et al.*, 2021). All these practices were designed to help retain or attract customers to accommodation facilities in the short run (Awan *et al.*, 2020; Atadil and Lu, 2021; Yu *et al.*, 2021). In addition, advanced technologies, such as robots, may also be used to reduce human contact and improve the sanitation condition of the hotels (Shin and Kang, 2020; Kim *et al.*, 2021). Hotel restaurants may expand their services to the food delivery market to guarantee a steady and stable income (Yang *et al.*, 2021).

In the wake of the pandemic, the CSR activities of hotels have taken different shapes. For instance, many hotels are giving free rooms to medical staff and discounted room rates to those who must quarantine (Gürlek and Kılıç, 2021; Qiu *et al.*, 2021; Ramírez-Cervantes *et al.*, 2021). These community-oriented CSR practices have been effective in maintaining stock returns and stakeholder attention on hotels during the pandemic (Qiu *et al.*, 2021). However, employees at quarantine hotels may suffer from symptoms of anxiety, stress and depression since they are directly exposed to the risk of cross-infection while serving quarantined guests (Teng *et al.*, 2020). Quarantined guests may be burdened by the cost of their unwanted long-term stays yet are only allowed limited use of facilities and services in quarantine hotels (Ramírez-Cervantes *et al.*, 2021). Thus, the risk of a double crisis (one internal with the employees and guests, one external with the pandemic) is apparent if management plans and counteractive measures are not carefully prepared.

Hotel business and CSR in Vietnam

Generally, hotel business activities in Vietnam are similar to those in other countries. The ultimate purpose of these activities is to satisfy guests and create loyalty (Le *et al.*, 2018). Concurrently, an increasing number of hotels in Vietnam are also undertaking relatively rudimentary forms of internal CSR, such as waste reduction, separation and recycling, and water conservation (Han *et al.*, 2018; Pham Phu *et al.*, 2019). However, certain external CSR initiatives, such as the establishment of marine protected areas for the benefit of tourists as well as the biophysical environment, may not be appreciated by residents because these can potentially affect their livelihood (ocean-based) in the long term (Svensson *et al.*, 2010). A balance between business and CSR activities, although not easy to achieve, must be thoughtfully considered by hotel businesses, not only in Vietnam but also in other contexts.

Under the global influence of the COVID-19 pandemic, the routine operations of hotels in Vietnam have been largely disrupted. After the first case of infection was detected in early 2020, the Vietnamese government ordered the whole country to implement a strict preventive measure approach, including the immediate closure of its borders leading to a restriction on travel to Vietnam (Huynh, 2020; Nguyen *et al.*, 2021). The border closure policy has significantly decreased international tourist arrivals and reduced the occupancy rates of hotels targeting this market. Losing a major source of revenue, many hotels have transformed their lodgings into paid quarantine facilities for overseas Vietnamese and foreigners who are specifically permitted to enter the country. The quarantine hotels must be approved and supervised by local governments to ensure health and safety procedures and regulations, including appropriate sanitation capabilities, good video surveillance systems, and proper employee training programmes, among others (Hoang *et al.*, 2021). In addition, hotel operations must be aligned with rescue or charter flights, which are only scheduled periodically at appointed airports (e.g. Cần Thơ, Đà Nẵng, Nội Bài – Hà Nội, Tân Sơn Nhất – Hồ Chí Minh city, and Vân Đồn – Quảng Ninh) (Kyodo News, 2020; Truc Anh, 2020). Typically, hotels near an appointed airport may gain a larger benefit from the operation. Under this uncertainty, the operation of quarantine hotels may temporarily be suspended when domestic COVID-19 infectious cases are on the rise, leading to a complete or partial lockdown of a whole area or certain sections of it (Gia Chinh, 2020; Thuy Dung, 2021). Hotels may also be removed from the approved list if they cannot ensure safe operation. Despite all these difficulties, a package of accommodation, food and health control services for 14–21-days' quarantine can generate a considerable revenue for the said hotels during this critical period.

Although the operation of quarantine hotels was initially justified as CSR activities (Vietnam Investment Review, 2020) there are, however, counter opinions suggesting that the income-directed approach of the current period might have weakened the notion of CSR. A study of both the CSR and business aspects of quarantine hotels in Vietnam is thus timely and necessary to document emerging issues and the questionable approach to CSR in the wake of the pandemic.

Study Methods

Internet-based media sources and user-generated content

Considering the exploratory nature of this study, a qualitative method of data collection was deemed appropriate. The method used was to collect secondary data on public opinion from various media sources. On the one hand, this approach was necessary due to the strict rules for social distance and the limitation of face-to-face contact to prevent the spread of the disease to communities. On the other hand, the social distance and lockdown rules have prompted people to utilize media and online platforms for communication and expression of their opinions. Therefore, the authors opted to use the content contributed by social media and their users in the online environment.

Press (e.g. news articles) and user-generated content (e.g. news readers' comments) has been widely used in academic research for decades as an important source of data (Carpenter, 2010; Lu and Stepchenkova, 2015). This type of data is constantly available and easy to access. In addition, the topics covered by this information are abundant. Moreover, the use of a secondary source of data such as this, which is readily available, will reduce the risk of exposure to COVID-19 that researchers must face if they choose to collect a primary set of data, especially when the subjects are the quarantine hotels.

In general, press and user-generated content has been used by many researchers during the COVID-19 pandemic. Salem *et al.* (2021), for example, analysed newspapers, magazines and other public information sources to reveal the initiatives and practices adopted by the Egyptian government concerning the tourism and hospitality industry in that country. In another study, Park *et al.* (2020) examined posts on Reddit, a popular social-networking site in the USA, to find out what employees in the tourism and hospitality industry were saying about COVID-19. Alternatively, Mehta *et al.* (2021) assessed guest reviews at a global scale to extract the reasons that made them dissatisfied with hotel products and services in the first 9 months of 2020. The most recent study on strategic responses of the hotel sector to COVID-19 highlight the need to use local language to examine the reactions of hotel businesses reported in media (Le and Phi, 2021).

In this study, writer-created content (the news articles) and reader-generated content (the comments on the news articles) were used to determine the impressions that the public in Vietnam had regarding the business and CSR activities of quarantine hotels there. The topic and the undertaking, thus, differed from those of previous studies on the hotel industry during the COVID-19 pandemic both in Vietnam and in other countries.

Data collection and analysis

The authors collected the database for this research in the middle of May 2021. Two methods were employed. First, we used the search engines provided by the news websites displayed on the 'Top Sites in Vietnam' list (Alexa, 2021) to locate the articles and the attached comments. Second, we utilized Google Search to find articles published on other websites or those missed in the previous attempt.

In the search process, the authors adopted several combinations of 'hotel' (*khách sạn*) and 'quarantine' (*cách ly*) in the Vietnamese language as the keywords. We carefully read the title and the content of each article to decide if they were suitable for the purpose of this study. Specifically, we focused on articles which reported the user-based activities of quarantine hotels and ignored the pieces that only mentioned administrative operations or similar information. A database of relevant articles and the comments was compiled into a sample of 46 articles and 409 comments for the current study. The database included the articles published on 18 websites between March 2020, when the first quarantine hotels started their operations in Vietnam, through to May 2021, when this research was conducted. The comments were mostly contributed by readers of thanhnien.vn, tuoitre.vn and vnexpress.net. The commenters (i.e. people who post comments on blogs and news websites), as they informed us in their comments, included prospective and actual quarantine hotel guests, people who were familiar with the hotel business, and people in the public, among others.

Researchers adopted the thematic analysis method (Braun and Clarke, 2006) to extract from the articles and the comments the major elements of their impressions. In our study, each theme was treated as an image or impression. Each image portrayed represented a cognitive impression (i.e. a thought) that the public had of the quarantine hotels. We also implemented an incident analysis (Gremler, 2004) to identify the events that triggered these impressions, and to divide the periods of time over which commenters developed their impressions about quarantine hotels. It should be noted that these periods slightly differed from the actual waves of COVID-19 observed in Vietnam so far (first wave: January–April 2020; second wave: July–December 2020; third wave: January–March 2021; fourth wave: April 2021 onwards).

For data analysis, the database was examined by a reading and theming process. Two researchers independently examined the themes and compared themes to reach a consensus, before finalizing the themes as suggested by Kassarjian (1977). Later, we grouped the themes in two general categories: (i) positive impressions; and (ii) negative impressions. The themes in one category shared a similar kind of thought about either the good or the bad of the quarantine hotels. Unfortunately, the number of the negative images (n = 10) seemed to exceed this of the positive ones (n = 7). Some of the images were even two sides of a coin, such as those of service quality, quarantine procedures and regulations, and burdens on quarantined people. Details of the themes and categories are displayed in Table 11.1. Finally, relevant quotes from the articles and comments demonstrating the themes were selected for illustration.

Table 11.1. Themes that emerged from data analysis. Compiled by the authors.

Themes	Frequency
Positive impressions	
Participating in and supporting the national COVID-19 prevention campaign	17
Guarantee of quarantine procedures and regulations	15
Reducing burdens for quarantined people	7
Satisfying the needs of quarantined people	7
Good service quality	6
Increasing workload and income for hotel employees	6
Increasing revenues for hotels	4
Negative impressions	
Expensive	13
Bad service quality	9
Increasing pressures on hotel employees	6
Increasing the spread of COVID-19 to the communities	6
Taking advantage of quarantined people	6
Violation of quarantine procedures and regulations	6
Unpublicized costs and cost structure	4
Inappropriate cost structure	2
Increasing the burdens of quarantined people	2
Reducing the meaningfulness of supportive activities	2

Findings

The analysis of the articles and their comments showed that the quarantine hotels were attached to a variety of images, both positive and negative (see Table 11.1). However, these images seemed to evolve over time, differing among the four stages. In addition, they were also affected by the main orientation of the hotels in each period, whether it was CSR-oriented, business-oriented, or a combination of both.

The first period – March and April 2020

On 20 March 2020, the Ministry of Health released its first guideline for quarantine at business hotels in Vietnam (Ministry of Health, 2020). Under this decision, hotels in big cities, such as Đà Nẵng, Hạ Long (Quảng Ninh province), Hà Nội and Hồ Chí Minh city, allowed the local governments to use their facilities as quarantine stations. The quarantine services were free thanks to hotels volunteering their services and also the subsidies from the local governments. Among themes, there is sight of resentment and disagreement with the free-quarantine service policy imposed by the government.

> When asked to accept guests returning from [a South Asian country] where COVID-19 was spreading, Ms. [name of a hotel manager] was worried. Other hotel managers shared this feeling because their future business might be affected if these guests were COVID-19 positive. However, after a thorough search of information, they realised that [it would be okay] if they were careful and implemented proper preventive measures.
>
> (laodong.vn writer, 17 March 2020)

The impressions of the quarantine hotels are: (i) 'participating in and supporting the national COVID-19 prevention campaign'; (ii) 'guarantee of quarantine procedures and regulations'; (iii) 'reducing burdens for quarantined people'; (iv) 'satisfying the needs of quarantined people'; and (vi) 'good service quality'.

> Many hotels in Đà Nẵng let [the local government] use their hotels as quarantine facilities amid the complicated progress of the COVID-19 pandemic.
> (baogiaothong.vn writer, 19 March 2020)

> This hotel [a hotel in Hạ Long City, Quảng Ninh province] not only gave [quarantined guests] free rooms and services but also provided them with free meals.
> (kenh14.vn writer, 19 March 2020)

> I think [the government] should let people to choose paid quarantine service. It would reduce stresses for quarantined people on the one hand, and financial burdens for the government on the other.
> (a quarantined guest quoted by halongcity.gov.vn writer, 26 March 2020)

The images that the public had of the quarantine hotels in the first period were thus mainly positive. That the hotels volunteered to participate in the government quarantine scheme partly contributed to develop these impressions.

The second period – September and October 2020

In contrast to the previous period, the policy of free accommodation was suspended in the second period. More hotels of the mid- to high range of three to five stars had joined the operation in addition to a limited number of facilities previously nominated or approved by local governments. Without government subsidy, quarantine hotels requested their guests pay at commercial rates for the use of facilities and meals. That, in turn, resulted in customers' complaints about being overcharged. For example, a group of Vietnamese people returning from South Korea publicly expressed their concerns with the price arrangements in the press and social media. Consequently, the negative image of 'expensive' became very apparent. Other negative images included, for example, 'hidden costs and cost structure', 'inappropriate cost structure', 'taking advantage of quarantined people' and 'bad service quality'. All these seemed to be the direct result of the 'expensive' impression.

> Hundreds of guests returning from Korea got angry when informed that their daily quarantine fee had increased from 1.3 million Vietnamese dong (VND) per day to 5 million VND per day.
> (tuoitre.vn writer, 30 September 2020)

> We must participate in labor export programs because we are poor. But unlucky for us, our current incomes are not better than in Vietnam. Now, with COVID-19, we lost our jobs and must go home. But we cannot afford such high quarantine fees. Please understand our situation. We are afraid that we may die in a foreign country, either due to COVID-19 or to our lack of money. We have just left Vietnam recently.
> (tuoitre.vn reader, 1 September 2020)

> They [the hotels] need to separate the costs for rooms, meals and medical services and inform their guests. They cannot only provide a price package then increase it without consideration. The government must control these hotels so as to not let them take advantage of the situation to cover their lost.
> (tuoitre.vn reader, 30 September 2020)

> On 26 October 2020, Ha Long City's People Committee received a claim from a guest regarding bad meal quality at [the hotel's name]. This guest returned from Japan through Vân Đồn International Airport and is quarantining in this hotel. When giving a costly meal (150.00 VND) with only meager dishes, this guest became irritated and reported the incident on the Facebook page of Vietnam's Embassy to Japan and other sites.
> (thanhnien.vn writer, 28 October 2020)

While the dominant themes emerging in the analysis of public opinion on the quarantine hotels in the second period were about their business aspects and were generally negative, positive impressions about their CSR also existed. Certain themes such as 'participating in and supporting the national COVID-19 prevention strategy', 'guarantee of quarantine procedures and regulations' and 'satisfying the needs of quarantined people' were still mentioned. Another positive business-linked image, 'increasing work load and income for hotel employees', was also talked about more frequently.

> Let's request all provinces to prepare their utmost capacities to receive guests returning from overseas. This is a multi-purpose undertaking: easing the entries for experts and tourists and creating jobs for local people.
> (tuoitre.vn reader, 1 September 2020)

The second period marked the 6-month operation of quarantine hotels in Vietnam. The quarantine service had gradually been commercialized. As a result, a mixed compilation of both the CSR- and the business-related images is observed.

The third period – November 2020 to February 2021

In this period, press releases did not cover a lot of news on the quarantine hotels, when it became a regular service with an increasing number of hotels participating and quarantined guests had been staying in these facilities. However, the press had started reporting some misbehaviour observed at quarantine hotels, such as 'violation of quarantine procedures and regulations' of hotel employees and administrative personnel.

> Two staff at [the hotel's name] became F1s [i.e. individuals who had close contacts with COVID-19 patients] after taking phone calls and changing money for a COVID-19 patient.
> (laodong.vn writer, 4 November 2020)

> Many quarantine hotels in Hồ Chí Minh City do not have security gates.
> (laodong.vn writer, 5 December 2020)

Overall, the images of the quarantine hotels, both positive and negative, were only sporadically portrayed in the third period. The mixed aggregation of the CSR-related and business-related images were still observed, but their boldness was less apparent.

The fourth period – April and May 2021

In this period, the public opinions redirect from the guests back to the hotel facilities. Questions on safety measures, facilities and the infrastructure of quarantine hotels were addressed in the media. The first one involved a case of COVID-19 among the quarantine hotel staff in Yên Bái province. Noticeably, this case has a direct connection with the Delta (Indian) variant, which has been found to be very active (Thiagarajan, 2021) and could spread to the guests who stayed there, and to the local communities. The second incident was related to bad food served at a hotel in Khánh Hòa province, and the third one was about the inappropriate cost structure found in a hotel in Hà Nội.

Driven by these incidents, the impressions about quarantine hotels became mostly negative, including 'increasing the spreading of COVID-19 to the communities', 'violation of quarantine procedures and regulations', 'bad service quality', 'expensive', 'taking advantage of quarantined people' and 'increasing the burdens of quarantined people', among other less bold ones. Positive impressions, if they were reported elsewhere ('guarantee of quarantine procedures and regulations' and 'reducing burdens for quarantined people'), were overshadowed.

> Yên Bái province must recheck the procedures implemented at their quarantine hotels, identifying the mistakes that led to the transmission of COVID-19 to hotel staff. Such inaccurate procedures are very alarming.
> (an epidemiologist quoted by laodong.vn writer, 27 April 2021)

> According to the guideline decreed by the Ministry of Health, all quarantine hotel staff must follow the quarantine procedures and regulations. So why could such a transmission happen? How many people did this staff meet in the last 10 days? We need to deliberately invest this case. It will be very deadly if we let the disease to spread in the communities.
> (vnexpress.net reader, 27 April 2021)

After more than a year of operation, quarantine hotels had been largely transformed from public service facilities (free) to commercialized service units (paid for). The initial positive and CSR-related images of the quarantine hotels were fading. On the other hand, the negative and business-related images were rapidly increasing. Quarantine at hotels was found to be not as efficient and effective as it was in the earlier periods in controlling the virus and its infection in the communities. Perhaps a change of the quarantine hotel model, and even a renewal of the overall prevention policy, must be considered.

Discussion

The findings of the study reported in this chapter showed that the quarantine hotels in Vietnam have attracted both CSR- and business-related images. On the one hand, the 'participating in and supporting the national COVID-19 prevention campaign', 'reducing burdens for quarantined people', and 'increasing workload and income for hotel employees' image elements clearly showed that the public did recognize and appreciate the contributions that the quarantine hotels were offering to society, the guests, and the employees, respectively. These images were mostly positive. On the other hand, the cost-, price- and service quality-related images demonstrated that the public were also aware of the business aspect of the quarantine hotels. These images could either be positive or negative. When they were negatively perceived, they did affect the whole operation of the hotels and weakened the images and CSR practices. In summary, participation in the quarantine operation has created new images of the hotels. These images, whether they are CSR related or business related and positive or negative, have doubtlessly affected the public's opinions about the hotels, at least during the COVID-19 pandemic.

The structure of the image components has seemed to change over time (see Fig. 11.1). When the hotels began their quarantine operations in the first period, positive CSR-related impressions dominated, such as 'participating in and supporting the national COVID-19 prevention campaign' and 'reducing burdens for quarantined people'. The voluntary and altruistic nature of this activity, supported by the fact that it was free, helped create these images. However, after hotels started charging fees, adverse business-related impressions increased, such as 'expensive' and 'bad service quality'. The relationship between the price and the quality has been increasingly scrutinized, and the mismatches between the two discovered and talked about. These mismatches have made the impression of the quarantine hotels deteriorate.

These findings have further demonstrated that the operation of quarantine hotels in Vietnam is very much in line with that in other countries, such as China and Spain (Qiu *et al.*, 2021; Ramírez-Cervantes *et al.*, 2021). Thus, the CSR activities of these hotels have a universally common value. This, together with other business-oriented tactics (Lai and Wong, 2020; Smart *et al.*, 2021), have provided the hotel industry with an additional strategic tool to survive the unprecedented pandemic (Le and Phi, 2021).

Theoretical contributions

Previous studies have observed that CSR can help improve the performance of hotels in general (Benavides-Velasco *et al.*, 2014; Ghaderi *et al.*, 2019). However, they often treated CSR and business activities as different domains and overlooked the overlaps between them. This study identified that CSR could bring in business opportunities, on the one hand, but on

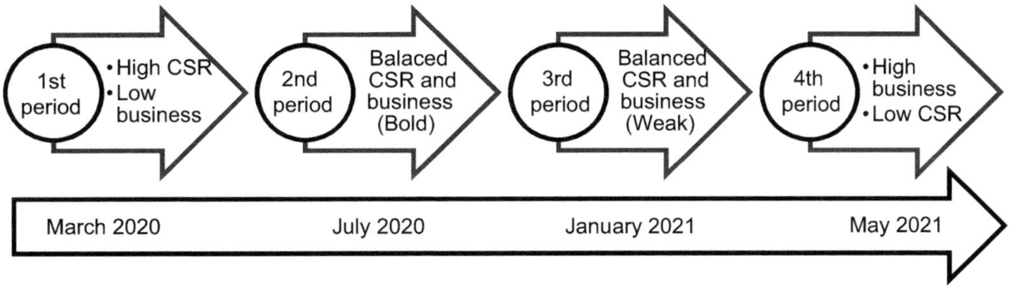

Fig. 11.1. Public perception of corporate social responsibility (CSR) in quarantine hotels. Compiled by the authors.

the other hand, such activities could weaken or even destroy the former if they are done without much consideration, especially regarding current social situations. Being a quarantine facility may not dilute the image or reputation of a hotel. None the less, being a quarantine facility to make money at all costs from people who are in a weak position is negatively affecting their reputation.

In addition, this study, implemented during the COVID-19 pandemic, has found that CSR activities adopted by hotels in Vietnam are not only limited to the consideration of environmental issues (Han *et al.*, 2018; Pham Phu *et al.*, 2019). The activities may be extended to include those involving social matters. By being outwardly responsible to the society at large, hotel businesses can also display and ensure their responsibilities to their employees and guests (Gürlek and Kılıç, 2021; Qiu *et al.*, 2021; Ramírez-Cervantes *et al.*, 2021). However, if the business aspect of the CSR activities is emphasized, the meaningfulness of these responsibilities will be diminished. A balance between business and CSR activities, therefore, must be searched for and respected by hotel business owners and operators.

Practical implications

The hotel sector in Vietnam is struggling to cope with the COVID-19 pandemic. Receiving guests for quarantine purposes will help to partially relieve the existing problems. However, as responsible business units, hotels should not pass the burdens on to their guests. An ethical approach should be adopted and followed in order not to mar the original positive purpose of the hotels, and hence, their image.

As the findings of this study suggest, hotels must correctly calculate and publicly inform their guests and other stakeholders about the costs of their accommodation and its components. Also clear policy directions from the authorities are important to ensure more transparency. The suitability and reasonability of the costs must be carefully considered. For example, additional and extra costs for medical and quarantine purposes may be acceptable, but those for public administrative costs are not. Charging an expensive and unreasonable price will bring in immediate financial benefits but will harm potential business opportunities in the long term.

In addition, hotels must be aware that guests who must pay a high price for involuntarily chosen products and services might be irritated from the very beginning. Therefore, the guarantee of service quality, especially that of the food and quarantine products (e.g. sanitisers and temperature-measuring devices), must be implemented with utmost care. When there are claims, proper crisis management methods (Lai and Wong, 2020) must be carried out to reduce guest dissatisfaction and public criticism. Again, short-term benefits cannot be ignored, but long-term benefits must also be properly regarded.

Concluding Remarks

Operating a hotel business during COVID-19 is certainly an ambiguous task. By accepting guests with high risk to stay for 2 weeks or more on the premises, many hotels in Vietnam are receiving a stable income to partly help with their financial issues and can project the image of a socially responsible business unit. In addition, the employees of these hotels can continue working and earning money, while certain groups of quarantined guests have their needs met. However, a balance between business and CSR activities must be suitably calculated and observed. The image of a greedy business should be avoided to better cultivate and improve the success of future operations.

This study, however, has some inherent limitations. First, only a limited collection of secondary data could be used. Consequently, the compilation of the images attached to the quarantine hotels is incomplete. In addition, the differences in the images of quarantine hotels in different geographical areas at different stages of the COVID-19 pandemic, if there were any, could not be identified. Second, changes in the images could only be determined in a qualitative manner. Statistical evidence is not produced. Third, the opinions of hotel managers and staff were not considered. Their comments on their hotel's perceived image, as well as their reactions towards this image, were also not identified.

Considering these limitations, future studies may want to collect primary data from hotel guests and other stakeholders, especially beyond May 2021, to better understand the impression that the public has of quarantine hotels. They may also want to use a quantitative text-mining method (Mehta *et al.*, 2021) to grasp the evolution of the public impression in a statistical manner. In addition, future attempts may involve the stakeholders on the supply and administration sides. They may also want to explore the supply chain of the domestic quarantine process, including the individual operations of the airlines (transporting the guests), the airports (receiving and scrutinizing the guests), the hotels (servicing the guests), and other supporting agencies (e.g. providing catering and healthcare services to the guests). These efforts will help create a fuller picture of both business- and CSR-related activities of hotels and other related entities in Vietnam, not only during COVID-19 but also the following period.

References

Alexa (2021) Top Sites in Vietnam. Available at: https://www.alexa.com/topsites/countries/VN (accessed 17 May 2021).

Alexander, M., Chen, C.C., MacLaren, A. and O'Gorman, K.D. (2010) Love motels: oriental phenomenon or emergent sector? *International Journal of Contemporary Hospitality Management* 22(2), 194–208.

Anguera-Torrell, O., Aznar-Alarcón, J.P. and Vives-Perez, J. (2021) COVID-19: hotel industry response to the pandemic evolution and to the public sector economic measures. *Tourism Recreation Research* 46(2), 148–157.

Aristovnik, A., Keržič, D., Ravšelj, D., Tomaževič, N. and Umek, L. (2020) Impacts of the COVID-19 pandemic on life of higher education students: a global perspective. *Sustainability* 12(20), 8438.

Atadil, H.A. and Lu, Q. (2021) An investigation of underlying dimensions of customers' perceptions of a safe hotel in the COVID-19 era: effects of those perceptions on hotel selection behavior. *Journal of Hospitality Marketing & Management* 30(6), 655–672.

Awan, M. I., Shamim, A. and Ahn, J. (2020) Implementing 'cleanliness is half of faith' in re-designing tourists, experiences and salvaging the hotel industry in Malaysia during COVID-19 pandemic. *Journal of Islamic Marketing* 12(3), 543–557.

Benavides-Velasco, C.A., Quintana-García, C. and Marchante-Lara, M. (2014) Total quality management, corporate social responsibility and performance in the hotel industry. *International Journal of Hospitality Management* 41, 77–87.

Boğan, E., Dedeoğlu, B.B. and Dedeoğlu, S.B. (2020) The effect of residents' perception of hotel social responsibility on overall attitude toward tourism. *Tourism Review*. Available at: https://doi.org/10.1108/TR-08-2019-0353

Braun, V. and Clarke, V. (2006) Using thematic analysis in psychology. *Qualitative Research in Psychology* 3(2), 77–101.

Calveras, A. (2015) Corporate social responsibility strategy in the hotel industry: evidence from the Balearic islands. *International Journal of Tourism Research* 17(4), 399–408.

Carpenter, S. (2010) A study of content diversity in online citizen journalism and online newspaper articles. *New Media & Society* 12(7), 1064–1084.

CBRE (2020) *CBRE Hotels Insights: Implications of COVID-19 on the Vietnam Hospitality Market*. CBRE, Ho Chi Minh City.

Cvelbar, L.K., Grün, B. and Dolnicar, S. (2021) 'To clean or not to clean?' Reducing daily routine hotel room cleaning by letting tourists answer this question for themselves. *Journal of Travel Research* 60(1), 220–229.

Ghaderi, Z., Mirzapour, M., Henderson, J.C. and Richardson, S. (2019) Corporate social responsibility and hotel performance: a view from Tehran, Iran. *Tourism Management Perspectives* 29, 41–47.

Gia Chinh (2020) Da Nang stops receiving repatriation flights. *VnExpress*, 24 July. Available at: https://e.vnexpress.net/news/news/da-nang-stops-receiving-repatriation-flights-4135777.html (accessed 20 May 2021).

Gremler, D.D. (2004) The critical incident technique in service research. *Journal of Service Research* 7(1), 65–89.

Gürlek, M. and Tuna, M. (2019) Corporate social responsibility and work engagement: evidence from the hotel industry. *Tourism Management Perspectives* 31, 195–208.

Gürlek, M. and Kılıç, İ. (2021) A true Friend becomes apparent on a rainy day: corporate social responsibility practices of top hotels during the COVID-19 pandemic. *Current Issues in Tourism* 24(7), 905–918. DOI: 10.1080/13683500.2021.1883557.

Han, H., Lee, J.-S., Ho Le, T.T. and Kim, W. (2018) Water conservation and waste reduction management for increasing guest loyalty and green hotel practices. *International Journal of Hospitality Management* 75, 58–66.

Hanoi Department of Health (2021) Danh sách các khách sạn được phép làm cơ sở cách ly phòng chống COVID-19 và thông tin về danh mục hồ sơ đề nghị cho người nhập cảnh và cách ly tại Hà Nội [Hotel list and immigration procedure]. Hanoi Department of Health, 15 April. Available at: https://soyte.hanoi.gov.vn/tin-tuc-su-kien-noi-bat/-/asset_publisher/4IVkx5Jltnbg/content/thong-tin-tren-website-cua-so-y-te-ve-danh-muc-ho-so-e-nghi-cho-nguoi-nhap-canh-va-cach-ly-tai-ha-noi (accessed 14 May 2021).

Ho Chi Minh City Department of Health (2021) Danh sách các khu cách ly y tế tập trung tại khách sạn trong phòng, chống dịch COVID-19 do người được cách ly tự nguyên chi trả trực thuộc Ủy ban nhân dân quận, huyện [Hotel list]. Ho Chi Minh City Department of Health, 24 March. Available at: http://www.medinet.hochiminhcity.gov.vn/chuyen-muc/danh-sach-cac-khu-cach-ly-y-te-tap-trung-tai-khach-san-trong-phong-chong-dich-c-cmobile2-42190.aspx (accessed 14 May 2021).

Hoang, T.G., Truong, N.T. and Nguyen, T.M. (2021) The survival of hotels during the COVID-19 pandemic: a critical case study in Vietnam. *Service Business* 15, 209–229.

Holcomb, J.L., Upchurch, R.S. and Okumus, F. (2007) Corporate social responsibility: what are top hotel companies reporting? *International Journal of Contemporary Hospitality Management* 19(6), 461–475.

Hung, K., Wang, S. and Tang, C. (2015) Understanding the normative expectations of customers toward Buddhism-themed hotels: a revisit of service quality. *International Journal of Contemporary Hospitality Management* 27(7), 1409–1441.

Huynh, T.L.-D. (2020) The COVID-19 containment in Vietnam: what are we doing? *Journal of Global Health* 10(1), 010338. DOI: 10.7189/jogh.10.010338.

Kassarjian, H.H. (1977) Content analysis in consumer research. *Journal of Consumer Research* 4(1), 8–18.

Kim, H., Rhou, Y., Topcuoglu, E. and Kim, Y.G. (2020) Why hotel employees care about corporate social responsibility (CSR): using need satisfaction theory. *International Journal of Hospitality Management* 87, 102505.

Kim, S., Kim, J., Badu-Baiden, F., Giroux, M. and Choi, Y. (2021) Preference for robot service or human service in hotels? Impacts of the COVID-19 pandemic. *International Journal of Hospitality Management* 93, 102795.

Kyodo News (2020) 1st Japan–Vietnam flight takes off after virus restrictions eased. Kyodo News, 25 June. Available at: https://english.kyodonews.net/news/2020/06/15da8f785978-1st-japan-vietnam-flight-takes-off-after-virus-restrictions-eased.html (accessed 20 May 2020).

Lai, I.K.-W. and Wong, J.W.-C. (2020) Comparing crisis management practices in the hotel industry between initial and pandemic stages of COVID-19. *International Journal of Contemporary Hospitality Management* 32(10), 3135–3156.

Latif, K.F., Pérez, A. and Sahibzada, U.F. (2020) Corporate social responsibility (CSR) and customer loyalty in the hotel industry: a cross-country study. *International Journal of Hospitality Management* 89, 102565.

Le, D. and Phi, G. (2021) Strategic responses of the hotel sector to COVID-19: toward a refined pandemic crisis management framework. *International Journal of Hospitality Management* 94, 102808.

Le, V.H., Nguyen Pham, T.H., Pham, L. and Berry, R. (2018) Technology readiness and satisfaction in Vietnam's luxury hotels. *International Journal of Management and Decision Making* 18(2), 183–208.

Lin, I.Y. (2009) The combined effect of color and music on customer satisfaction in hotel bars. *Journal of Hospitality Marketing & Management* 19(1), 22–37.

Ling, L., Guo, X. and He, L. (2012) Optimal pricing strategy of hotel for long-term stay. *International Journal of Services Technology and Management* 17(1), 72–86.

Liu, Y., Teichert, T., Rossi, M., Li, H. and Hu, F. (2017) Big data for big insights: investigating language-specific drivers of hotel satisfaction with 412,784 user-generated reviews. *Tourism Management* 59, 554–563.

Lu, W. and Stepchenkova, S. (2015) User-generated content as a research mode in tourism and hospitality applications: topics, methods, and software. *Journal of Hospitality Marketing & Management* 24(2), 119–154. DOI: 10.1080/19368623.2014.907758.
Lu, X. (2020) E-sports hotel, a brand new personalized hotel in the age of e-sports. In: Markopoulos, E., Goonetilleke, R., Ho, A. and Luximon, Y. (eds) *Advances in Creativity, Innovation, Entrepreneurship and Communication of Design. AHFE 2020. Advances in Intelligent Systems and Computing*. vol 1218. Springer, Cham, Switzerland, pp. 213–218.
McKibbin, W. and Fernando, R. (2021) The global macroeconomic impacts of COVID-19: seven scenarios. *Asian Economic Papers* 20(2), 1–30.
Mehta, M.P., Kumar, G. and Ramkumar, M. (2021) Customer expectations in the hotel industry during the COVID-19 pandemic: a global perspective using sentiment analysis. *Tourism Recreation Research* 167, 1–18. DOI: 10.1080/02508281.2021.1894692.
Ministry of Health (2020) *Quyết định 1246/QĐ-BYT ngày 20/3/2020 Về việc ban hành 'Hướng dẫn tạm thời cách ly y tế tập trung tại khách sạn trong phòng, chống dịch COVID-19 do người được cách ly tự nguyện chi trả'*, [Temporary guideline for paid quarantine hotels]. Ministry of Health, Hanoi.
Nguyen, T.V., Tran, Q.D., Phan, L.T., Vu, L.N., Truong, D.T. *et al.* (2021) In the interest of public safety: rapid response to the COVID-19 epidemic in Vietnam. *BMJ Global Health* 6(1), e004100. DOI: 10.1136/bmjgh-2020-004100.
Palacios-Florencio, B., del Junco, J.G., Castellanos-Verdugo, M. and Rosa-Díaz, I.M. (2018) Trust as mediator of corporate social responsibility, image and loyalty in the hotel sector. *Journal of Sustainable Tourism* 26(7), 1273–1289.
Park, E., Kim, W.-H. and Kim, S.-B. (2020) Tracking tourism and hospitality employees' real-time perceptions and emotions in an online community during the COVID-19 pandemic. *Current Issues in Tourism* 83, 1–5. DOI: 10.1080/13683500.2020.1823336.
Pham Phu, S.T., Fujiwara, T., Hoang, M.G., Pham, V.D. and Tran, M.T. (2019) Waste separation at source and recycling potential of the hotel industry in Hoi An city, Vietnam. *Journal of Material Cycles and Waste Management* 21, 23–34.
Qiu, S., Jiang, J., Liu, X., Chen, M.-H. and Yuan, X. (2021) Can corporate social responsibility protect firm value during the COVID-19 pandemic? *International Journal of Hospitality Management* 93, 102759.
Ramírez-Cervantes, K.L., Romero-Pardo, V., Pérez-Tovar, C., Martínez-Alés, G. and Quintana-Diaz, M. (2021) A medicalized hotel as a public health resource for the containment of COVID-19: more than a place for quarantining. *Journal of Public Health* 43(1), 89–97. DOI: 10.1093/pubmed/fdaa129.
Salem, I.E., Elkhwesky, Z. and Ramkissoon, H. (2021) A content analysis for government's and hotels' response to COVID-19 pandemic in Egypt. *Tourism and Hospitality Research* 7(1). DOI: 10.1177/14673584211002614.
Shin, H. and Kang, J. (2020) Reducing perceived health risk to attract hotel customers in the COVID-19 pandemic era: focused on technology innovation for social distancing and cleanliness. *International Journal of Hospitality Management* 91, 102664. DOI: 10.1016/j.ijhm.2020.102664.
Smart, K., Ma, E., Qu, H. and Ding, L. (2021) COVID-19 impacts, coping strategies, and management reflection: a lodging industry case. *International Journal of Hospitality Management* 94, 102859.
Sohrabi, B., Vanani, I.R., Tahmasebipur, K. and Fazli, S. (2013) An exploratory analysis of hotel selection factors: a comprehensive survey of Tehran hotels. *International Journal of Hospitality Management* 31(1), 96–106.
Stringam, B.B. (2008) A comparison of vacation ownership amenities with hotel and resort hotel amenities. *Journal of Retail & Leisure Property* 7, 186–203.
Sun, J., Chen, P.-J., Ren, L., Shih, E.H.-W. and Ma, C. (2021) Place attachment to pseudo establishments: an application of the stimulus-organism-response paradigm to themed hotels. *Journal of Business Research* 129, 484–494.
Svensson, P., Rodwell, L.D. and Attrill, M.J. (2010) The perceptions of local fishermen towards a hotel managed marine reserve in Vietnam. *Ocean & Coastal Management* 53(3), 114–122.
Tanford, S., Raab, C. and Kim, Y.-S. (2012) Determinants of customer loyalty and purchasing behavior for full-service and limited-service hotels. *International Journal of Hospitality Management* 31(2), 319–328.
Teng, X., Teng, Y.-M., Wu, K.-S. and Chang, B.-G. (2021) Corporate social responsibility in public health during the COVID-19 pandemic: quarantine hotel in China. *Frontiers in Public Health* 9, 620930. DOI: 10.3389/fpubh.2021.620930.

Teng, Y.-M., Wu, K.-S., Lin, K.-L. and Xu, D. (2020) Mental health impact of COVID-19 on quarantine hotel employees in China. *Risk Management and Healthcare Policy* 13, 2743–2751. DOI: 10.2147/RMHP.S286171.

Thiagarajan, K. (2021) Why is India having a COVID-19 surge? *The BMJ* 373, n1124. DOI: 10.1136/bmj.n1124.

Thuy Dung (2021) Viet Nam suspends inbound flights for fear of new COVID-19 variants. Vietnam National Administration of Tourism, 6 January. Available at: https://vietnamtourism.gov.vn/english/index.php/items/15530 (accessed 20 May 2021).

Truc Anh (2020) Vietjet stages COVID-19 rescue flights. *Vietnam Investment Review*, 21 August. Available at: https://vir.com.vn/vietjet-stages-covid-19-rescue-flights-78706.html (accessed 20 May 2021).

Vietnam Investment Review (2020) Business community expresses their social responsibility. *Vietnam Investment Review*, 6 April. Available at: https://vir.com.vn/business-community-expresses-their-social-responsibility-75373.html (accessed 14 May 2021).

Vietnam National Administration of Tourism (2020) *Vietnam Tourism Annual Report 2019*. Labour Publishing House, Hanoi.

Wang, L., Wang, X.-K., Peng, J.-J. and Wang, J.-Q. (2020) The differences in hotel selection among various types of travellers: a comparative analysis with a useful bounded rationality behavioural decision support model. *Tourism Management* 76, 103961.

Wong, A.K.-F. and Kim, S. (2020) Development and validation of standard hotel corporate social responsibility (CSR) scale from the employee perspective. *International Journal of Hospitality Management* 87, 102507.

Wu, C.H.-J. and Liang, R.-D. (2009) Effect of experiential value on customer satisfaction with service encounters in luxury-hotel restaurants. *International Journal of Hospitality Management* 28(4), 586–593.

Yang, F. X. and Lau, V.M.-C. (2015) 'LuXurY' hotel loyalty – a comparison of Chinese Gen X and Y tourists to Macau. *International Journal of Contemporary Hospitality Management* 27(7), 1685–1706.

Yang, F. X., Li, X., Lau, V.M.-C. and Zhu, V.Z. (2021) To survive or to thrive? China's luxury hotel restaurants entering O2O food delivery platforms amid the COVID-19 crisis. *International Journal of Hospitality Management* 94, 102855.

Yu, J., Seo, J. and Hyun, S.S. (2021) Perceived hygiene attributes in the hotel industry: customer retention amid the COVID-19 crisis. *International Journal of Hospitality Management* 93, 102768.

Zhou, L., Ye, S., Pearce, P.L. and Wu, M.-Y. (2014) Refreshing hotel satisfaction studies by reconfiguring customer review data. *International Journal of Hospitality Management* 38, 1–10.

12 Policies and Stakeholder Engagement in Cultural Events

Phuong-Anh Dang[1]*, Le-Anh Trinh[1], Giang T. Phi[2] and Huong T. Bui[3]
[1]Vietnam National University, Hanoi; [2]VinUniversity, Hanoi, Vietnam; [3]Ritsumeikan Asia Pacific University (APU), Japan

Introduction

The development of event tourism has accelerated in the last two decades at many locations across Vietnam, however, the study of the event sector in Vietnam is still in the early stages (Bui, 2010; Cao, 2014). Event tourism is universally important for its social and cultural roles, as well as its critical function in place marketing and destination branding (Andersson and Getz, 2008). Studies on event tourism in the Vietnamese context have examined cultural festivals organized by the local communities (Tran, 1986; Duong, 2004; Bui, 2009; Dang, 2017). Beyond their cultural significance, festivals and cultural events are also frequently held for enhancing the local image, attracting tourists and bringing economic benefits to localities. The characteristics, the process and the subsequent impacts of these events depend on the policies issued by different levels of the government, along with the diverse stakeholders' roles and engagement. This chapter analyses policies and stakeholders' engagement in the planning and management of festivals and cultural events.

The chapter starts with the key literature on planned and cultural events, event tourism, and stakeholder theory within the context of event policy making, before moving on to the key context of Vietnam. Research methods and findings are then presented through the illustration of four representative vignettes: (i) Gióng Festival (Hà Nội city); (ii) Thành Tuyên Festival (Tuyên Quang province); (iii) Tịch Điền Festival (Hà Nam province); and (iv) the Tea Festival (Thái Nguyên province). By comparing and contrasting these four events, the authors deepen insight on connections and interactions among stakeholders for event planning and management, in turn providing recommendations for practical implications and future research.

Literature Review

Classification of planned events

Events have been steadily gaining interest in the research community (Stokes, 2008; Patterson and Getz, 2013; Connell *et al.*, 2015). Planned events are divided into four main categories associated with their main venues and the facilities required (Getz and Page, 2016):

- Business events (or the MICE (meetings, incentives, conferences and exhibitions)

*Corresponding author: dangphuonganh@vnu.edu.vn

 Vietnam Tourism: Policies and Practices
(eds Huong T. Bui *et al.*)
DOI: 10.1079/9781789242782.0012

sector) require convention and exhibition centres, and include numerous smaller private parties and functions held in restaurants, hotels or resorts;

- Sport events also require special-purpose facilities including athletic parks, arenas and stadia;
- Entertainment events such as concerts are generally provided by the private sector and utilize many types of venues;
- Festivals and other cultural celebrations are less dependent on facilities and can use any relevant public or private venue such as parks, streets, theatres and concert halls.

Another classification of planned events is based on a portfolio of events by type, season, target market and value. Within the portfolio model, two terms are notable: 'mega events' and 'hallmark events'. Mega events are those that have a large cost and attract a large number of visitors, have significant media reach and impact on the built environment and local populations (Müller, 2015). Such events include the Olympics, the Asian Games and the World Expo. On the other hand, a hallmark event in the Ritchie (1984) definition is a major one-time or recurring event, developed primarily to enhance the awareness, appeal and profitability of a tourism destination. A primary function of hallmark events is to provide the host community with an opportunity to secure high prominence in the tourism marketplace (Hall, 1989). Accordingly, hallmark events often vary in scale, being either grassroots or state-led. Festivals and cultural events are often used as examples of hallmark events, given their strong links to specific tourism destinations and local communities.

Classification of cultural festivals

Festivals are often referred to as cultural events, which are celebrations that focus on specific themes and/or special characteristics of the host community's culture (Yeoman *et al.*, 2012). Traditional festivals are based on long-term folk traditions, religious beliefs, celebration of seasons and agricultural harvests, or other cultural aspects of indigenous communities (Hitchcock and Nuryanti, 2000). In contrast, modern festivals tend to have stronger focus on performing and visual arts, as well as food and drink. Many modern festivals are developed with the intention of driving event tourism and can have an international focus, as opposed to the local or domestic focus of traditional festivals (Yeoman *et al.*, 2012). However, the boundary between traditional and modern events is not always clear. The term 'reinvention of tradition' or 'reinvented tradition' (see e.g. Beiner, 2007) is derived from anthropology and signifies a creative process that enables the adoption, revival or reconstruction of traditional cultural practices that can be applied to identify traditional and modern festivals. The extent to which a cultural festival relies on local tradition or reinvented tradition may have implications for stakeholder engagement with event planning and management, as well as subsequently impacting event outcomes.

Event tourism: positive and negative impacts

The use of events as tourist attractions is widely recognized in the event literature (e.g. Mules and Faulkner, 1996; Getz, 1997; McDonnell *et al.*, 1999). Recent research also demonstrates the role of hallmark events in filling the gap of a seasonal drop in tourism demand by foreign tourists in the off-peak season, as local residents and domestic visitors provide a substitute form of demand stimulated by events (Connell *et al.*, 2015). Positive economic impacts of events are not constrained to tourists' expenditure at the event but are spread throughout the local economy, with tourism-related services such as transportation, accommodation, restaurants and shops being the major beneficiaries (McDonnell *et al.*, 1999). All levels of government also receive taxes generated from this expenditure, which constitutes one of the biggest benefits of event tourism (Turco, 1995).

Events may not only serve to attract tourists, but also assist in the development or maintenance of community or regional identity (Hall, 1992; Getz *et al.*, 1998). Getz and Page (2016), for instance, found that events can create positive images for a destination and help to brand or reposition cities, as well as contributing to place marketing by making cities more liveable and attractive to visit and revisit. In addition, event tourism can act as a catalyst for other forms of desired development such as heritage preservation, urban renewal, community capacity building, and volunteerism, thereby generating a long-term or permanent legacy (Getz and Page, 2016).

While successful events contribute significantly to the positive image and economic growth of a destination, poorly developed and managed events can also lead to many negative impacts. For example, bad media reports about problems associated with an event might devastate the host destination's image (Getz, 2002; Boo and Busser, 2005). Another potential negative impact of event tourism is 'leakage', the amount of tourist expenditure that leaks out of the local economy (Dwyer *et al.*, 2007).

In terms of cultural impacts, various concerns have been expressed that events might bring commercialization and commodification of local culture to meet the needs of visitors (Getz, 1997; Whitford, 2009). Moreover, the exposure of the host community to global homogenization, brought about by both a given event and the out-of-region visitors that the event attracts, may have adverse impacts on the community's traditional culture and values, leading to a gradual loss of identity (Douglas *et al.*, 2001).

Event tourists, local communities and committees, entrepreneurs, private companies or government officials all play certain roles in event planning and management, and would be affected either positively or negatively by these activities. Moreover, many of these impacts last well beyond the duration of the event itself. In order to maximize benefits and reduce negative impacts, event tourism needs to be approached from a stakeholders' perspective, with a link to event planning and policy making.

Stakeholder engagement in cultural events

Stakeholders are defined as groups or individuals who can affect, or be affected by the achievement of an organization's objectives (Freeman, 1984). Event stakeholders are understood as 'those people and groups with a stake in the event and its outcomes, including all that [are] participating in the event production, sponsors and grant-givers, community representatives, and everyone impacted by the event' (Getz, 1997, p. 15). The context of event tourism is complex, involving a diverse range of stakeholders concerned with achieving their own goals (Phi *et al.*, 2014). It is essential that the planners and managers involved in event tourism recognize the existing relationships among stakeholder groups as they interact with events in the respective destination (Getz and Page, 2016). Hence, organizational stakeholder theory is also applicable in event and festival settings (Kristiansen *et al.*, 2016).

With particular reference to event context, Getz *et al.* (2006) contend that the planning and management of festivals and events result from a set of managed stakeholders' relationships. They proposed a typology of event stakeholders based on functional roles, including: (i) facilitators providing cash grants, sponsorships and in-kind support; (ii) co-producers such as performers and associations who are involved through voluntary activities in creating the event experience; (iii) suppliers acting as service providers inside the festival value chain; (iv) the audience; and (v) those impacted, including the local community and special interest groups.

In addition, stakeholders can also be classified into 'primary' and 'secondary' groups, based on their saliency and reciprocal relationships with a given event's organization (Clarkson, 1995). Primary stakeholders are those who are most involved and engaged with an event; they have a direct impact on the survival and continued success of an organization or an event. In contrast, although secondary stakeholders may still be important and can assume multiple roles in the organization or event, their participation usually has less direct impact on the outcome (Todd *et al.*, 2017).

An event can only occur as a result of the interactions of key stakeholders with the managing organization (Clarkson, 1995). Accordingly, certain attributes have to be present in order to identify the primary stakeholders who play the most important roles in the survival and success of an event. Mitchell *et al.* (1997) defined the characteristics of those primary stakeholders as follows: (i) their power could be coercive, utilitarian or normative; (ii) their legitimacy can be discerned when a stakeholder's claim is perceived as being appropriate, socially acceptable, and expected based on individual, organizational or social norms; and (iii) urgency would be felt when the stakeholder's claim is perceived as time sensitive and critical or highly important to event management. They further underline that power is transitory; it can be acquired as well as lost, and these attributes are not static but rather dynamic. Importantly, the attributes such as power, legitimacy and urgency depend very much on event policies shaped by the managing organization and/or the authorities, defined as both the formal rules, regulations, strategies and guidelines, as well as the operational planning and management decisions made in relation to the event (Phi, 2010). Event stakeholders and their roles are thus often specified or influenced by event policies.

In the past, many hallmark and cultural events had long histories and often evolved organically from the local community, and therefore represented a grassroots or bottom-up approach to event policy making (Todd *et al.*, 2017). In recent years, the public sector (e.g. general public administration and specific tourism-related government bodies) has increasingly become involved in the planning and management of cultural events, be it through institutionalizing and reviving mature events or creating new events with the aspiration of becoming hallmark events (Getz *et al.*, 2012). With significant resources at its disposal (e.g. funding, legal authority and access to public areas), the public sector's involvement often represents a top-down approach to event planning and policy making (Phi, 2010). The success of cultural events and festivals, however, relies very much on sustained support and resources from local stakeholders where the events are based, and requires that the public sector gradually take on a more flexible approach to better engage a diverse range of stakeholders in event planning and management. Consequently, the roles and engagement of stakeholders are not fixed but are constantly changing as an event evolves over time (Todd *et al.*, 2017).

Policy Context of Events and Cultural Festivals in Vietnam

A system of legal documents and policies contributes to establishing the broader institutional structure that enables Vietnamese cultural events and festivals to be run. In particular, Decree Number 5, issued by the Central State on July 1998, states: 'Culture is the mental foundation for our society, being both the goal and driver for social-economic development' (Vietnam Communist Party, 2013). The central state's cultural policy has enabled a range of traditional practices, including traditional cultural events and festivals, to be revived and to flourish (Ngo, 2013). In 2001, the *Law on Cultural Heritage* officially legalized the rights and responsibilities of the various levels of government and communities in preserving and promoting Vietnam's cultural heritage and traditions. In particular, Article 25 of this law states that festivals are considered intangible heritage and are under the support of the state governments for their continuation and for the revitalization of cultural values, while measures should be taken to minimize superstitious practices and excessive commercialization of these events (National Assembly of the Socialist Republic of Vietnam course X, 2001). In 2009, four major amendments were made to Article 25 of the *Law on Cultural Heritage*, which further strengthened the supporting role of state government. These include to: (i) establish favourable conditions for planning festivals; (ii) encourage the organization of traditional cultural activities and traditional arts as part of the festivals; (iii) revitalize selectively traditional rituals in festivals; and (iv) encourage the promotion of the origins, contents and unique value of traditional festivals (National Assembly of the Socialist Republic of Vietnam course XII, 2009).

More specific regulations and policies on events have also been issued and amended periodically to meet different social, economic and political objectives. The *Regulations on Opening Ethnic Traditional Festivals*, issued in 1989, and the *Regulations on Festivals*, promulgated in 1994, define festivals as traditional events, characterized by history, revolutionary history and religion. The 2001 *Regulations on Event Organisation* also include 'events originated from overseas' as an event category (Ministry of Culture and Information, 1989; 1994; 2001). However, the major objectives of these cultural events remain relatively unchanged, which are 'to educate the population on traditions, cultural heritage conservation and meeting cultural and religious needs' (Bui, 2011).

The coalition between cultural events and tourism destination branding has featured in the newest regulation, *Decree on Event Management and Organisation*, number 110/2018/NĐ-CP, promulgated in 2018. A new definition for 'cultural festival' has been introduced as 'activities which introduce and promote unique values of culture and sport, and tourism potential in relation to the landscapes and people of Vietnam' (Vietnam Government, 2018). Subsequently, the institutionalization of event planning and management has been established on par with the recognition of tourism for its contribution to development (Tran, 2005) through the promulgation of the *Law on Tourism* 2005 and its amendment in 2017 (National Assembly of the Socialist Republic of Vietnam course XI, 2005, 2017).

For the purpose of promotion of local identity, the concepts for cultural festivals centre on local core values in three formats: (i) revitalization or scale-up traditional community festivals to be touristic cultural festivals, for example Hội Gióng (Hà Nội), Bà Chúa Xứ (An Giang) and Tịch Điền (Hà Nam); (ii) establishment of new festivals based on unique local resources, for example Thái Nguyên Tea Festival (since 2011), Buôn Ma Thuột Coffee Festival (since 2005), Nha Trang Ocean Festival (since 2010), Đà Lạt Flower Festival (2005), etc.; and (iii) establishment of new festivals from foreign concepts, such as Hạ Long Carnival (2007) and Đà Nẵng Firework Festival (2008). These festivals all recur annually or biennially to strengthen their local tourism profile.

Study Methods

The purpose of this study is to apply stakeholder theory to the practical setting of event tourism in Vietnam. Therefore, the issues concerning event stakeholders should emerge from the reality of multiple contexts and cases. A qualitative research methodology was thus adopted for this research using a case-study design conducted at four different locations in northern Vietnam.

Background of the four festivals

Four cases were selected based on the following key criteria: (i) geographically dispersed throughout the north; (ii) taking place during one of the two main festival seasons in Vietnam (spring and autumn); and (iii) representative of two main formats in event categories (i.e. grassroots and state-led) with the planning and management varied with reference to key stakeholders. Four events that suit these criteria are Gióng Festival in Hà Nội, Tịch Điền Festival in Hà Nam, Thành Tuyên Festival in Tuyên Quang and the Tea Festival in Thái Nguyên (see Table 12.1).

The four festivals have different organizational structures, ranging from grassroots (Gióng Festival), to local state-led (Thành Tuyên Festival and Thái Nguyên Tea Festival) and central state-led (Tịch Điền Festival). The involvement of diverse stakeholders at different levels of administration substantially illustrates the complexity of cultural events in Vietnam.

Gióng Festival

Gióng Festival is one of the most significant and long-lasting traditional folk festivals in the Red River Delta of northern Vietnam, organized to honour and glorify the hero-turned-saint Gióng (Thánh Gióng). The legend of Thánh Gióng is a symbol for the thousand-year-old Vietnamese tradition of fighting against foreign invaders. The Gióng Festival has continued to be held to date in different suburban areas of Hà Nội. The two locations which host this event on a significant scale are Sóc Temple (Sóc Sơn Commune)

Table 12.1. Profile of the festivals. Compiled by the authors.

Festival name	Category	Length (days)	Visitor numbers
Gióng Festival	Traditional folk festival	3	10,000
Thành Tuyên Festival	Modern folk festival	2	5,000
Tịch Điền Festival	Revived traditional festival	3	1,000
Thái Nguyên Tea Festival	Modern international festival	3	4,000

and Phù Đổng (Gia Lâm Commune). The festival is held annually for 3 days in January in the lunar calendar, attracting more than 10,000 visitors while over 1000 people voluntarily participated in its planning and implementation.

Thành Tuyên Festival

Thành Tuyên Festival is a modern folk festival, initiated in 2008 on the occasion of the mid-autumn festival. During this occasion, local people of Tuyên Quang province make large, colourful and creative lanterns for a Light Parade along major streets. The parade immediately attracts many local participants and visitors from nearby areas. The Light Parade is a ritual rooted in the folk belief that the sacred Gods and secular people are connected during the brightest full moon of the year. Besides this underpinning folk belief, the festival attracts and retains visitors thanks to its colourfulness, festive environment and communal spirit, all of which connect people.

Tịch Điền Festival

Tịch Điền Festival is a revived traditional festival with more than a thousand years of history. The festival originates in the rituals of tribal society, performed by tribal leaders on the ground of rice fields at the beginning of the year to pray for a prosperous harvest season. Tịch Điền Festival is held at the beginning of January in the lunar calendar. The first ritual took place as far back as 987 in Đọi Sơn (present-day Hà Nam province), led by King Lê Đại Hành. The festival was revived during the Nguyễn Dynasty in the 19th and early 20th centuries. After a long period of suspension (since the mid-1940s), in 2009 the Vietnamese Ministry of Culture, Sport and Tourism decided to revive and reorganize this festival for the celebration of 'A Thousand Years of Thang Long – Hà Nội in 2010' (Trinh, 2018). Over the last decade, Tịch Điền Festival has continued to be a success and expanded into a major hallmark event of Hà Nam province.

Thái Nguyên Tea Festival

Thái Nguyên Tea Festival is a biennial modern international festival, initiated in 2011 by the Thái Nguyên government. Tea is a unique product and a major driver for economic activities of Thái Nguyên province. Thái Nguyên tea has long been promoted and is well perceived in both domestic and international markets, yet many local tea producers were struggling after the global financial crisis of 2007–2008. The Festival was thus created to facilitate and coordinate the production and consumption of tea and its derivative commodities such as foods and drinks in domestic and international markets (Thái Nguyên Provincial People's Committee, 2015). There is no fixed date for the festival, and it has been previously arranged in February, March and, most recently, April. Besides being a meeting venue for tea companies to connect and expand their markets, the festival also promotes the intangible heritages of Thái Nguyên through various festive activities such as tea ceremonies, tea roads and tea tours to traditional villages (Thái Nguyên Newspaper, 2010).

Data collection and analysis

The researchers identified key stakeholders by their multiple roles in event planning and management of the four chosen festivals. Respondents were further identified through a snowball sampling technique. In-depth interviews are a common practice in qualitative research; they are useful for gathering detailed information about a phenomenon or to explore insights into new issues (Boyce and Neale, 2006). Interviews were undertaken with respondents at a convenient location within each festival, ensuring that respondents were comfortable within their natural setting, allowing for more accurate representations of the phenomena to emerge (see Guba and Lincoln, 1989).

A team researcher examined reports of the festivals issued by respective event organizations to identify 22 primary stakeholders and 32 secondary stakeholders, including local residents, visitors and media agencies, with whom to conduct in-depth interviews. Equal distribution of respondents across the stakeholders across case sites and groups ensured the contribution of diverse insights. By site, 13 in-depth interviews were conducted at Gióng Festival, 15 interviews at Tịch Điền Festival, 12 interviews at Thành Tuyên Festival and 14 interviews at

Thái Nguyên Tea Festival. A semi-structured interview is usually guided by certain themes with a prepared set of questions, yet still ensuring a high level of flexibility by allowing room for new emerging questions and issues to be addressed (Boyce and Neale, 2006). These interviews were in between 40 and 150 minutes and utilized a semi-structured approach with a list of questions derived from the literature, including the following main questions: (i) What are your roles and how do you contribute to the event?; (ii) How were resources mobilized for events?; (iii) How were the events regulated?; and (iv) What are the relationships between involved stakeholders?

The interview data was subjected to thematic analysis, which involves the identification and reporting of key patterns and themes that 'capture something important about the data in relation to the research question' (Braun and Clarke, 2006, p. 82). The analysis was guided by stakeholder theory, as well as the key approaches to event planning and policy making mentioned in the literature. In addition, categories for thematic coding are pre-described into drivers for stakeholder involvement, the traditional aspect of the events and different layers of administration to show the dimension of top-down and bottom-up in decision making and governance. All of these themes are derived from the literature and examination of secondary resources, such as general reports by event organizations and outcomes issued by relevant administrators, obtained before conducting the fieldwork and analysis of the interview data.

Findings and Discussion

Endogenous drivers for stakeholder engagement

This section discusses endogenous drivers for stakeholder engagement in event planning and management by analysing two cases – Gióng and Thành Tuyên Festivals. Although the traditions on which the two festivals are based are both from within and from outside the folk beliefs of the locals, the festival that is rooted in communal life has longevity and viability.

Gióng Festival has been a grassroots cultural event by locals and for locals; a form of cultural practice symbolizing the local community's pride. The locals are the primary stakeholders, who have designed the concept, planned and managed the festival over the long history of the community. Prior to its institutionalization in 2010, this festival was performed on a discrete, small scale, utilizing budgets from local communities, contributed on a voluntary basis. The ritual and festive elements of the festival were performed based on local oral history for generations. The 73-year-old resident, keeper of Phù Đổng temple said:

> The key attraction of Hội Gióng is the Assembled Fight performance created based on the legend of Gióng's fight against the invaders. This performance has taken place every year since I was a child. The scale of the performance was very small back then, only including five to six people.
>
> (keeper of Phù Đổng temple)

It was the year when the Gióng Festival at Sóc Temple (Sóc Sơn, Hà Nội) and Phù Đổng Temple (Gia Lâm, Hà Nội) was designated by UNESCO as intangible heritage and the festival was included as part of the activities for the landmark event of the 1000-year anniversary of Thăng Long – Hà Nội. The designation has scaled up this local festival to the level of being a national hallmark event, thus its significance has expanded beyond the boundaries of a grassroots event. Subsequently, the role and involvement of stakeholders have been transformed in four ways. First, the purpose of the festival has changed from performing local cultural rituals to being a showcase of national cultural tradition linking to the history of a thousand-year-old Hà Nội. Second, by scaling up this event, it attracts domestic and international tourists who are external to the locals. Third, the role of event planning and management has shifted from the local community to the local province. Fourth, the budget for the event, originally mobilized from the community, is now supported by the central state. As the festival has grown in scale and its objectives have changed, the stakeholder profile has also changed, which now includes state media agencies and tourism organizations.

Despite the change in scale, the local community and local authorities remain the

Fig. 12.1. Gióng Festival – performance preparation. Photo courtesy of Phuong-Anh Dang.

primary stakeholders of this event, representing a bottom-up approach to event planning and policy making. In particular, the community maintains ownership over the local traditions and rituals being practised during the event period. The locations of the event are original, linking to the long-existing oral history and social memory of the locals (see Fig. 12.1). The festival has been managed through administrative divisions such as the Department of Culture and the Centre for Heritage Conservation. The local authorities, in this case acts as coordinator, connecting the local community to the central state to achieve the multiple objectives of protecting intangible heritage, attracting state budget and fulfilling the socio-economic objectives of the festival. For example, the Head of Sóc Sơn Department of Culture, Vice-chairman of the Festival's organizing committee stated:

> The representatives of the People's Committee at the commune and village levels participated throughout the planning of the festival, in order to ensure security, safety and cleanliness for locals and tourists during the festival. The main activities of the festival are still actively planned and implemented by the local residents, and the state created an official plan to be approved by the higher level of the government. The marketing and promotion of the festival is primarily carried out by the organising

> committee (i.e. the local state), as the local residents do not have existing relationships with the media.
>
> (Head of Sóc Sơn Department of Culture)

However, along with the recognition of Gióng Festival by UNESCO as intangible heritage, the same respondent also indicates that stronger intervention of the state and the academic community in 'standardizing' the rituals performed in the festival:

> The festival plan is proposed annually by the festival's organising committee to higher levels of government based on the discussions and group meetings with representatives of the villages and communes surrounding the heritage sites. Initially, local people's traditions were all integrated into the plan. When the festival was recognised by UNESCO, the festival's organisation was standardised by scientists, and we [now] follow these standards.
>
> (Head of Sóc Sơn Department of Culture)

In contrast to Gióng Festival, Thành Tuyên Festival is a modern one with a reinvented tradition created and managed by the local community that has proven its viability since its inauguration in 2008. Over the years, the local festival has grown to become a Disney-like Light Parade (see Fig. 12.2). The festive element of the event on the night of the full moon is perhaps the key factor attracting visitors to come and return every year, as in the visitor's opinion below:

> This is the fourth year I am coming to this festival, often on the weekend prior to the mid-autumn full moon night. The Light Parade is very colourful with many large and beautifully decorated lights marching along main streets. The illuminated vehicles are diverse in shapes, sizes and functions. People sing, dance, and splash water to each other.
>
> (visitor to the Light Parade)

Having noticed the growing popularity and attention of the events beyond the community scale, the local state institutionalized and controlled the format, performance and budget of the festival. The local state plays a coordinating role connecting the central state to the local community. For example, a resident who participated in the Light Parade said:

> There are rules for the lantern models that are registered into the lantern contest organised by the government. For example, these models are not allowed to have sound, flashing light or water spraying features that are too strong. However, we still made the lanterns our way and

Fig. 12.2. Thành Tuyên Festival – the Light Parade. Photo courtesy of Phuong-Anh Dang.

just did not use these features during the main parade. Outside of the parade, these features are important to create the most festive atmosphere for visitors to enjoy.

(resident and participant in the Light Parade)

In summary, there are two key findings in relation to stakeholder engagement in grassroots festivals. First, the grassroots festivals present an organic endogenous driver, where the existence and success of the festival are underpinned by the pre-existing or invented traditions by locals. Second, when the festivals are scaled up, the transitional role of the local state is significant. On the one hand, local state authorities may represent the local community when the festivals have deep roots in local cultural traditions and rituals. On the other hand, the local state may represent the voice and function of the central state to control the festival when the local community is not the owner of the rituals. While the local communities are the main participants and play essential roles, the local state has increased its influence and administrative functions in relation to the festivals.

Exogenous drivers for stakeholder engagement

In the case of Tịch Điền Festival, the central state-led policy on reinvented tradition required substantial scientific evidence from historical records of the event coming from both written and oral histories of senior community members. The festival's core concept originates in the rice-field grounding memorial ceremony first performed by King Lê Đại Hành in the 10th century, a ritual to honour the local gods who are enshrined to protect humans and lands from disasters. The festive elements are re-enacted through traditional popular games and sports, traditional performances such as a dragon dance and creative new traditions such as graffiti painting on buffalos (see Fig. 12.3). The national level of the festival is envisaged through participation from various levels of government, from the highest-level state officials to high-profile provincial officials. For example, the grounding ritual is now carried out by the current president of Vietnam (Communist Party

Fig. 12.3. Tịch Điền Festival – buffalo painting. Photo courtesy of Le-Anh Trinh.

online newspaper, 2010). The involvement of the central state has been expanded on the vertical axis (from central to provincial and local authorities) as well as the horizontal axis by the relevant administrative office (e.g. from the national Ministry of Information, Culture and Tourism to the local Department of Culture). The chairman of the festival management board said:

> In 2009, after nearly 100 years of suspension, this important national ritual was officially reconstructed at Đọi Sơn, Hà Nam province [on] 5–7 January. From 2010, the festival also had the participation of Vietnam President Nguyễn Minh Triết and [subsequent] President Trương Tấn Sang, along with representatives from other countries and other high-level government officials.
>
> (Chairman of the festival management board)

Alongside the ultimate role of the central state, a group of scholars who research, establish and operate the concept of the festival are also considered to be primary stakeholders. The scholars operationalize the state's mission and vision into plans and strategies to be implemented by local authorities and communities. A lead scholar of the 2019 project stated:

> The Tịch Điền Festival in Đọi Sơn is reconstructed based on the Tịch Điền rituals throughout different imperial dynasties of Vietnam, especially the rituals performed during the Nguyễn dynasty at Huế ancient capital. Based on both written historical books and the oral history of the elders we chose Đọi Sơn as the main location for the festival to reflect the historical location of King Lê Đại Hành's first ritual.
>
> (lead scholar of the 2019 festival)

Although being driven by state policy backed by the academic community, the local communes perform the key activities in the festival. However, due to the long suspension, the social memory of the rice-field grounding ceremony no longer exists among the locals of Đọi Sơn (Duy Tiên, Hà Nam). Instead, the researchers reinvented the tradition mainly from their desk research, and then implanted it into a selected locality. Accordingly, the locals were instructed to perform rituals, popular cultural ceremonies and activities under the guidance of academic culturalists. On the one hand, the performances of the local communities have been well prepared and appeared to be professional to outsiders. On the other hand, since the cultural performance is not rooted in local social and spiritual lives, they are performed more or less as procedures without much belief or commitment. When the planning and management are directed entirely from outside the community, the festival becomes a 'cultural performance'. Interviews with the locals reveal that the participation in and commitment to the festival is largely driven by their financial gains from participation in the event as performers, as stated by a local community leader:

> In the first few years, each of us was paid VND 50,000 per day of rehearsal and VND 100,000 per day of performance. We were very excited back then because our hometown was chosen to host a national festival. However, the pay remained unchanged after many years and many no longer want to take a day off work to participate.
>
> (local community leader)

Despite these drawbacks, Tịch Điền Festival has grown to be a hallmark event in the last 10 years, partly because the destination Hà Nam is relatively close to the main tourist market of Hà Nội. Naturally, tourism and service providers and tourists have also gained more prominent roles in the festival's planning and management, in turn further reducing the community's ownership of the festival. With the goal of driving event tourism, new entertainment activities and contemporary art performances have also been added to the festival programme, in order to increase the festival's attractiveness beyond the repetition of the core rituals through time (Trinh, 2018).

While also strongly led by the state, the concepts behind Thái Nguyên Tea Festival are developed from unique local features. Tea has long been the prime agricultural product of Thái Nguyên, being a springboard for the integrated economic activities of the province to develop. The local authorities play a key role in the planning and management of this festival. At the planning stage, the state draws up the concepts, calls for participation and manages budgets. At the implementation stage, the local authorities act as gatekeepers and monitors of participant and activity selection through licensing and admission. These different roles played by the local authorities are demonstrated in the quote by a representative of Thái Nguyên's Department of Culture, Sport and Tourism:

> We developed the Tea Festival in Thái Nguyên based on the central state's policy, which encouraged each province to develop a hallmark festival that brings out local characteristics. The festival's plan is assigned by the Thái Nguyên local government to the Department of Culture, Sport and Tourism, in coordination with Department of Agriculture and Rural Development and the Vietnam Tea Association ... The exhibition of the local tea business is an important part of the festival, and we have different policies each year to either encourage or limit the number of exhibitors. ... While we encourage donations from the tea businesses, the main source of funding for this festival is still from the province's public funds.
>
> (representative of Thái Nguyên's Department of Culture, Sport and Tourism)

Owing to the main economic objective of the festival, local businesses contribute financially through donation and paying fees, which directly makes up part of the festival budget. The festival in fact functions more or less like an exhibition of local products. For example, a local tea business said:

> We received an invitation from the Vietnamese Tea Association and were given a showroom and some money to cover expenses during the festival. When we registered, the businesses will usually donate to the management to show our support.
>
> (local tea business)

The active role of businesses in this case shows the propensity for public–private partnership in event tourism, an ongoing trend in the event industry in Vietnam. For example, some local governments organize bidding for hosting their hallmark event among large businesses operating in the area, such as the FLC Group, the main sponsor for the Hạ Long Carnival, and Sun Group, which partners as the main sponsor for the Đà Nẵng International Firework Festival. Through these partnerships, the local state plays a directing and hosting role, while local businesses might finance the event and organize some packages of the overall festive activities. This new trend of partnership for event hosting has been growing quickly and has attracted the participation of large businesses, in which context the Thái Nguyên Tea Festival is not exceptional.

As the festival is organized with a top-down approach, local residents (besides those with tea businesses) did not participate in the planning and management of the event. They were simply spectators and event-goers who watched and enjoyed the festive atmosphere without spending on accommodation and food and beverage services. However, as the event has expanded through the years and more activities have been added in different local areas, a larger portion of the Thái Nguyên community also directly participates as service providers or cultural performers to tourists (see Fig. 12.4). The strong administrative role of local government makes this event different both from conventional hallmark events and exhibitions hosted by professional associations, where the private sector directly plans and manages the event, and state governments only play a supporting role through regulation and policy making (Robertson and Wardrop, 2004).

Conceptualization of Stakeholder Engagement in Vietnam Cultural Events

The framework shown in Fig. 12.5 describes the engagement and functions of the stakeholders across the four different events discussed above. Within the four quadrants resulting from the interactions of the drivers, the four festivals under investigation are located along with the objectives set for each event to achieve. The endogenous-exogenous driver presented on the vertical axis refers to the approaches to event planning and management, whether from within the community (endogenous) and bottom up or from outside (exogenous) and top down. There are three levels of stakeholder engagement – macro-level (central state), meso-level (local state) and micro-level (local community) – and these and their respective roles are shown in the centre of the framework diagram.

The horizontal axis refers to the function of cultural tradition. 'Local tradition' is that which is deeply rooted in the social and religious life of the locals. When the tradition is not organic to the locals, having been recently invented by

Fig. 12.4. Thái Nguyên Tea Festival – performance of 'The Beauty of the Land of Tea'. Photo courtesy of Le-Anh Trinh.

locals or imposed on them by external actors, it is termed a 'reinvented tradition'. It is important to note that the use of reinvented tradition at festivals does not necessarily equate to less local engagement. For example, in the case of Thành Tuyên Festival, although the event has been recently created and is not originally rooted in local cultural lives, the acts of making lanterns and the Light Parade are universal traditions shared by many countries in Asia, an essential ritual during the mid-autumn festival. Therefore, this reinvented tradition has grown and gradually became organic to the locality where it was reinvented. The tradition has grown organically because it was reinvented from within the community, performed by the community, and first and foremost, it enriched the local community's social and spiritual life before expanding its influence to the outside and attracting the local government's attention and involvement. Thus, the Thành Tuyên Festival illustrates organic reinvented traditions, while the Tịch Điền Festival shows certain fragility as the revived traditional festival was reconstructed by scholars and implanted into the community under state command.

While the categories of the above-chosen festivals vary widely, the leading and coordinating roles of authorities at different levels are essential. At the macro-level, the central state's top-down involvement is shown in the case of Tịch Điền. At the meso-level, the local state maintains their primary role in all forms of festivals. If the festival is imposed by the central state, the local state monitors the operation of the event. The Thái Nguyên Tea Festival is a case where the local state coordinates a public–private partnership. When the festivals are initiated by the local communities (as with the Thành Tuyên Festival), the local state institutionalizes the activities and operations through regulations. And when the festival has its roots in community life (as with the Gióng Festival), the local state represents the locals and coordinates activities and operations with the higher level. At the micro-level, the local communities have multiple roles to play. They implement the reinvented tradition in the case of Tịch Điền, or actively invent and practise the tradition for the festival in the case of Thành Tuyên Festival. They are the owners of the Gióng Festival where their role is central and primary, while in the

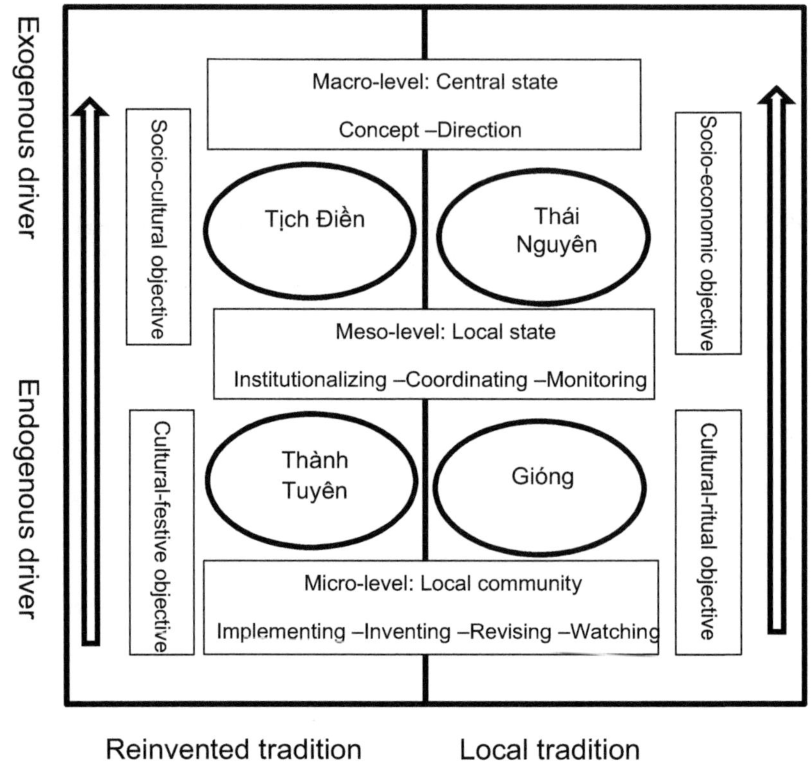

Fig. 12.5. The framework of stakeholder engagement in Vietnam cultural festivals. Approaches to event planning and management and key event objectives are shown on the vertical axis. Details of stakeholder engagement and their roles are indicated in the central rectangles. Compiled by the authors.

case of Thái Nguyên, locals' engagement is still mainly limited to being event attendees and the audience, watching the event take place on their own land. Stakeholder engagement in events in Vietnam is different from conventional practice internationally, where 'states are now juggling two potentially competing roles in: (1) facilitating the bidding for and staging of events; and (2) protecting broader public interests and managing negative impacts' (Phi *et al.*, 2014, p. 406). In Vietnam, the governments have already played primary, indispensable roles and are involved much deeper in all processes of event planning and management of cultural festivals.

As shown in Fig. 12.5, each type of festival has slightly different primary objectives: socio-cultural objectives in the case of Tịch Điền Festival; socio-economic objectives in the case of Thái Nguyên Tea Festival; cultural-festive objectives in the case of Thành Tuyên Festival; and cultural-ritual objectives in the case of Gióng Festival. It is important to note that these primary objectives are not mutually exclusive or static, but often change along with the event's long-term evolution as new stakeholders emerge and gain power, legitimacy and urgency in the event planning and management process. With the involvement of a diverse and expanding number of stakeholders, it is normal for a cultural festival to simultaneously seek to achieve a range of economic, social, cultural and political goals. Many of these goals align with the literature on positive impacts of planned events mentioned earlier, such as enhancing national, regional and local identity, enriching locals' social and spiritual lives, and more recently, attracting tourists, increasing tourism expenditure, and creating future business opportunities.

The event planners and managers, however, need to be mindful of potential negative impacts that may affect the primary groups of stakeholders and hinder the event's achievement of key objectives. This may include economic internal 'leakage', where a large portion of the event budget may be used to hire famous performers from outside for new entertainment activities, while the pay for local performers remains the same (e.g. the case of Tịch Điền Festival). With sustained expansion of cultural festivals to cater to event attendees from outside the community (including foreign tourists) and sustained intervention of the state over a long period, many (revived) traditional festivals (e.g. Gióng Festival) also run the risk of being gradually reconstructed and commodified, with authentic landscapes and rituals gradually replaced by 'official' ones, often those that have more appeal to tourists (Chen and Tao, 2017).

Conclusions

Events and festivals as catalysts for tourism have grown substantially in Vietnam. The flourish of cultural events has been backed by local community's initiatives and the state's policies and regulations connecting culture and tourism. Under these conditions, stakeholders have been actively involved in the creation of modern festivals, or the revitalization of traditional festivals to upscale them into hallmark events. Underpinned by stakeholders and cultural theories, the authors explore the topic of stakeholder engagement at four different festivals in Vietnam, in relation to the event policy and policy-making context. Four key findings are noted.

First, the key stakeholders who play primary roles in the survival of cultural festivals are local communities. Many cultural festivals, either rooted in local tradition or having been invented from local uniqueness and beliefs, are the product of communities, reflecting the communal characteristics of rural Vietnam. These cultural festivals originate from cultural capital, social memory, and legends long existing in the beliefs of the local community and have been practised and performed with modifications for the modern context. Thus, many grassroots events lay foundations for expansion and upscaling to become national or provincial hallmark events.

Second, local governments are perhaps the most important primary stakeholder in most cases. Institutionalizing and upscaling local festivals to become hallmark events often requires strong engagement from local administrators. This key stakeholder connects the local community to the central state by aligning the community events with state policies and objectives. The local states play dual roles, to represent the communities but also to transmit state policies to locals. The role of the local state is therefore rather complex, ranging from institutionalizing to coordinating and monitoring. The local state also plays a connecting and coordinating role in event public–private partnerships.

Third, at the macro-level, the central state creates legal and regulatory frameworks within which events are initiated, operate and evolve. The hallmark event on a national scale might have participation of high-ranking government officials for political and symbolic purposes. The central state, however, tends to be less involved with direct event management issues. In some cases, the central state might appoint particular stakeholders such as academia or private-sector actors to coordinate respective events to ensure their sustainability.

Fourth, stakeholders' roles in cultural festivals are not fixed but constantly evolving, together with the long-term evolution of the event itself. From the case studies, it is noted that there is a constant dialogue between different levels of administration and the local community. The central role of community is indispensable in the planning and management of cultural festivals; therefore, flexibility and negotiation with the community should always be part of stakeholder interactions and implementation of these events.

Under the current COVID-19 pandemic situation, the event industry has been hard hit and many scheduled events have been cancelled or delayed. Even though Vietnam emerged as a success in controlling the early spread of the disease, the strict social distancing and lockdown policies since early 2020 have put most of the event industry on hold. With the global rollout of COVID vaccines and the potential return of tourism, the event industry has opportunities to thrive and bounce back after the COVID-19 pandemic. In this context, the coordination role of various levels of the state remains essential to ensuring health and safety standards and to quickly react to changes in the 'new normal'.

References

Andersson, T. and Getz, D. (2008) Stakeholder management strategies of festivals. *Journal of Convention & Event Tourism* 9(3), 199–220. DOI: 10.1080/15470140802323801.

Beiner, G. (2007) *Remembering the Year of the French: Irish Folk History and Social Memory*. University of Wisconsin Press, Madison, Wisconsin.

Boo, S. and Busser, J.A. (2005) Impact analysis of a tourism festival on tourists destination images. *Event Management* 9(4), 223–237. DOI: 10.3727/152599506776771562.

Boyce, C. and Neale, P. (2006) *Conducting In-depth Interviews: a Guide for Designing and Conducting In-depth Interviews for Evaluation Input*. Path Finder International, Watertown, Massachusetts.

Braun, V. and Clarke, V. (2006) Using thematic analysis in psychology. *Qualitative Research in Psychology* 3(2), 77–101. DOI: 10.1191/1478088706qp063oa.

Bui, Q.T. (2010) Tổ chức lễ hội như l tổ chức sự kiện. Tạp chí *Văn hóa nghệ thuật*, số 318.

Bui, S.H. (2009) *Quản lý lễ hội truyền thống của người Việt*. Nxb Văn hóa dân tộc, HàNội.

Bui, S.H. (2011) Quản lý lễ hội truyền thống của người Việt ở châu thổ Bắc Bộ – nhìn từ các văn bản quản lý từ sau đổi mới đến nay. *Tạp chí Di sản văn hóa, số* 1(34), 17–24.

Cao, D.H. (2014) *Quản lý lễ hội v sự kiện*. Nxb Lao động, Hà Nội.

Chen, H. and Tao, W. (2017) The revival and restructuring of a traditional folk festival: cultural landscape and memory in Guangzhou, South China. *Sustainability* 9(10), 1767. DOI: 10.3390/su9101767.

Clarkson, M.E. (1995) A stakeholder framework for analyzing and evaluating corporate social performance. *Academy of Management Review* 20(1), 92–117. DOI: 10.2307/258888.

Communist Party online newspaper (2010) Chủ tịch nước Nguyễn Minh Triết dự Lễ hội Tịch điền Đọi Sơn 2010. Available at: https://dangcongsan.vn/chinh-tri/chu-tich-nuoc-nguyen-minh-triet-du-le-hoi-tic h-dien-doi-son-2010-332909.html (accessed 20 November 2020).

Connell, J., Page, S. and Meyer, D. (2015) Visitor attractions and events: responding to seasonality. *Tourism Management* 46(3), 283–298. DOI: 10.1016/j.tourman.2014.06.013.

Dang, A.T.P. (2017) The relationship between intangible cultural heritage and community development: a case study of kite playing in Vietnam's Red River Delta. *Journal of Social Sciences and Humanities* 3(2), 203–220.

Douglas, N., Douglas, G. and Derrett, R. (eds) (2001) *Special Interest Groups: Context and Cases*. Wiley, Brisbane, Queensland, Australia.

Duong, S. V. (2004) *Lễ hội Việt Nam trong sự phát triển du lịch*. Trường Đại học Văn hóa, Hà Nội.

Dwyer, L., Forsyth, P. and Spurr, R. (2007) Economic impact of sport events: a reassessment. *Tourism Review International* 10(4), 207–216. DOI: 10.3727/154427206779367118.

Freeman, E.R. (1984) *Strategic Management: a Stakeholder Approach*. Pitman, Boston, Massachusetts.

Getz, D. (1997) *Event Management and Event Tourism*. Cognizant Communication Corporation, New York.

Getz, D. (2002) *Event Management and Event Tourism*, 2nd edn. Cognizant Communication Corporation, New York.

Getz, D. and Page, S.J. (2016) Progress and prospects for event tourism research. *Tourism Management* 52(2), 593–631. DOI: 10.1016/j.tourman.2015.03.007.

Getz, D., Anderson, D. and Sheehan, L. (1998) Roles, issues, and strategies for convention and visitors' bureaux in destination planning and product development: a survey of Canadian bureau. *Tourism Management* 19(4), 331–340. DOI: 10.1016/S0261-5177(98)00035-1.

Getz, D., Andersson, T. and Larson, M. (2007) Managing festival stakeholders: concepts and case studies. *Event Management* 10(2/3), 103–122. DOI: 10.3727/152599507780676689.

Getz, D., Svensson, B., Peterson, R. and Gunnervall, A. (2012) Hallmark events: definition, goals and planning process. *International Journal of Event Management Research* 7(1), 47–67.

Guba, E. and Lincoln, Y. (1989) *Fourth Generation Evaluation*. Sage, Newbury Park, California.

Hall, C.M. (1989) The definition and analysis of hallmark tourist events. *Geo Journal* 19(3), 263–268. DOI: 10.1007/BF00454570.

Hall, C.M. (1992) *Hallmark Tourist Events: Impacts, Management and Planning*. Belhaven, London.
Hitchcock, M. and Nuryanti, W. (eds) (2000) *Building on Batik: the Globalization of a Craft Community*. Ashgate, Aldershot, UK.
Kristiansen, E., Strittmatter, A. and Skirstad, B. (2016) Stakeholders, challenges and issues at a co-hosted youth Olympic event: lessons learned from the European Youth Olympic Festival in 2015. *The International Journal of the History of Sport* 33(10), 1152–1168. DOI: 10.1080/09523367.2016.1238822.
McDonnell, I., Allen, J. and O'Toole, W. (1999) *Festival and Special Event Management*. Wiley, Milton, Queensland, Australia.
Ministry of Culture and Information (1989) *Regulations on Opening Ethnic Traditional Festivals*, No 54 VHQC. Publisher of the Law Library, Hanoi.
Ministry of Culture and Information (1994) *Regulations on Festivals*, with the decision No 636 QĐ-QC. Publisher of the Law Library, Hanoi.
Ministry of Culture and Information (2001) *Regulations on Event Organisation*, with the decision No 39/2001/QĐ-BVHTT. Publisher of the Law Library, Hanoi.
Mitchell, R.K., Agle, B.R. and Wood, D.J. (1997) Toward a theory of stakeholder identification and salience: defining the principle of who and what really counts. *Academy of Management Review* 22(4), 853–886. DOI: 10.2307/259247.
Mules, T. and Faulkner, B. (1996) An economic perspective on special issue. *Tourism Economics* 2(2), 107–117. DOI: 10.1177/135481669600200201.
Müller, M. (2015) What makes an event a mega-event? Definitions and sizes. *Leisure Studies* 34(6), 627–642. DOI: 10.1080/02614367.2014.993333.
National Assembly of the Socialist Republic of Vietnam course X (2001) *Law on Cultural Heritage*, No 28/2001/QH10. Publisher of the Law Library, National Assembly of the Socialist Republic of Vietnam, Hanoi.
National Assembly of the Socialist Republic of Vietnam course XI (2005) *Law on Tourism, No 44/2005/ QH11*, the 7th meeting. Publisher of the Law Library, National Assembly of the Socialist Republic of Vietnam, Hanoi.
National Assembly of the Socialist Republic of Vietnam course XII (2009) *The Law on Amending and Supplementing a Number of Articles of the Law on Cultural Heritage*, the 5th meeting. Publisher of the Law Library, National Assembly of the Socialist Republic of Vietnam, Hanoi.
National Assembly of the Socialist Republic of Vietnam course XIV (2017) *Law on Tourism*, No 09/2017/ QH14. Publisher of the Law Library, National Assembly of the Socialist Republic of Vietnam, Hanoi.
Ngo, D.T. (2013) Giữ gìn nét đẹp của lễ hội. Tạp chí *Du lịch Việt Nam*, số 3/2013, tr.21.
Patterson, I. and Getz, D. (2013) At the nexus of leisure and event studies. *Event Management* 17(3), 227–240. DOI: 10.3727/152599513X13708863377836.
Phi, G.T. (2010) Problem identification in event planning and management: the 2009 Australian World Rally Championship. Honours dissertation, Southern Cross University, Gold Coast, Queensland, Australia.
Phi, G., Dredge, D. and Whitford, M. (2014) Understanding conflicting perspectives in event planning and management using Q method. *Tourism Management* 40(3), 406–415. DOI: 10.1016/j.tourman.2013.07.012.
Ritchie, J.R.B. (1984) Assessing the impacts of hallmark events: conceptual and research issues. *Journal of Travel Research* 23(1), 2–11. DOI: 10.1177/004728758402300101.
Robertson, M. and Wardrop, K. (2004) Events and the destination dynamic: Edinburgh Festival, entrepreneurship and strategic marketing. In: Yeoman, I., Robertson, M., Ali-Knight, J., Drummond, S. and McMahon-Beattie, U. (eds) *Festivals and Events Management: an International Arts and Culture Perspective*. Routledge, London, pp. 115–129.
Stokes, R. (2008) Tourism strategy making: insights to the events tourism domain. *Tourism Management* 29(2), 252–262. DOI: 10.1016/j.tourman.2007.03.014.
Thái Nguyên Newspaper (2010) Thái Nguyên đăng cai Festival Tr quốc tế. Available at: https://baothaingu yen.vn/tin-tuc/thai-nguyen-de-nhat-danh-tra/thai-nguyen-dang-cai-festival-tra-quoc-te-47229-46228.html (accessed 20 November 2020).
Thái Nguyên Provincial People's Committee (2015) *Decision on approving the script of Tea Culture Festival at the third Thai Nguyen Tea Festival*, No 2738/QĐ-UBND. Thái Nguyên Provincial People's Committee, Thái Nguyên, Vietnam.
Todd, L., Leask, A. and Ensor, J. (2017) Understanding primary stakeholders' multiple roles in hallmark event tourism management. *Tourism Management* 59(1), 494–509. DOI: 10.1016/j.tourman.2016.09.010.

Tran, D.T. (2005) *Nhap mon khoa hoc du lịch [Introduction to Tourism]*, 4th edn. Vietnam National University (VNU), Hanoi.

Tran, Q.V. (1986) Lễ hội một cái nhìn tổng thể. Tạp chí Văn hóa dân gian, số 1, tr.3–6.

Trinh, A.L. (2018) *Quản lý lễ hội truyền thống gắn với phát triển sản phẩm du lịch*. Viện Văn hóa Nghệ thuật Quốc gia, Việt Nam.

Turco, D.M. (1995) Measuring the tax impacts of an international festival: justification for government sponsorship. *Festival Management and Event Tourism* 2(3–4), 191–195. DOI: 10.3727/106527095792315585.

Vietnam Communist Party (2013) 'Báo cáo chính trị của Ban Chấp hành Trung ương Đảng khóa VII tại Đại hội đại biểu to n quốc lần thứ VIII của Đảng', *Văn kiện Đại hội Đảng thời kỳ đổi mới và hội nhập (Đại hội VI, VII, VIII, IX, X, XI)*, Nxb. Chính trị quốc gia Sự thật, Hà Nội, 2013, tr. 356–357.

Vietnam Government (2018) *Decree on Event Management and Organisation*, No 110/2018/NĐ-CP. Publisher of the Law Library, Vietnam Government, Hanoi.

Whitford, M. (2009) A framework for the development of event public policy: facilitating regional development. *Tourism Management* 30(5), 674–682. DOI: 10.1016/j.tourman.2008.10.018.

Yeoman, I., Robertson, M., Ali-Knight, J., Drummond, S. and McMahon-Beattie, U. (eds) (2012) *Festival and Events Management*. Routledge, London.

13 Human Resources Management Practices in the Tourism Industry

Thu G. Hoang[1]*, Huong H. Do[2] and Andrew Le[3]
[1]Hanoi University, Hanoi, Vietnam; [2]Thai Binh Duong University, Nha Trang, Vietnam; [3]International College of Management Sydney (ICMS), Sydney, Australia

Introduction

Tourism has been an important economic driver and employment generator for Vietnam in the last 10 years. The exponential growth of tourism from 2016 to 2019 exceeded the range of major forecasting indicators, demanding a reappraisal of the development strategy. In 2018, the Prime Minister of Vietnam approved the new *Master Plan for Việt Nam Tourism Development to 2030 with Vision to 2050* (VNAT, 2018a), to ensure that responsible tourism development contributes to the overall socio-economic development, environmental protection and poverty reduction. However, Vietnam's tourism development faces various challenges, namely the rapid visitor growth, especially in the low-paying segment, and the increase in accommodation supply. These, in turn, have increasingly put strains on tourism infrastructure, transportation systems, human resources (HR), the natural environment and local communities (Dinh *et al.*, 2019). Among these challenges, those associated with HR are the most pressing and hold the most promise as a more capable workforce will be at the forefront of the solutions to the industry's other challenges.

Acknowledging the necessity to help organizations deal with these challenges from within, this chapter reviews the common HR practices of businesses in the tourism industry in Vietnam, focusing on the issues related to these practices. Commencing with a brief review of the HR practices in the tourism and hospitality sector contexts, the chapter then draws an overall picture of the HR for Vietnam tourism and discusses the widespread concerns about the quantity and quality of tourism workers. Finally, based on the review of extant studies and reports concerning HR management in the tourism sector, specific recommendations are provided to the industry's key stakeholders.

HR Practices in the Tourism and Hospitality Industry

Kramar *et al.* (2015) argue that HR management covers three of the five crucial business strategy implementation aspects (i.e. task design, selection, training, development of people, and reward systems). Companies need to consider the following HR practices to implement their strategy: (i) job analysis and design; (ii) employee recruitment and selection; (iii) employee learning and development; (iv) performance management; (v) pay structure, incentives and benefits; and (vi) industrial relations (see Table 13.1).

*Corresponding author: thuhg@hanu.edu.vn

(eds Huong T. Bui *et al.*)

DOI: 10.1079/9781789242782.0013

Table 13.1. A menu of human resource (HR) practices. Adapted from Kramar *et al.*, 2015.

	1. Job analysis and design	
Few tasks	←→	Many tasks
Simple tasks	←→	Complex tasks
Few skills required	←→	Many skills required
Specific job description	←→	General job descriptions
	2. Employee recruitment and selection	
External sources	←→	Internal sources
Limited socialization	←→	Extensive socialization
Assessment of specific skills	←→	Assessment of general skills
Narrow career paths	←→	Broad career paths
	3. Employee learning and development	
Focus on current job skills	←→	Focus on future job skills
Individual orientation	←→	Group orientation
Train few employees	←→	Train all employees
Spontaneous unplanned	←→	Planned, systematic
	4. Performance management	
Behavioural criteria	←→	Results criteria
Development orientation	←→	Administrative orientation
Short-term criteria	←→	Long-term criteria
Individual orientation	←→	Group orientation
	5. Pay structure, incentives and benefits	
Pay weighted towards salary and benefits	←→	Pay weighted towards incentives
Short-term incentive	←→	Long-term incentives
Emphasis on internal equity	←→	Emphasis on external equity
Individual incentives	←→	Group incentives
	6. Industrial relations	
Collective bargaining	←→	Individual bargaining
Top-down decision making	←→	Participation in decision making
Formal due process	←→	No due process
View employees as an expense	←→	View employees as assets

Although of utmost importance in strategic implementation, HR management issues have been considered the most difficult challenge, according to hotel managers and corporate executives (Enz, 2001, 2009). A study involving a sample of 170 hospitality managers from over 25 different countries shows that the 'thorniest' issues managers face are using human capital or how to care for and get the most from hospitality employees. The most frequent concerns are attracting, retaining, motivating, training and developing the current and future workforce (Enz, 2001). Eight years later, the same author duplicated her study with a larger sample size of 243 individuals from over 60 countries. Interestingly, HR issues remained at the top of the list (Enz, 2009).

Furthermore, HR practices are argued to play a central role for a successful service

Table 13.2. Eight good HR practices. Adapted from Fáilte Ireland, 2005, cited in Baum, 2007.

Practice	Description
Flexibility	Preparedness on the part of the enterprise to match their demands on staff time as far as possible with the available supply of time from workers and to recognize the work–life balance that must be managed in contemporary society.
Participation	Preparedness to encourage staff to be involved in local decision making and to exercise some influence over the patterns of their working day.
Performance management	The regular systems of performance review to ensure that individual and team performance is delivering the service standard and business results expected of it.
Recognition	The attribution of credit to high achievers within the peer group and the celebration of the achievements of high-performing individuals and teams.
Rewards	The payments (both monetary and non-monetary) that are linked to performance and that are clearly understood by staff to track the profitability of the enterprise.
Communication	The routine dialogue with staff, in the openness and accessibility of management, and in regular feedback loops and follow-through actions.
Learning and development	Readiness to access on-site learning opportunities, support for learning progression through attendance at off-site courses, and a continuous emphasis on on-site team development.
Empowerment	The active support and guidance of staff so as to enable them to participate in the workplace in the manner described above.

firm (Schneider and White, 2004) and have a strong relationship with hotel service quality and employees' attitudes (Tiago *et al.*, 2019). Another study of good practices in tourism and hospitality HR was conducted by Fáilte Ireland (2005, cited in Baum, 2007). Eight recurring themes identified by that study are listed in Table 13.2.

Overview of Vietnam Tourism HR

The tourism industry in 2019 in Vietnam employed approximately 3,070,000 people in total, with 1,023,000 being direct tourism workers (ITDR, 2020). Most tourism workers are employed in the accommodation sector (70%), while tour operators and transportation attract 10%, and 20% work in other tourism services. The total number of people working in the accommodation sector in 2019 was 502,759, of which 12.57% were in the front office, 18.34% in restaurants, 31.96% in housekeeping, and 37.13% in other departments. In addition, inbound tour guides account for the majority of tour guides (64.4%) with 17,080 people, while 8579 and 860 are the number of tour guides for the domestic market (32.4%) and on-site tour guides (3.2%), respectively. Thus, the total number of tour guides is 26,519. In terms of gender, more females (55.6%) than males (44.4%) work in tourism. Moreover, 90% of tourism workers are aged under 45, which is advantageous for tourism in the country in meeting its HR development goals.

From an education and training perspective, 192 tourism institutes across the country comprise 62 schools of tourism in universities, 126 colleges, and four training centres responsible for providing formal tourism education and training (ITDR, 2020). In terms of vocational tourism certification, the international cooperation programmes funded by Luxembourg and the European Union (EU) have successfully established 13 Vietnam Tourism Occupational Standards (VTOS), a system of skills standards, and Vietnam Tourism Certification Board (VTCB). In addition, 14 accreditation centres are placed in 14 tourism schools and more than 47 temporary accreditation centres in hotels for housekeeping, cooking and hotel security. To date, 1806 VTOS trainers are accredited. In addition, VTOS materials have been used for internal training in large tourism organizations (Nguyen, 2019). Selected trainers have been sent to the Association of South-east Asian

Nations (ASEAN) for training and accreditation after the Mutual Recognition Agreement on Tourism Professionals (MRA-TP) was established. Currently, there are 14 ASEAN trainers and 13 accreditors for six tourism occupations. However, the Vietnam National Administration of Tourism (VNAT) has to improve the VTOS to be recognized as equivalent by the MRA-TP. In addition, MRA-TP also requires the establishment of two separate national boards: a National Tourism Professional Board and a Tourism Professional Certification Board, instead of the current single national tourism certification board (Le, 2020). Therefore, in complying with the ASEAN requirements, Vietnam will truly immerse itself in the free regional movement trend of tourism workers inside South-east Asia.

However, the number of tourism workers is far from meeting the demand from the tourism industry. For example, only 80% of accommodation jobs are filled. Moreover, there is a critical concern about the quality of the human tourism resources, with only 42% receiving formal training in tourism, 38% with training in other disciplines, and 20% without any form of training (Thao An, 2019). During 2015–2019, among the 42% with formal tourism training, 24% were university graduates, 0.37% were postgraduates while 13% held diplomas and 14% earned intermediate vocational training. The remaining 48.63% relied on short course training and primary vocational training. In 2018, 60% of the tourism employees had some foreign language skills. The most popular language is English, followed by Chinese and French. In terms of information and communications technology (ICT) skills, the number of capable tourism workers is 68% (VNAT, 2018b). In 2019, the World Economic Forum Travel and Tourism Competitiveness Report showed that Vietnam's tourism HR and labour market index was ranked 47th of 140 countries measured, falling from 36th of 136 countries in 2017 (Calderwood and Soshkin, 2019). Additionally, it is worth noting that Vietnam was one of the two countries in South-east Asia that experienced a fall in rank during this period while almost all other countries enjoyed a rise.

There is a lack of industry-education linkage in the hospitality sector (Le *et al.*, 2018). University graduates are challenged to compete with students from vocational schools in satisfying the immediate needs of the workplace due to limited occupational skills, weak problem-solving capacity, critical thinking and, again, lack of English proficiency (Nguyen and Robinson, 2010; Le *et al.*, 2018). Other VNAT publications show a need to retrain 60–70% of tourism university graduates since their skills are mismatched with the requirements of employers (VNAT, 2007).

Study Methods

Various methods were employed to collect data, including desk study research, secondary data from past research, and in-depth interviews. First, a desk study was conducted to gain a general understanding of the topic. Various sources of information were used including reports on international tourism (e.g. the Annual Tourism Report 2013–2019 by the United Nations World Tourism Organization) and reports by the Vietnam tourism industry (e.g. ESRT Programme, 2013, 2015; VNAT, 2013; the *Vietnam Tourism Annual Report* for 2013–2019 (VNAT, 2013–2019); Bodewig and Badiani-Magnusson, 2014). The review of these sources and relevant data from the Hospitality and Tourism Career Cluster in the Occupational Information Network (O*NET) provides a thorough understanding of the literature, particularly on the current trends and challenges and industry outlook. Moreover, the literature review also provides vital tourism statistics that help shape the focus of the chapter, such as tourism growth, revenue and job demand.

Second, secondary data was derived from the Manpower Survey (MPS) 2015 (ESRT Programme, 2015) commissioned by the Environmentally and Socially Responsible Tourism (ESRT) Capacity Development Programme in Vietnam. The first author participated in the MPS as a project leader and was responsible for the literature review, methodology, training for surveyors and in-depth interviewers, analysing data, and preparing a part of the report. The MPS was conducted in the Central Coast provinces of Vietnam, consisting of Thừa Thiên Huế, Đà Nẵng and Quảng Nam. These provinces are among the most visited destinations in Vietnam and have experienced significant increases in both international and domestic visitors consistently over the last 10 years (Dinh *et al.*, 2019). The MPS surveyed 213 respondents consisting of 115 accommodation facilities, 79 travel enterprises, and 19 tourism vocational training colleges from June to October 2015. The questionnaires covered essential information regarding the workforce, including: (i) labour statistics (e.g. number of employees, gender, education and training, and work arrangement); (ii) labour performance; and (iii) key HR practices (e.g. hiring, training and development, employee separation, and compensation). The MPS limited the sample to three–five-star hotels and significant travel companies in the region, expecting these larger organizations would archive HR records and have a standard of operations to be examined. Instead of collecting data on all job titles, the MPS focused on those that had high impact on the operation of the businesses. Based on 32 ASEAN job titles in six common labour divisions, the MPS chose to focus on 11 job titles in the hotel services (i.e. chef, commis pastry, executive chef, executive housekeeper, food and beverage manager, front desk agent, front office manager, front office supervisor, floor manager, head waiter, laundry manager) and five job titles in the travel services (i.e. product manager, sales and marketing manager, travel consultant, tour guide, travel operator).

Finally, the authors also conducted a series of ten in-depth interviews with industry leaders, business owners, managers and employees to gain better insight into the tourism industry's current practices, challenges and opportunities. The findings derived from the analysis of multiple sources are presented by critical themes that emerged from the document analysis and in-depth interviews. Based on this analysis, key findings and recommendations for industry improvement in management are outlined below.

Findings

Tourism businesses in Vietnam face a severe labour shortage due to competition from within the industry and other fast-growing parts of the economy (Hoang *et al.*, 2020). VNAT estimated that the tourism industry would need about 40,000 new employees/year from 2017 to 2020 to catch up with the growth in HR demand, while the capacity of tourism-related training institutions is only 15,000 graduates/year nationwide. This serious labour gap is forcing organizations to be more strategic in attracting qualified applicants. This section reviews critical HR practices and discusses relevant HR issues within the tourism industry.

Recruitment

A critical element of an organization's recruitment strategy is its decisions about where to look for applicants (Noe *et al.*, 2020). Concerning recruitment practices, the MPS findings indicate that internal recruitment, referral from current employees, and specialized job-search websites (e.g. https://www.hoteljob.vn/ accessed 27 March 2021) were the most popular recruitment sources for hotels and travel businesses when seeking to identify qualified candidates for vacant positions. College recruitment and company websites were not popular among surveyed businesses.

Manpower Vietnam 2011 (ESRT Programme, 2013) suggested that the skills shortage at the management level is very high. More specifically, middle managers in Vietnam lack generic managerial skills. Additionally, most businesses in tourism and hospitality were small and medium-sized enterprises (SMEs), which might limit their ability to tap into qualified staff that often requires cost and relationships (ESRT Programme, 2015). Tourism businesses indicated that they do not have specific written HR policies guiding their recruitment activities. Recruitment seems to be less challenging among travel agencies for several reasons, such as their small size, a lower turnover rate (for larger travel agencies), and a better understanding of the market's labour supply situation due to frequent contact among employees of competing businesses.

To evaluate the effectiveness of recruitment sources, organizations often based recruitment on post-hire criteria such as cost, the applicant to hire ratio (the number of applicants hired divided by the number of applicants), and the success ratio (the number of hires who perform well divided by the total number of applicants hired). However, recruitment sources are ineffective in helping organizations to attract the right kind of candidates (ESRT Programme, 2015). This weakness further amplifies the labour shortage issue because organizations are not able to reach out to the already limited applicant pool. For larger hotel groups, job fairs, college recruitment and specialized job-search websites are more effective for entry-level positions, while internal recruitment and referral from current employees are more suitable for higher-ranking positions. In the interviews with larger hotel groups, a respondent revealed that:

> Partnerships with universities, job fairs, and specialised job-search websites are among the most popular recruitment sources. The size and resources available for recruitment in larger hotels might allow them to do more elaborate recruitment.
>
> (interviewee from a large hotel group)

Selection

Biodata (e.g. résumés) comparison, interviews and work samples were the most frequently used selection procedures in hotel and travel businesses (ESRT Programme, 2015). From our interviews with both HR specialists and heads of departments, the use of biodata and interviews seems to be the standard in the industry and applies to most positions. Most hotels and travel agencies collect biodata via applications, where applicants must provide required information such as work experience, education and training, and other qualifications. In addition, semi-structured interviews and the use of situational interview questions helped recruiters to learn how applicants might respond to work-related situations. Also, candidates are usually required to perform a miniature replica of the work activities such as cooking a sample dish, setting tables or pricing a tour. Despite the popularity of biodata, work samples and interviews, the MPS results indicated that they are not highly effective selection procedures (rated just above 3 on a 5-point scale).

There are a few noted variations in the selection practices among tourism businesses in Vietnam. First, the selection process is more formally structured and reinforced in the hotel business than the travel business. Travel agencies are more willing to hire less-qualified candidates for the job, possibly because they can pair them with more experienced and senior employees without sacrificing the quality of the services. Many travel agencies recruit their employees directly from college as interns and will offer them formal positions if they show potential for the job (ESRT Programme, 2013). However, it is commonly believed that the best way to know if the candidates are qualified is to require them to go through a probation period which usually lasts for 1–3 months. For more experienced candidates, interviewers are more interested in assessing the person–organization fit in terms of organizational culture, compensation expectation, and other related terms of employment.

Second, the selection process in internationally managed hotels is often more scientifically designed and supported by better training for those involved in selection decision making than those using the common selection procedures in domestically managed hotels. However, it is essential to note that although international hotel operators have a robust HR management system with well-designed HR policies, processes, training and tools in place, they do not systematically apply them in all HR activities (e.g. recruitment, selection, performance management and training). Interviewees suggested that there are cases where they have to accept less-qualified candidates due to high labour demand and low labour supply.

Third, the selection process for higher-ranking positions in the tourism business is less structured than those at the lower levels. Applicants, for example, are not usually required to complete a generic application form; instead, they usually submit their résumé, which is believed to be a better way to communicate their qualifications. Interview questions are also less structured and tailored to the qualifications and experience of the applicants. Finally, there is a more substantial interest for the investors to participate in the selection process for hotels under management contracts, especially for the higher-ranking positions. Both hotel and travel businesses found it was more challenging to staff their managerial positions than their non-managerial positions (Manpower Group, 2011; Bodewig and Badiani-Magnusson, 2014).

Performance appraisal and labour performance

Performance appraisal consists of identifying, measuring and managing human performance in organizations (Gomez-Mejia *et al.*, 2016). Performance appraisal is a formal HR function carried out by most businesses in the tourism industry. However, there are considerable variations. In many big businesses, the performance appraisal is more intentionally designed as a strategic HR tool to assist the management in making HR decisions (e.g. training, promotion or termination). The employee's performance is based on specific key performance indicators (KPIs) derived from the job requirements. One respondent from the hotel industry confirmed that:

> Supervisors and managers are required to conduct performance appraisals more frequently, ranging from one every month to one every 6 months, to provide more immediate feedback.
>
> (interviewee from the hotel industry)

In smaller businesses, the performance appraisal is less structured in that an employee's performance is more likely to be evaluated on overall performance without specific performance feedback. Feedback from management tends to be more spontaneous but is not consistently communicated, making it difficult for employees to make timely adjustments. Therefore, the information generated from the performance appraisal is often used for administrative purposes such as rewards, promotions or terminations.

The quality of HR is a significant concern for the tourism industry. The MPS (ESRT Programme, 2015) indicated that although a large portion of the tourism workforce in the region was rated as being able to meet the expectations of the jobs, at least 10–20% of the employees were identified as poor performers, failing to meet the minimum job requirements. In addition, Dinh *et al.* (2019) reported that the annual growth rate of labour productivity in the hotel and food services since 2010 had only been 1.4%, compared with 4.7% for the whole economy. As a result, the productivity level of the tourism sector is only 23% of the economy-wide average in 2017 and is among the lowest in South-east Asia.

Another indicator for the lack of quality HR is the increasing gap between real wage growth relative to actual labour growth in the industry. Annual Hotel Survey data (Grant Thornton, 2012–2018) reveal a consistent increase in average payroll expenses per employee from 2015 to 2018. According to Dinh *et al.* (2019), the tourism industry's wage/productivity index ratio between 2012 and 2017 was approximately 20% higher than that of the whole economy. The growing wage-productivity gap is due to the labour shortage, the report argued, and businesses are bidding up wages to lure qualified workers.

Further investigation indicated that some areas of performance need an immediate solution. Therefore, the MPS investigated specific performance dimensions of the surveyed positions to better understand how employees perform in specific aspects of the job. The results indicate that the performance problems were significantly more concerning in several dimensions, such as *core competency* (basic and essential skills for anyone to do the job), *managerial competency* (managing, supervising or influencing the work of others) and *responsible competency* (specific skills required for the enhancement of services and products towards a sustainable tourism development). These performance issues identified by the MPS were congruent with past research. Several other reports (e.g. Manpower Group, 2011; Bodewig and Badiani-Magnusson, 2014) also discussed the challenges facing organizations in Vietnam in filling positions due to serious skill gaps in core and management skills.

Training and development

Due to the shortage of skilled labour in the tourism industry, most hotel and travel services rely on internal training as an immediate solution for labour performance issues. However, not all businesses are prepared to meet the challenges due to limited resources, lack of adequate training programmes, or the unwillingness to invest in training because their trained employees might leave for better job opportunities, which are abundant in the market. The MPS results (ESRT Programme, 2015) indicated that:

> Hotel businesses are better prepared for this challenging task, with almost 80% of the hotels having a dedicated budget for training instead of the travel services, where only 30% of the businesses had a budget for training.
>
> (ESRT Programme, 2015)

Further analysis of the in-depth interviews with the travel sector confirmed that travel businesses have little motivation to train their employees because they were afraid of losing trained individuals to competitors or having to compete with their employees who quit and start a new company in the same market. Most of the businesses that participated in the interviews indicated that on-the-job and in-house training was the primary approach to training. However, they also indicated that the training quality is often compromised because of a lack of funding or qualified trainers.

Over the last decade, various efforts led by both the public and the private sectors focused on improving the quality of the tourism workforce. For example, the ESRT Programme (2011–2016) is a €12.1 million project funded by the EU, co-managed by the Ministry of Culture, Sports and Tourism and the delegation of the EU to Vietnam. The programme has organized a wide range of HR development activities for the owners and managers of SMEs, VNAT and Department of Culture, Sport and Tourism (DCST) administrators, training institutions, tourism workers in disadvantaged areas and local tourism communities. The training opportunities offered by the ESRT Programme are diverse, including responsible tourism, business management, vocational competencies and training of trainers. Likewise, large hotel groups such as Vinpearl, Thiên Minh Group and Sun Hospitality have invested heavily in the training and development of their managers to meet their future demand for highly qualified employees.

Compensation

Compensation in the tourism industry is generally perceived to be lower than that of competing industries in the economy (Dogru *et al.*, 2019). In Vietnam, compensation varies significantly across positions and sectors (i.e. hotel and travel businesses). While some managerial positions (e.g. executive chefs, front office managers, food and beverage managers or executive housekeeper) receive attractive compensation, many other non-managerial positions (e.g. receptionists, comis pastry and travel consultants) are paid just above the minimum legal wage. According to a recent study conducted by the Vietnam General Confederation of Labour, the minimum legal wage only covers 65–70% of the minimum living expenses. Another important finding of the study relating to compensation was the surprisingly low salaries for product managers and sales and marketing managers. According to the results, these positions receive an average salary of just above VND 7 million. This level of salary is not equitable with the responsibilities and duties of the two positions. Research (e.g. Colquitt *et al.*, 2001) has found that the lack of internal equity and external equity might lead to undesirable outcomes such as poor performance, lack of organizational commitment and job satisfaction, and high turnover of personnel.

Employees' separation

The shortage of skilled labour in the tourism industry encourages businesses to attract qualified employees from competitors more aggressively. Indeed, the MPS (ESRT Programme, 2015) reveals that 'alternative job opportunities from competitors were the primary reason for people to turn over in both hotel and travel businesses'. Although most organizations in the MPS did not have severe problems with employment separation, in-depth interviews with industry leaders indicated that:

> The talent war is much more severe for high-ranking positions and those in the hotel business. It is estimated that the current turnover rate for the hotel business is around 20–30% making it extremely challenging for hotels to fill their vacancies.
>
> (interviewee from a large hotel group)

As a result, it is not uncommon for tourism businesses, especially hotels, to double the compensation to attract critical talents from competitors. On the other hand, major travel businesses reported a near-zero turnover rate due to their ability to maintain more sustained revenue sources.

HR management in the wake of pandemic

While this chapter was being written, the tourism workforce has been in crisis due to COVID-19. The border closure and social-distancing policies during 2020 caused the number of international visitors to drop by 79.5% and domestic visitors to fall by 34.1% compared with 2019. During 2020, nearly 60% of the tourism jobs were cut due to 95% of inbound travel companies closing down and only approximately 10-15% of available hotel rooms being occupied (Trang Linh, 2021). A survey conducted by the Tourism Advisory Board in 2021 revealed that 18% of tourism firms had fired all their staff while 48% had discharged 50–80% of their total employees. The number of tourism companies that provided different types of financial support for COVID-19-affected employees is 75% (Hoa Quynh, 2021). The impact of two national promotional campaigns, '*Vietnamese people travel to Vietnam destinations*' and '*Viet Nam tourism – a safe and attractive destination*', has contributed to the tourism industry's temporary recovery. The notion of 'flexibility' in HR management (Fáilte Ireland, 2005, cited in Baum, 2007) is applicable in this context and perhaps requires each tourism firm to be creative and government financial support to match the low tourist demand with staff time. In the long term, how to retain the workforce so that firms will be ready when COVID-19 is over seems to be the greatest challenge for tourism firms because once labourers move to other sectors, it is likely to be hard to attract them back to tourism (Tien Minh, 2021).

Recommendations

Establishment and utilization of job databases

Developed and recently revised by the EU-funded ESRT Programme, VTOS is one of the most updated job databases for the tourism industry in Vietnam. The revised VTOS have been developed and benchmarked against international occupational standards and ACCSTP (ASEAN Common Competency Standards for Tourism Professionals) Standards. The revised VTOS, following the Guidelines from Decision No. 09/2008/QĐ-BLĐTBXH (27 March 2008) of the Minister of Labor, Invalids and Social Affairs (MOLISA, 2008), set out the principles and processes for the development of National Occupational Skill Standards. VTOS covers two significant sectors within the tourism industry (hospitality and travel), with six key and four specialist occupational areas.

In addition, O*NET is a unique, powerful source for continually updated occupational information and labour market research. By using a contemporary, interactive skills-based database and a common language to describe worker skills and attributes, O*NET transforms mountains of data into precise, focused occupational intelligence that anyone can understand quickly and efficiently. Currently, the information in O*NET is available for over 900 occupations. This is a comprehensive source of job information that the US Department of Labor maintains and it is available free of charge to individual job seekers, businesses and HR professionals. This is particularly important because more than 65% of the hotels and most travel businesses that participated in the current study are privately and independently owned. The lack of funding and expertise for job-analysis activities has often been cited as a major barrier to conducting HR management tasks more professionally.

To maximize the benefits of these job information databases, specific recommendations are given as follows. First, it is vital to set up a task force and allocate resources to local job analysis for crucial positions in specific sectors (i.e. hotel and travel) based on VTOS and O*NET information. Second, in addition to local job analysis, provincial destination management organizations (DMOs) should provide training to local businesses on applying job information in their HR activities (i.e. how to use the job information in HR decision making). Priorities can be given to critical areas of HR such as selection, training, performance appraisal and job redesign. Lastly, we suggest that DMOs develop regulations and policies at the provincial level to reinforce job information in all HR decisions at local businesses. While it is common in many countries (e.g. in Europe and North America) to have laws and regulations overseeing and reinforcing how organizations conduct their HR activities, the current Vietnam labour law and other HR regulations do not directly deal with equal employment opportunities. Therefore, it is totally up to the business to ensure their decisions concerning employment opportunities are fair.

Recruitment and selection

It is recommended that recruitment of managers should come from both internal and external sources. Additionally, businesses should make better use of college recruitment and their website and social media to attract entry-level jobs such as receptionists and tour guides. Furthermore, provincial government agencies (e.g. DCST, Department of Labor, Invalids and Social Affairs and DMOs) and educational institutions can collaborate with businesses in their recruitment efforts. Finally, educational institutions are encouraged to collaborate with businesses in creating internship opportunities and allow businesses to interact directly with students to attract potential employees.

With that being said, several specific recommendations to improve selection systems are as follows. First, biodata is popular but using it effectively requires careful consideration. For lower-ranking positions, it is recommended that organizations should develop their application forms to compare the credentials and qualifications across candidates more accurately and efficiently. For higher-ranking positions where job applicants are fewer, organizations can consider a customized résumé because it tells more about the candidates. Second, ability and personality-based assessments are becoming more popular selection tools and are used by many organizations across industries. They are very effective in handling a large number of candidates for lower-ranking positions and have shown excellent validity in helping organizations choose the best candidates. Ability tests (or intelligence tests) are efficient for complex jobs, such as managers, and personality inventories can allow organizations to identify candidates with suitable traits for the jobs. Third, interviews can be compelling only if they are conducted correctly. Recent research evidence suggests that structured and behavioural interviews are the most effective in the selection. Structured interviews refer to the kind of interviews whose questions are based on job-analysis information. Structured interviews use the same questions for all candidates and follow the same scoring system evaluated by multiple judges in each interview. Behavioural interviews focus on past behaviours of the candidates which are relevant to the current job. This focus allows interviewers to understand better how candidates might behave in a relevant situation in the future. Fourth, we recommend that organizations continue to use work samples for jobs that they can meaningfully create as 'miniature replicas' of the jobs. Work samples are beneficial to measure job skills by taking samples of behaviours under realistic, job-like conditions. Lastly, assessment centres are recommended for the hiring of medium- and top-level people for positions in the business. Assessment centres are collections of standardized procedures used for the evaluation of candidates for managerial positions.

Training and development

It is recommended that organizations provide trainers (usually the most skilled employees or supervisors) with training knowledge and skills, including communication skills and a solid understanding of how people learn, for on-the-job training delivery. In addition, succession planning and promotion-ability forecasting can develop future supervisors' and managers' competencies in leadership, motivation skills, delegation, planning and controlling.

Compensation

Equity is the single most crucial aspect of a compensation system. Achieving equity requires organizations to pay attention to: (i) internal equity (fairness of the pay structure within a firm); (ii) external equity (fairness in comparison to the same job with other employers); and (iii) individual equity (fairness in pay decisions for employees holding the same job). Organizations are recommended to evaluate their compensation systems using job-analysis information. By identifying all compensable factors (efforts, skills, responsibility and working conditions) for all jobs and evaluating jobs using a predetermined evaluation system, organizations can make sure they compensate fairly within the firm.

Fair compensation requires fair performance appraisal. To facilitate the equity in compensation, it is suggested that organizations should utilize VTOS's competency units in performance appraisal because this task requires raters to differentiate individual performance across different aspects of a job. In addition to fairness, this will facilitate performance feedback by allowing managers to give more specific responses to their employees, and organizations can design more appropriate training for their employees.

Conclusions

This chapter identifies a range of HR management in the context of Vietnam tourism. Preferred recruitment channels are internal recruitment, a referral from current employees, and specialized job-search websites. In larger hotel groups, firms pay more attention to partnerships with universities, job fairs and specialized job-search websites. Attending job fairs organized by universities and training institutions increases recruitment costs and applicants' hiring ratios. To perform an essential task in HR practices, selection methods frequently used in both hotel and travel businesses are résumés, interviews and work samples. More scientifically proven tests such as job knowledge tests, personality tests and assessment centres are not widely used, especially in locally managed firms. International hotels tend to adopt a more scientifically designed selection process with training sessions for selection decision makers. A more challenging issue is selection for managerial positions, as found in both tourism and hotel businesses.

Evidence in this study also revealed that performance appraisal is poorly conducted in small businesses with less structured assessment on overall performance. The employees find it difficult to adjust their work performance from non-specific and spontaneous feedback and inconsistent communication. Considerably recognized as an HR strategic tool to assist decisions in other HR areas, more prominent organizations use KPIs and more structured performance appraisal forms. It is worth noting that compensation significantly varied across different positions and sectors (hotel or travel). There is a big gap between the attractive salaries for managerial positions and the just-above-the-minimum legal wage of various non-managerial positions. In addition, training findings indicated that most hotel and travel businesses utilize internal training to improve their staff performance, although limited resources, lack of adequate training programmes and lack of qualified trainers are some challenges they face. Retaining employees remains the reason that inhibits firms from investing more in training since they are afraid of employees quitting their jobs after upgrading their knowledge and skills.

Findings from this study suggested that job-analysis and job-design tools such as VTOS and O*NET might be beneficial for the firms to reach international standards and continually update occupational tourism information and labour market research (Le, 2020). Close industry-higher education linkages in recruiting and improving the higher education curriculum and immersing students in the real world before graduation imply a positive change in the quality of the workforce for more complex tasks (Le *et al.*, 2018). In addition, there is an opportunity for tourism businesses to introduce pre-employment testing to select applicants with the necessary knowledge, skills, abilities and personal characteristics that underpin firms' success in achieving their strategic HR management goals (Kramar *et al.*, 2015). As the free movement of the labour market inside ASEAN allows tourism firms to attract more candidates for managerial positions from other regional countries, this study posits a heightened need to meet MRA-TP requirements.

Finally, some limitations need to be considered. First, the chapter mainly focuses on the HR practices in tourism and accommodation businesses. Second, the empirical data was only sourced from the MPS of the central provinces (i.e. Thừa Thiên Huế, Đà Nẵng and Quảng Nam) before COVID-19. Further empirical research is needed to understand the HR practices in the digital age post-COVID-19 and how it works in other provinces.

References

Baum, T. (2007) Human resources in tourism: still waiting for change. *Tourism Management* 28(6), 1383–1399. DOI: 10.1016/j.tourman.2007.04.005.

Bodewig, C. and Badiani-Magnusson, R. (2014) *Skilling up Vietnam: Preparing the Workforce for a Modern Market Economy*. World Bank, Hanoi. Available at: https://doi.org/10.1596/978-1-4648-0231-7

Calderwood, L.U. and Soshkin, M. (2019) *The Travel & Tourism Competitiveness Report 2019: Travel and Tourism at a Tipping Point*. World Economic Forum. Available at: http://www3.weforum.org/docs/WEF_TTCR_2019.pdf (accessed 26 May 2021).

Colquitt, J., Conlon, D., Wesson, M., Porter, C. and Ng, K. (2001) Justice at the millennium: a meta-analytic review of 25 years of organizational justice research. *Journal of Applied Psychology* 86(3), 425–445. DOI: 10.1037/0021-9010.86.3.425.

Dinh, V.T., De Kleine Feige, A. I., Pham, D.M., Eckardt, S. and Vashakmadze, E.T. (2019) *Taking Stock: Recent Economic Developments of Vietnam (Special Focus: Vietnam's Tourism Developments – Stepping Back from the Tipping Point – Vietnam's Tourism Trends, Challenges, and Policy Priorities)*. No. 138475. World Bank, Hanoi, pp. pp. 1–62.

Dogru, T., McGinley, S., Line, N. and Szende, P. (2019) Employee earnings growth in the leisure and hospitality industry. *Tourism Management* 74(3), 1–11. DOI: 10.1016/j.tourman.2019.02.008.

Enz, C.A. (2001) What keeps you up at night: key issues of concern for lodging managers. *Cornell Hotel and Restaurant Administration Quarterly* 42(2), 38–45.

Enz, C.A. (2009) Human resource management: a troubling issue for the global hotel industry. *Cornell Hospitality Quarterly* 50, 578–583.

ESRT Programme (2013) *Manpower and Training Needs Analysis of Vietnam Tourism Industry*. Environmentally and Socially Responsible Tourism (ESRT) Capacity Development Programme. Vietnam National Administration of Tourism, Hanoi.

ESRT Programme (2015) *Tourism Manpower Analyses in Central Coast of 3 Provinces*. Environmentally and Socially Responsible Tourism (ESRT) Capacity Development Programme. Vietnam National Administration of Tourism, Hanoi.

Gomez-Mejia, L.R., Balkin, D.B. and Cardy, R.L. (2016) *Managing Human Resources*, 8th edn. Pearson, New York.

Grant Thornton (2012–2018) *Vietnam Lodging Industry Hotel Survey*. Grant Thornton, Hanoi.

Hoang, L. V., Vu, H.M. and Ngo, V.M. (2020) Factors affecting job pursuit intention in hotel industry in Ho Chi Minh, Viet Nam. *The Journal of Asian Finance, Economics and Business* 7(11), 281–290. DOI: 10.13106/jafeb.2020.vol7.no11.281.

Hoa Quynh (2021) COVID-19 Pandemic: Tourism businesses are facing labour crisis. *Cong Thuong*, 10 May. Available at: https://congthuong.vn/dich-covid-19-doanh-nghiep-du-lich-doi-dien-khung-hoang-lao-dong-156774.html on 26/05/2021 (accessed 26 May 2021).

ITDR (Institute for Tourism Development Research) (2020) Introduction. In: *Examining Quality of Vietnam Tourism Workers Conference Proceedings*. ITDR, Hanoi, pp. 1–8.

Kramar, R., Bartram, T., De Cieri, H., Noe, R.A. and Hollenbeck, J.R. (2015) *Human Resource Management in Australia: Strategy, People, Performance*. McGraw Hill Education, New York.

Le, A.H., McDonald, C.V. and Klieve, H. (2018) Hospitality higher education in Vietnam: voices from stakeholders. *Tourism Management Perspectives* 27(4), 68–82. DOI: 10.1016/j.tmp.2018.05.002.

Le, A.T. (2020) Improving occupational skill standards in Vietnam tourism. In: *Examining Quality of Vietnam Tourism Workers Conference Proceedings*. Institute for Tourism Development Research, Hanoi, pp. 10–17.

Manpower Group (2011) *Building a High-Skilled Economy: the New Vietnam*. Manpower Group, Hanoi.

MOLISA (Minister of Labor, Invalids and Social Affairs) (2008) Principles and Procedures for Development and Issuance of National Occupational Skill Standards. MOLISA, Hanoi.

Nguyen, P. and Robinson, A. (2010) Managing continuous improvement in Vietnam: unique challenges and approaches to overcome them. *Quality Management Journal* 17(2), 27–41. DOI: 10.1080/10686967.2010.11918268.

Nguyen, T.B. (2019) Current situation and issues of skilled labourers in the Vietnam accommodation sector. In: *Examining Quality of Vietnam Tourism Workers Conference Proceedings*. Institute for Tourism Development Research, Hanoi, pp. 18–34.

Noe, R.A., Hollenbeck, J.R., Gerhart, B.A. and Wright, P.M. (2020) *Fundamentals of Human Resource Management*, 8th edn. McGraw-Hill/Irwin, New York.
Schneider, B. and White, S.S. (2004) *Service Quality: Research Perspectives*. Sage Publications, Thousand Oaks, California.
Thao An (2019) Enormous opportunity for tourism worker. *Nhan Dan* (Hanoi), 23 October. Available at: https://nhandan.com.vn/dien-dan-dulich/co-hoi-viec-lam-rong-mo-voi-lao-dong-nganh-du-lich-374739 (accessed 7 April 2021).
Tiago, F., Borges-Tiago, T. and Couto, J. (2019) Human resources' role in hospitality service quality. In: Kavoura, A., Kefallonitis, E. and Theodoridis, P. (eds) *Strategic Innovative Marketing and Tourism: the 8th ICSIMAT*. Springer, Northern Aegean, Greece, pp. 81–90.
Tien Minh (2021) Tourism workers overcome the challenges during the COVID-19 pandemic. *Ninh Binh*, 1 March. Available at: https://baoninhbinh.org.vn/lao-dong-nganh-du-lich-vuot-kho-trong-dich-benh-covid-19/d20210226140338526.htm (accessed 26 May 2021).
Trang Linh (2021) Vietnam Tourism 2020: Developing internal resources in COVID-19 crisis. *Nhan Dan*, 3 January. Available at: https://nhandan.vn/dien-dan-dulich/du-lich-viet-nam-2020-phat-huy-noi-luc-trong-bao-covid-19-630469 (accessed 26 May 2021).
VNAT (Vietnam National Administration of Tourism) (2007) *Vietnam Human Resources Development in Tourism Project*. VNAT, Hanoi.
VNAT (Vietnam National Administration of Tourism) (2013) *Vietnam Tourism Occupational Standards*. VNAT, Hanoi.
VNAT (Vietnam National Administration of Tourism) (2013–2019) *Vietnam Tourism Annual Report*. VNAT, Hanoi.
VNAT (Vietnam National Administration of Tourism) (2018a) *Master Plan for Việt Nam Tourism Development to 2030 with Vision to 2050*. VNAT, Hanoi.
VNAT (Vietnam National Administration of Tourism) (2018b) *Vietnam Tourism Annual Report*. VNAT, Hanoi.

Index

Note:- Page numbers in **bold** type refer to figures and page numbers in *italic* type refer to tables.